Introduction to

HUMAN
GEOGRAPHY

Edited by
David Dorrell Ph.D. and Joseph P. Henderson Ph.D

University System of Georgia
"Creating A More Educated Georgia"

UNG
UNIVERSITY *of*
NORTH GEORGIA™
UNIVERSITY PRESS

Blue Ridge | Cumming | Dahlonega | Gainesville | Oconee

This title is a product of an Affordable Learning Georgia Textbook Transformation grant.

Produced by the University System of Georgia

Cover Design and Layout Design by Corey Parson

Cover Photo by Atik Sulianami

Contents

CHAPTER 4: FOLK CULTURE AND POPULAR CULTURE   **64**

Dominica Ramírez and David Dorrell

CHAPTER 5: THE GEOGRAPHY OF LANGUAGE   **81**

Arnulfo G. Ramírez

CHAPTER 6: RELIGION   **107**

David Dorrell

1

Introduction to Geography

Joseph Henderson

STUDENT LEARNING OUTCOMES

By the end of this section, the student will be able to:

1. Understand: the importance of maps and some tools used to create them.
2. Explain: the concept of places and how they are characterized from a spatial perspective.
3. Describe: the various types of diffusion.
4. Connect: the discipline of geography with other academic disciplines.

CHAPTER OUTLINE

1.1 INTRODUCTION

Geography is a diverse discipline that has some sort of connection to most every other academic discipline. This connection is the spatial perspective, which essentially means if a phenomenon can be mapped, it has some kind of relationship to geography. Studying the entire world is a fascinating subject, and geographical knowledge is fundamental to a competent understanding of our world. In this chapter, you will learn what geography is as well as some of the fundamental concepts that underpin the discipline. These fundamental terms and concepts will be interwoven throughout the text, so a sound understanding of these topics is critical as you delve deeper into the chapters that follow. By the end of the chapter, you will begin to think like a geographer.

1.2 WHAT IS GEOGRAPHY?

The Greek word *geographos* from which geography is derived, is literally translated as writing (*graphos*) about the Earth (*geo*). Geography differs from the discipline of geology because geology focuses mainly on the physical Earth and the processes that formed and continue to shape it. On the other hand, geography involves a much broader approach to examining the Earth, as it involves the study of humans as well. As such, geography has **two major subdivisions, human (social science) and physical (natural science)**. This text focuses primarily on human geography, but because the physical aspects affect humans and vice versa, physical geography will not be completely excluded, but will receive less emphasis.

Geography is the study of the physical and environmental aspects of the world, from a spatial perspective. As geographers study the Earth, the one element that binds the discipline of geography and makes it unique is studying the Earth from a spatial perspective. The spatial perspective means that the phenomenon you are studying can be displayed on a map, so geography focuses on *places* around the world. **Geography**, then, is a physical (or natural) and social science that asks the fundamental questions, "What is *where*, and why?" Human geography is a social science that focuses on people, where they live, their ways of life, and their interactions in different places around the world. A simple example of a geographic study in human geography would be where is the Hispanic population concentrated in the U.S., and why? A physical geography research endeavor might ask where do most hurricanes strike the U.S. coastline, and why? In addition, because the Earth is dynamic, geographers also look at how places change through time, and why, so there is a natural connection with history.

1.2.1 Geography and its relationship to other disciplines

Not only is there a connection between geography and history, but geography is also related to a broad range of other academic disciplines (**Figure 1.1**). If you examine Figure 1.1, you may find your own major on the outside margin of the circle, with the corresponding subdiscipline in geography on the interior of the circle. Again, if a phenomenon can be depicted on a map and studied from a spatial perspective, it is geographical. A basic example would relate to the health sciences or medical geography, the subfield of geography that focuses on the spatial patterns of various aspects of health.   For example, when the spread of a disease from its source area is mapped, medical professionals can get a better idea of the causes of a disease and the mechanisms of its transmission. Often, the understanding of cultural practices or the environmental conditions (such as the habitat for a mosquito-borne disease) can shed light on the process of how the disease operates. Another example of how geography relates to other disciplines is in economic geography, the subfield that examines the different economic activities in various places, and how places interact economically. A fundamental concept in economic studies is that the location of a business is often important to the success of that business. If the business is located in close proximity to its clientele, for example, the customers might be more likely to visit that restaurant, store, etc. on a regular basis. A business owner would be wise to consult maps of both transportation networks as well as the population of the customers to which they intend to cater.

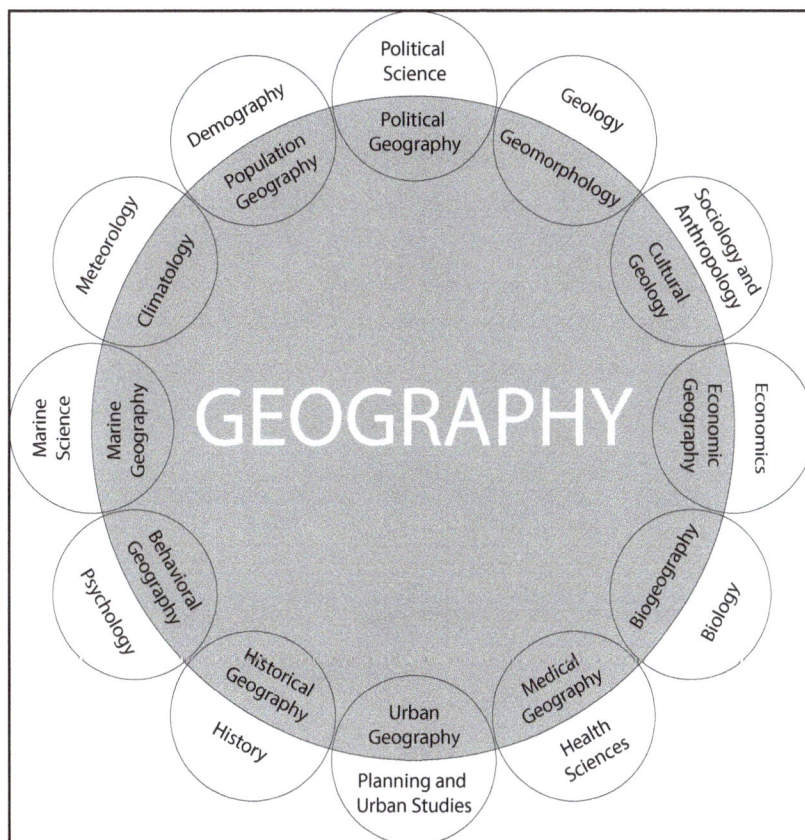

Figure 1.1 | Geography Relationships
Geography and its relationship to other disciplines.[1]
Author | Corey Parson
Source | Original Work
License | CC BY-SA 4.0

1.3 MAPPING THE WORLD

Maps are fundamental to the discipline of geography and have been used by humans since before 6,000 B.C. Today's maps are much more sophisticated, complex, and precise, and are used by many people who employ GPS mapping systems in their vehicles. This technology allows motorists to navigate from place-to-place with relative ease, but the process by which these digital and other maps are created is exceptionally complex.

Essentially, a map, which is a flat presentation of a place on Earth, is actually depicting a curved surface. The Earth, which looks like a sphere, is technically an oblate spheroid, which means that the "middle" of the Earth, around the equator, is slightly wider, and the north/south pole axis is slightly shorter, than a perfect sphere. When any curved surface is depicted on a flat surface, that process is known as **projection**, and many types of map projections exist. A fundamental characteristic of all maps is they involves projections, and all projections have some sort of distortion inherent in them. The size, shape, distance, and direction of objects are distorted to various degrees on maps. The reason this distortion occurs can be visualized by simply imagining peeling an orange, and trying to flatten the

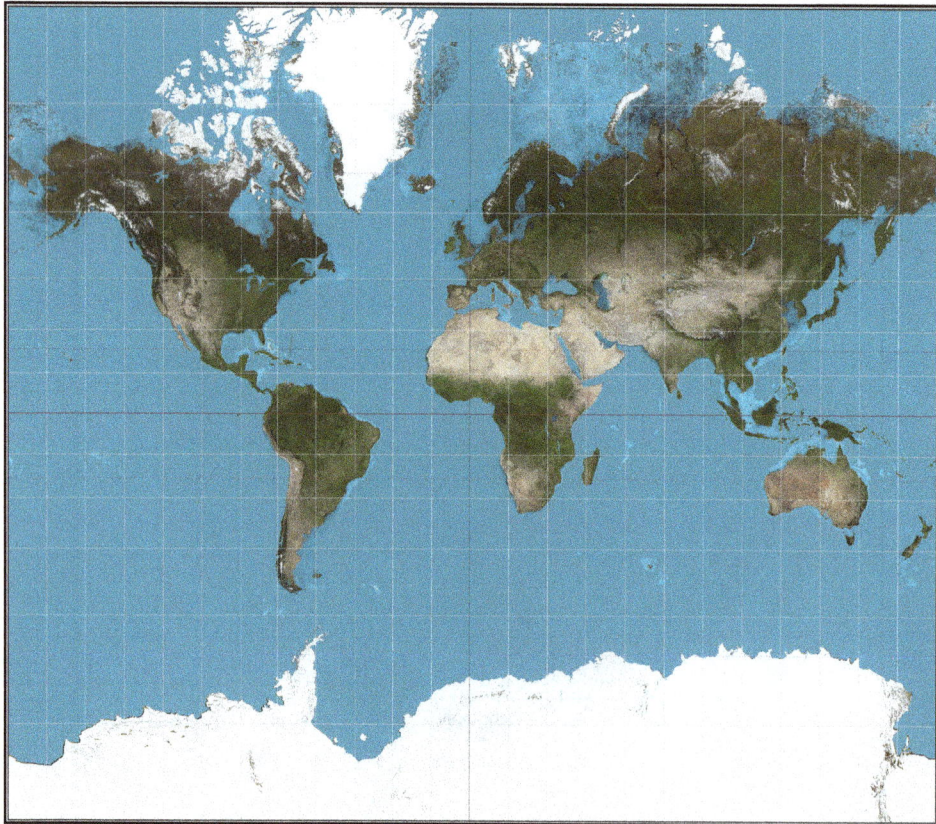

Figure 1.2 | World Map
World Map with Mercator Projection.
Author | User "Strebe"
Source | Wikimedia Commons
License | CC BY-SA 3.0

peel on a table. If you drew the continents on that orange before peeling it, the continents would most certainly be distorted when you try to flatten the peel on the table. This analogy does not precisely describe how projections are created; the process is much more involved. However, the underlying principle still applies. An example of distortion is shown on the map of the globe below (**Figure 1.2**). Note, for example, in this Mercator projection that Greenland appears to be larger than South America, although it is, in fact, much smaller.

Besides projections, another important characteristic of maps is the **scale**. The scale of a map is a ratio of the length or distance on the map versus the length or distance on the Earth or ground (actual). The amount of detail shown on a map will vary based on the scale. For example, a map with a scale of 1:100,000 (which means 1 in/cm on the map equals 1,000,000 in/cm on the ground) would show much less detail than a map at a scale of 1:10,000 (**Figure 1.3**). Besides showing scale as a ratio, it can also be presented as a bar graph or as a verbal statement. Scale can also mean the spatial extent of some kind of phenomena. For example, one could examine migration at the global, national, state, or local scale. By either definition, however, each refers to the level of detail about the place that the geographer is researching. Examining the world from different scales enables different patterns and connections to emerge.

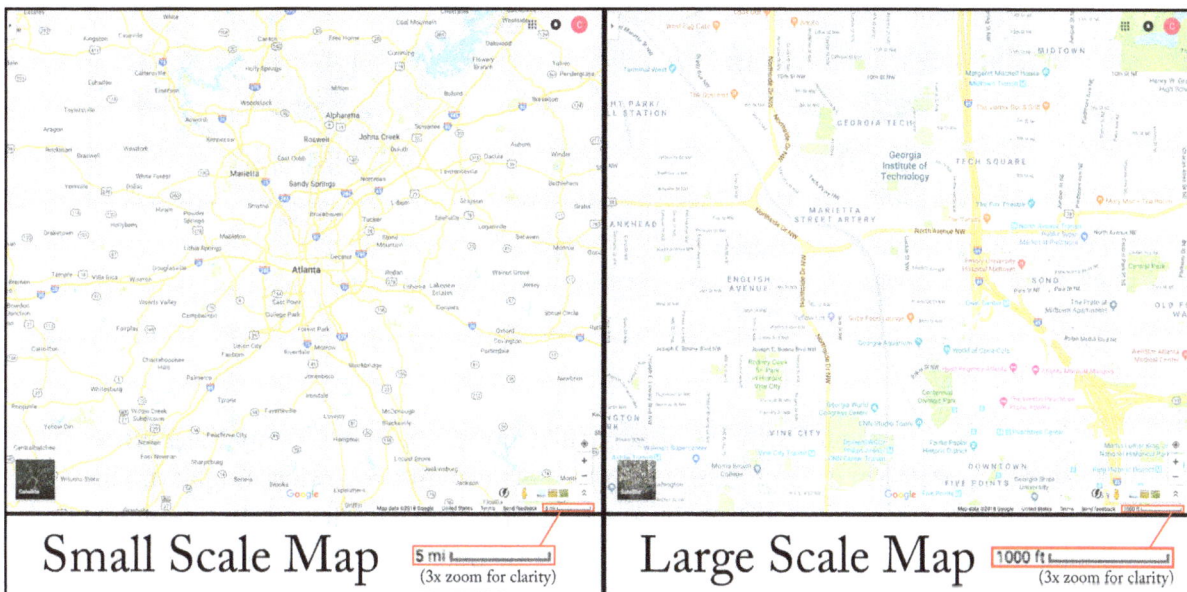

Figure 1.3 | Comparison of Map Scales
The map on the left is a small scale map, showing a larger area. The map on the right is a large scale map, showing a smaller area.
Author | Corey Parson, Google Maps
Source | Google Maps
License | © Google Maps. Used with Permission.

1.4 WHERE IN THE WORLD AM I?

One of the most important pieces of information that maps provide is location. Knowing precisely where a place is in the world is fundamental to geography.

While one can define a location simply by using a street address, not all places on Earth have such an address. Therefore, one of the basic ways to pinpoint a location on the Earth is using the geographic grid. The geographic grid is comprised of meridians and parallels, which are imaginary lines and arcs crisscrossing the Earth's surface. **Meridians** are half circles that connect the north and south poles, and **longitude** refers to the numbering system for meridians. **Parallels** are circles that encompass the Earth and are parallel to the equator, and the numbering system for these circles is known as **latitude (Figure 1.4)**. Where meridians and parallels intersect at precise locations (points) on the Earth on the geographic grid, a location can be known by its latitude and longitude.

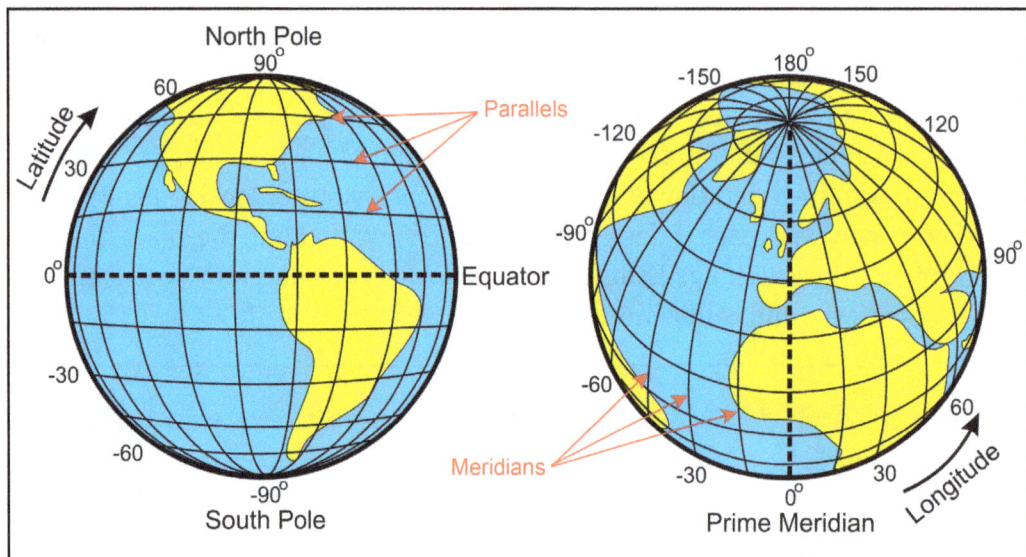

Figure 1.4 | Longitude and Latitude
The geographic grid comprised of meridians and parallels with longitude and latitude.
Author | User "Djexplo" and Corey Parson
Source | Wikimedia Commons
License | CC 0

A few meridians on Earth are of particular importance, one being the **Prime Meridian** located at 0° longitude, which passes through Greenwich, England. The other important meridian, called the **International Date Line**, follows roughly along 180° longitude, and this meridian is on the opposite side of the world from the Prime Meridian (**Figure 1.5**). When a traveler crosses the International Date Line, the day of the week instantaneously changes. When moving westward, the day moves forward, and when traveling eastward, the date jumps backward one day. Fortunately, the International Date Line is in the middle of the Pacific Ocean, so disruptions to the daily calendar are minimal for most people in the world. Moreover, the International Date Line does not precisely follow the 180° longitude line, and this accommodation allows countries and territories consisting of islands that straddle 180° longitude to share the same calendar date.

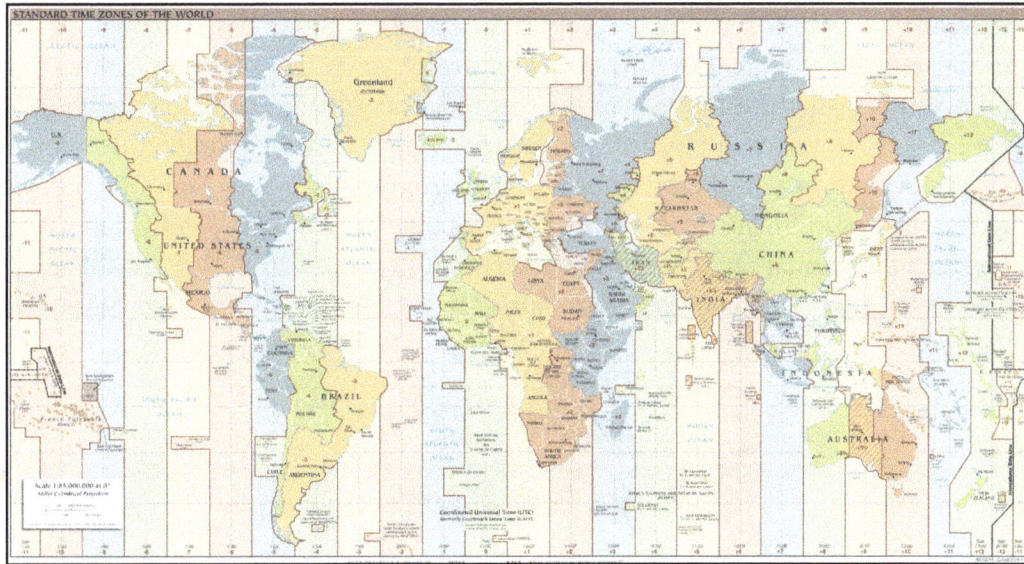

Figure 1.5 | Time Zones
This world map shows the international date line and global time zones.
Author | Central Intelligence Agency
Source | Wikimedia Commons
License | Public Domain

1.5 HOW DO I DESCRIBE WHERE I AM?

Defining a location by using the geographic grid is only part of the process of describing a place. Geographers are primarily concerned with two ways of describing a place: site and situation. **Site** refers to the physical characteristics, such as the topography, vegetative cover, climatic conditions, and the like. **Situation**, on the other hand, refers to the area surrounding the place, and is sometimes referred to as relative location. In other words, where is this place relative to other places, and how is it connected to its surroundings via transportation networks? New Orleans provides an excellent example of site versus situation. The site of New Orleans is not ideal for a city, as it lies below sea level and is prone to flooding. However, the situation of New Orleans is much better in that New Orleans is connected to large portion of the Mississippi River's network of navigable waterways while also being close to the Gulf of Mexico and convenient to coastal traffic. Hence, the situation of New Orleans is why the city has not long since been abandoned, despite catastrophic flooding such as during Hurricane Katrina in 2005. As we examine various places around the world, both site and situation are key considerations in determining the "why" of where a place is located.

1.5.1 Regions

While site and situation can help describe a place, a broader view of the world and the connections between places can be derived from the concept of regions. A **region** is an area that shares some sort of common characteristic that binds the area into a whole. Geographers use regions to help one understand the interconnections between places and simplify a complex world. Two major types of regions are formal and functional. **Formal regions** are characterized by homogeneity or uniformity in one or a number of different characteristics. These characteristics can be both human and physical-related, so regions could be defined by climate or vegetation types, in the sense of physical geography, or they could be defined by language or ethnicity, in the sense of human geography. One example of a map that includes formal regions would be a map of the states in the US. In this map, each state could be considered a formal region because each state is governed in a common or unique way, and hence portrays homogeneity (**Figure 1.6**).

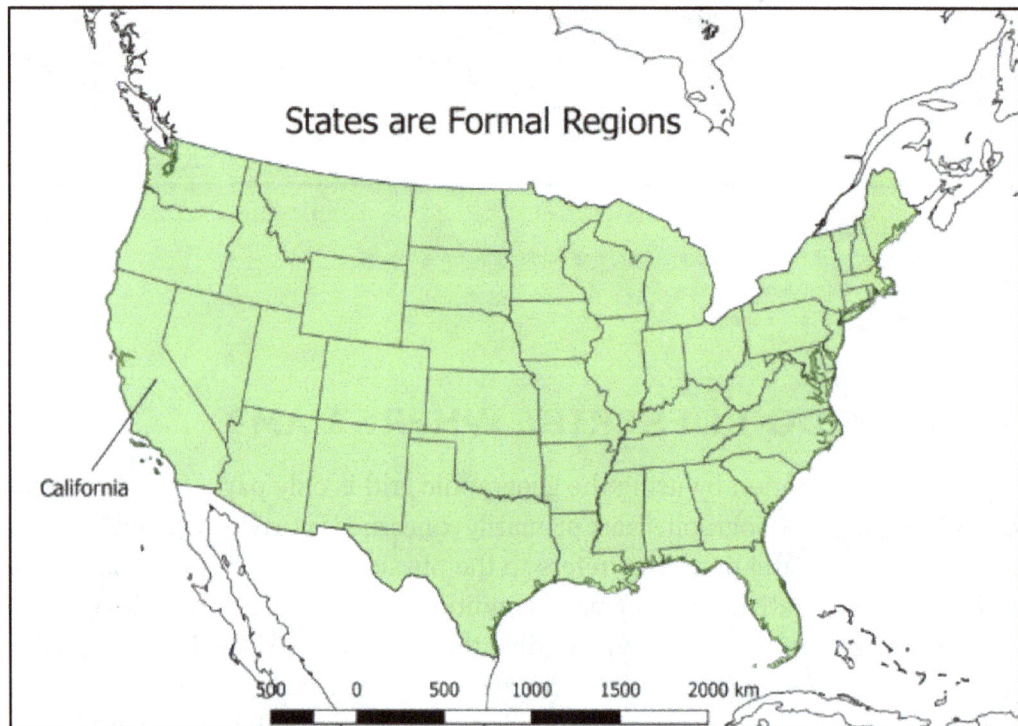

Figure 1.6 | United States
This map shows the formal regions of the states.
Author | David Dorrell
Source | Original Work
License | CC BY SA 4.0

A **functional region,** which is sometimes called a nodal region, is an area that contains a central node or focal point to which other places in the region are connected by some activity. Functional regions can be seen in cities where the central area of the city might serve as the focal point for the rest of the metropolitan area (**Figure 1.7**). At a smaller scale, a Wi-Fi hotspot could be considered the focal point of a functional region that extends to the range of the Wi-Fi signal. Even the delivery area for the local pizza restaurant would be a functional region with the restaurant as a central node.

Regions are devised and not absolute, so whether or not a particular place fits within a region is sometimes a matter of dispute. For example, scholars disagree

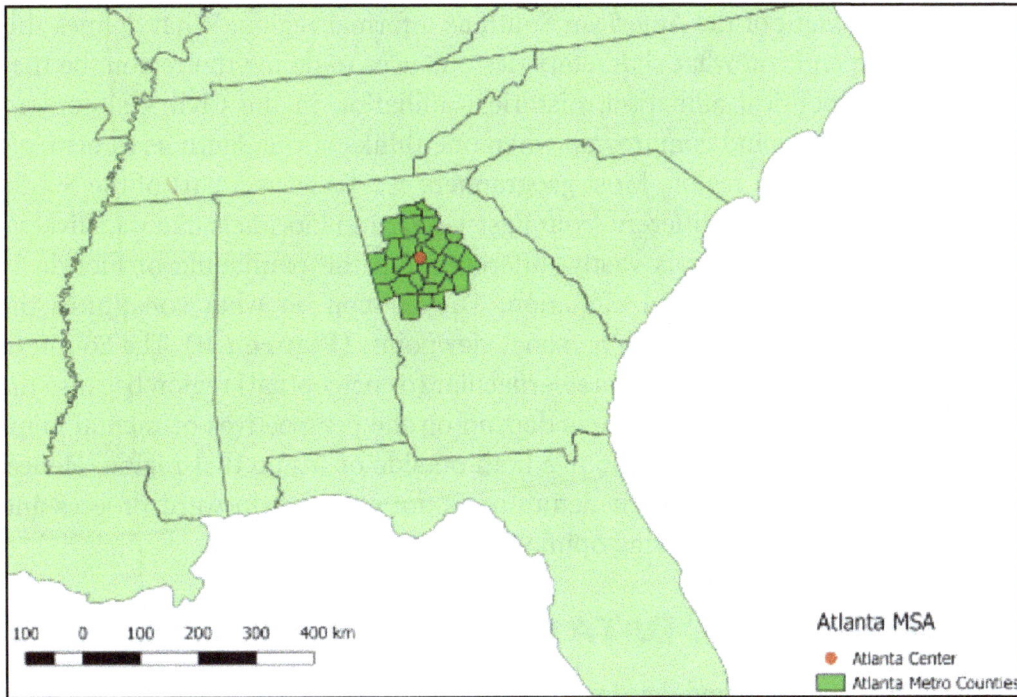

Figure 1.7 | Atlanta MSA
The Atlanta Metropolitan Statistical Area as defined by the United States Census Bureau.
Author | David Dorrell
Source | Original Work
License | CC BY SA 4.0

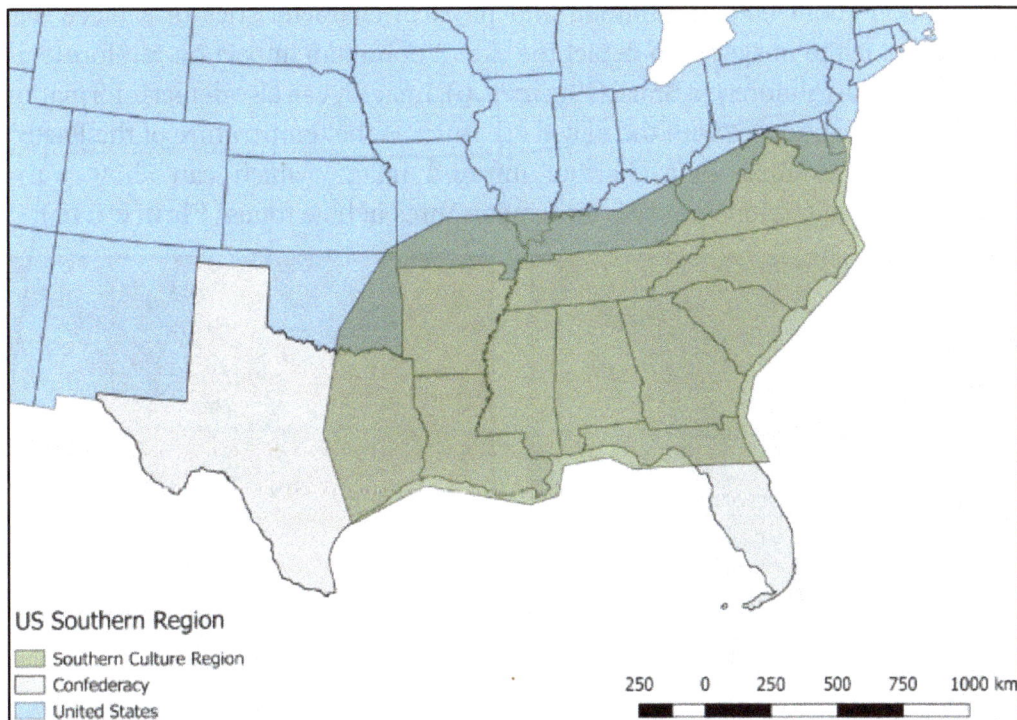

Figure 1.8 | The Fuzzy Boundaries of the American South
It's not exactly the old Confederacy, or the slave states. And it varies from one part to the next.
Author | David Dorrell
Source | Original Work
License | CC BY SA 4.0

on the exact extent of the American South as a formal region. What defines this region? It depends on what characteristics one uses to define the region, be they food, dialect, political affiliation, historical affiliation in the Civil War, or any other element the mind conjures up when one thinks of the South as a relatively homogeneous formal region. Most geographers see Texas as a part of the South, but West Texas is much different from East Texas, and Florida is likewise diverse. Extreme southern Florida is vastly different from the panhandle of Florida in ethnic make-up and political affiliation. The opinion on what constitutes the Southern region varies based on personal viewpoints (**Figure 1.8**). The Southern region, then, may be thought of as a vernacular (or perceptual) region because the boundaries of these types of regions depend on the perspectives or mental maps of different groups of people who live both outside or inside that region. Hence, the concept of regions and their definition is not a straightforward process and involves generalities and varying opinions.

1.6 GEOGRAPHIC DATA COLLECTION AND ANALYSIS

In order to analyze and develop regions, describe places, and conduct detailed geographic analysis, two important tools have been developed that are of particular value to geographers. The first is **remote sensing**, or the acquisition of data about the Earth's surface from aerial platforms such as satellites, airplanes or drones. Images taken from these airborne machines can provide a wealth of valuable information about both the human and physical characteristics of a place. For example, satellite imagery can depict the extent of human impact on rainforests in the Amazonian rainforest of Brazil (**Figure 1.9**). Imagery can also depict information that humans cannot see with the naked eye, such as the temperature of the Earth's surface. One example is a thermal infrared image, which can show warm temperatures in red tones and cooler temperatures in blue tones (**Figure 1.10**).

Figure 1.9 | Deforestation
Deforestation in the state of Rondônia, western Brazil.
Author | NASA
Source | Earth Observatory
License | Public Domain

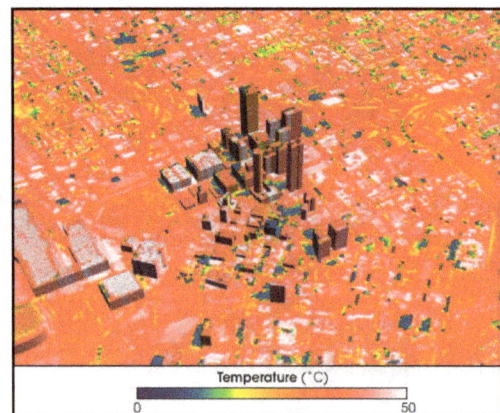

Figure 1.10 | Thermal Imaging
Thermal imagery of Atlanta, GA.
Author | NASA
Source | Wikimedia Commons
License | Public Domain

Digital imagery like the one in Figure 1.9 is in a format that can be entered into **Geographic Information Systems (GIS)**, the second important tool employed by geographers. GIS combines computer hardware and software in a system that stores, analyzes and displays geographic data with a "computer mapping" capability. Geographic data is stored in layers, and these layers of data can be queried in a number of sophisticated ways to analyze some aspect of an area (**Figure 1.11**). Each data point in a GIS is georeferenced to a precise location on the Earth's surface (latitude and longitude, for example), and these data points have different attributes corresponding to the data layer they are associated with. Data layers can represent a myriad of characteristics about that data point, such as elevation, soils, the presence of water, per-capita income, ethnicity, etc. Overlaying the data layers can provide incredible insights into the connections between characteristics/ factors in places, such as the connection between per-capita income and ethnicity or the links between soil types and vegetative cover. GIS also has a vast suite of other capabilities such as least-cost path for transportation, line-of-sight perspectives from a particular location, or 3-D models of urban areas. Because of their multi-faceted capacity to present geographic information, businesses and government agencies around the world use GIS to answer questions, plan development, chart delivery routes, and even monitor crime and first responder activity (**Figure 1.12**). It is not surprising that one of the fastest growing job markets is in GIS technology, as GIS jobs exist at the local, state, and national level as well as in many businesses in the private sector. Even the U.S. Census Bureau maintains an extensive GIS database known as Topologically Integrated Geographic Encoding and Referencing (TIGER).

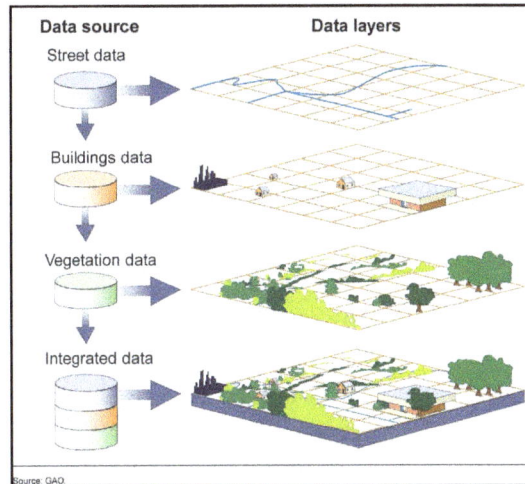

Figure 1.11 | Data Layers
Data layers in a Geographic Information System (GIS).
Author | US Government Accountability Office
Source | National Geographic
License | Public Domain

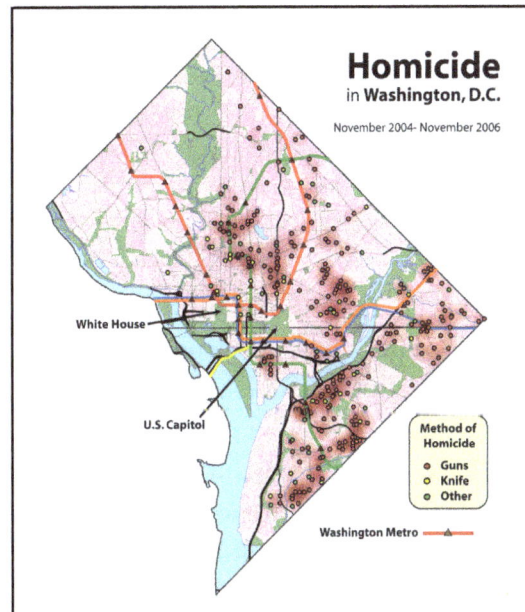

Figure 1.12 | Crime Analysis
A crime analysis of Washington DC.
Author | User "Aude"
Source | Wikimedia Commons
License | CC BY SA 2.5

1.7 CHANGES IN PLACES: DIFFUSION

Thus far, we have examined the Earth in a rather static fashion by learning about places and regions, how maps are created, and how geographic information is gathered and analyzed. However, the Earth is dynamic and constantly changing, and one of the reasons places change is because of diffusion. **Diffusion** is the spread of ideas, objects, inventions, and other practices from place to place. As people migrate or move to a new area, they bring their ideas, objects, and the like with them in a process call **relocation diffusion**. Another diffusion process involves the spread outward from a core area that contains the idea, cultural practice, etc. This type of diffusion is **expansion diffusion**, and this type of diffusion can occur from person-to-person contact (as with a contagious disease) or through a hierarchy, or stratified condition, where the idea might originate in a major city, spread to medium-sized cities, and so on to smaller cities (**Figure 1.13**). In Chapter 6, we will examine how religion has spread across the world through both relocation and expansion diffusion, and in Chapter 10, we will see that domesticated plants and animals have diffused extensively across the Earth.

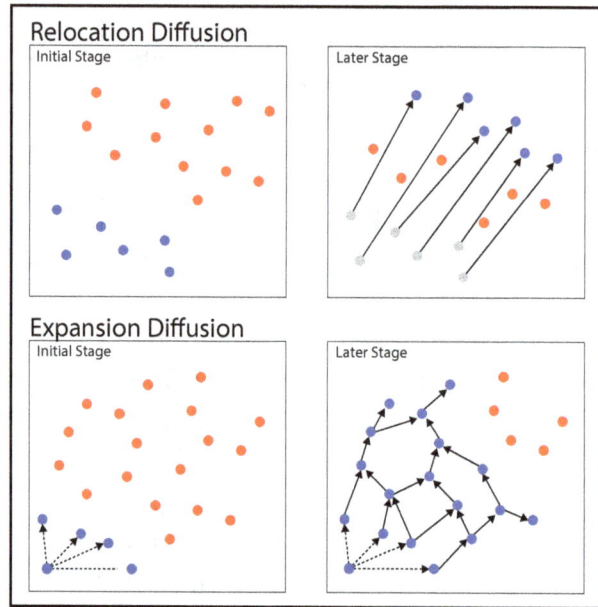

Figure 1.13 | Diffusion
Relocation and expansion diffusion.[2]
Author | Corey Parson
Source | Original Work
License | CC BY SA 4.0

1.8 THE HUMAN-ENVIRONMENT RELATIONSHIP

The process of spatial diffusion can be profoundly affected by the physical terrain, such as is the case with a mountain range. Because migration and transportation over mountain ranges can be limited, diffusion can be slowed or even stopped by these physical barriers. This example is but one instance of the relationship between humans and their environment. The environment can significantly affect human activities, and vice versa, humans can shape and change the Earth's surface and its atmosphere. Two major perspectives on the human-environment relationship in the field of geography are environmental determinism, which has been largely rejected, and possibilism. **Environmental determinism** is the idea that the natural or physical environment shapes and creates cultures; in other words, the environment essentially dictates culture. For example, environmental determinists in the 1920s thought that people who lived in the tropics were slothful and backward because finding food in the tropics was thought to be rather easy. In contrast, Europeans, who lived in "stimulating" climates with a sharp change in seasons were more industrious and inventive. The racist undertones of this sort of perspective is clear, but modern geographers still recognize the definitive impact of the environment on societies, as can be seen, for example, in the theorized demise of the Anasazi people in the American Southwest because of extended drought (**Figure 1.14**).

The Anasazi, who are believed to have inhabited in the Southwestern U.S. from 100 B.C. to 1300 A.D., were ill-equipped to deal with drought, compared to those who inhabit the modern-day states of New Mexico and Arizona. In the view of **possibilism**, people can adapt to their environmental conditions, despite the limitations they might pose, and if a society has better technology, the people are better able to adapt and develop their culture in a number of possible ways. The possibilities are greater, hence, the term possibilism. One excellent example of possibilism is found in Dubai, in the United Arab Emirates. Although snow skiing in the Middle East may seem preposterous, plans are in place to build the longest indoor ski slope in the world in this city, where one ski slope already exists (**Figure 1.15**). Even technological advanced

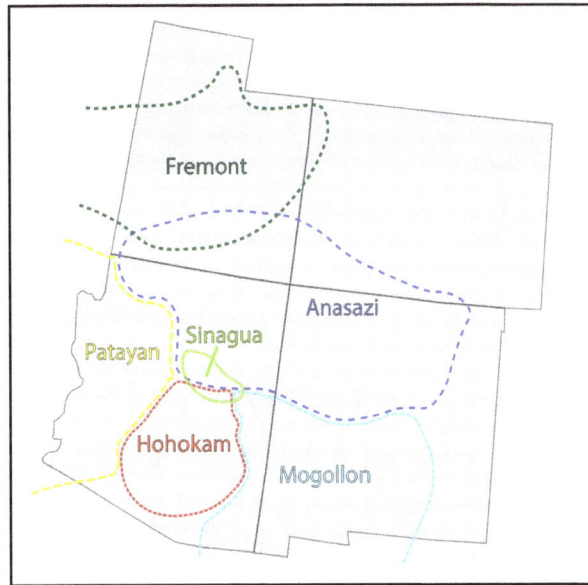

Figure 1.14 | Indigenous People of the Four Corners
Extent of Anasazi and other indigenous people, U.S. Southwest.[3]
Author | Corey Parson, User "Theshibboleth"
Source | Wikimedia Commons
License | CC BY SA 3.0

Figure 1.15 | Ski Dubai
Indoor ski slope at Ski Dubai located in the Mall of the Emirates.
Author | Filipe Fortes
Source | Wikimedia Commons
License | CC BY SA 2.0

societies, however, can still be tremendously affected by the environment and have little or no control over the power of nature. The devastating impact of hurricanes in the United States, tsunamis in Japan, and fire in the United States are but a few examples (**Figures 1.16** and **1.17**).

Figure 1.16 | Colby Fire, California, 2014
Colby Fire in the San Gabriel Mountains foothills.
Author | User "Eeekster"
Source | Wikimedia Commons
License | CC BY 3.0

Figure 1.17 | Tennessee Wildfires, 2016
An aerial shot of the wildfires in Sevier county.
Author | Tennessee National Guard
Source | Flickr
License | CC BY 3.0

1.9 KEY TERMS DEFINED

Diffusion – spread of ideas, objects, inventions, and other practices from place to place.

Environmental determinism – the idea that the natural or physical environment shapes and creates cultures; in other words, the environment essentially dictates culture.

Expansion diffusion – the type of diffusion involves the spread outward from a core area that contains the idea, cultural practice, etc. and can occur from person-to-person contact (as with a contagious disease) or through a hierarchy.

Formal region – a region defined by homogeneity in one or a number of different characteristics.

Functional region – a region that is define by a central node or focal point to which other places in the region are connected.

Geographic information systems – combines computer hardware and software in a system that stores, analyzes and displays geographic data with a "computer mapping" capability in a system of data layers.

Geography – literally, writing about the Earth; the study of the physical and environmental aspects of the world, from a spatial perspective.

International Date Line – roughly follows 180° longitude.

Latitude – the numbering system for parallels.

Longitude – the numbering system for meridians.

Meridian – half circles that connect the North and South poles.

Parallel – circles that encompass the Earth and are parallel to the equator.

Possibilism – the theory people can adapt to their environmental conditions and choose from many alternatives (possibilities), despite the limitations that the environment pose.

Prime Meridian – 0° longitude, passes through Greenwich, England.

Projection – the process of transferring locations from the Earth's curved surface to a flat map.

Region – an area that shares some sort of common characteristic that binds the area into a whole.

Relocation diffusion – the diffusion process in which people migrate or move to a new area, and bring their ideas, objects, and the like with them.

Remote sensing – acquisition of data about the Earth's surface from aerial platforms such as satellites, airplanes, or drones.

Scale – ratio of the length or distance on the map versus the length or distance on the Earth or ground (actual); can also refer to the spatial extent of some phenomenon.

Site – a way to describe a location; refers to the physical characteristics, such as the topography, vegetative cover, climatic conditions, etc.

Situation – a way to describe a location by referring to the area surrounding the place, and is sometimes referred to as relative location.

1.10 WORKS CONSULTED AND FURTHER READING

Bjelland, Mark, Daniel R. Montello, Jerome D. Fellmann, Arthur Getis, and Judith Getis. 2013. Human Geography: Landscapes of Human Activities. 12 edition. New York: McGraw-Hill Education.

Boyle, Mark. 2014. Human Geography: A Concise Introduction. 1 edition. Chichester, West Sussex ; Malden, MA: Wiley-Blackwell.

deBlij, Harm, and Peter O. Muller. 2010. Geography: Regions, Realms, and Concepts. Fourteenth edition. Hoboken: John Wiley & Sons.

Fouberg, Erin H., Alexander B. Murphy, and Harm J. de Blij. 2015. Human Geography: People, Place, and Culture. 11 edition. Hoboken: Wiley.

Knox, Paul L., and Sallie A. Marston. 2015. Human Geography: Places and Regions in Global Context. 7 edition. Boston: Pearson.

Malinowski, Jon, and David H. Kaplan Professor. 2012. Human Geography. 1 edition. New York: McGraw-Hill Education.

National Aeronautics and Space Administration. "Earth Observatory." Accessed August 16. https:earthobservatory.nasa.gov.

National Geographic Society. "Encyclopedia Entries." Accessed August 16. https://www.nationalgeographic.org/encyclopedia.

Rubenstein, James M. 2016. The Cultural Landscape: An Introduction to Human Geography. 12 edition. Boston: Pearson.

1.11 ENDNOTES

1. "Western Illinois University." Liberal Arts Lecture 2011 - University News - Western Illinois University. Accessed April 26, 2018. http://www.wiu.edu/news/lecture_archive/liberalArts2011.php.

2. Fellmann, Jerome D., Arthur Getis, Judith Getis, and Jon C. Malinowski. *Human Geography: Landscapes of Human Activities.* Boston: McGraw Hill Higher Education, 2005.

3. The Anasazi. Accessed April 27, 2018. http://sangres.com/features/anasazi.htm#. WuOBNITyuUl.

2 Population and Health

David Dorrell

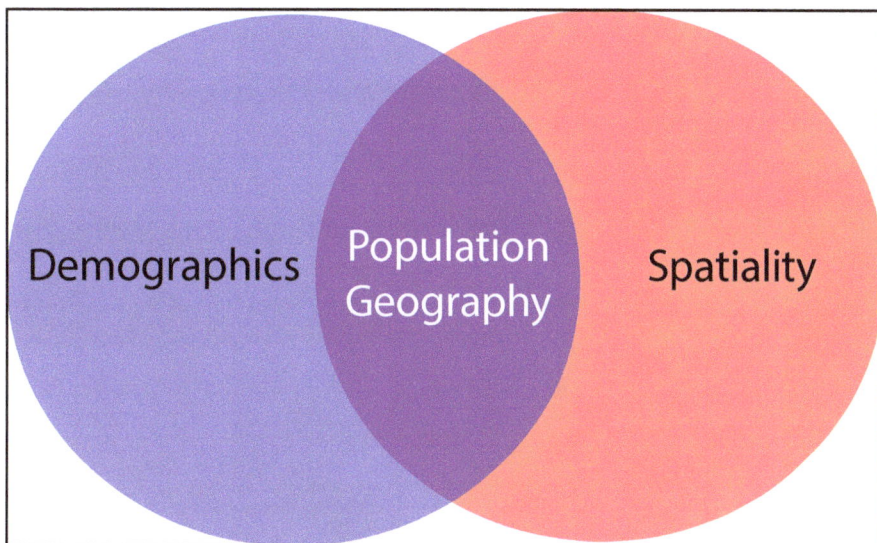

STUDENT LEARNING OUTCOMES

By the end of this section, the student will be able to:

1. Understand: the spatial organization of the human population
2. Explain: the dynamics of population as they are reflected in fertility, morbidity, and mortality
3. Describe: the relationship between population and other spatial phenomena such as living standards, agriculture, and health
4. Connect: development, migration and population as a fluid self-balancing system

CHAPTER OUTLINE

2.1 INTRODUCTION

"You are one in a million, there are 1700 people in China exactly like you."

In this chapter we will look at the human population. We'll look at the size of it, and whether it may be growing or shrinking. We'll explore the role of scale. We'll look at differences between countries. And we'll do all of this through the lens of spatiality.

The human population is at 7.5 billion, an all-time high. In the space of a few centuries it has gone from less than one billion to more than seven, with projections of several billions more in the relatively near future (**Figure 2.2**).

At the global level we can talk about population without consideration of migration, since the earth is a closed system in this regard, but when we discuss countries, it is useful to separate the **natural increase rate**—the rate of population change only accounting for births and deaths—from the effect of migration. A full discussion of migration occurs in Chapter Three. The following map shows a choropleth map of countries of the world categorized by population (**Figure 2.3**). Notice Bangladesh, the small country nearly surrounded by India. Now look at Russia.

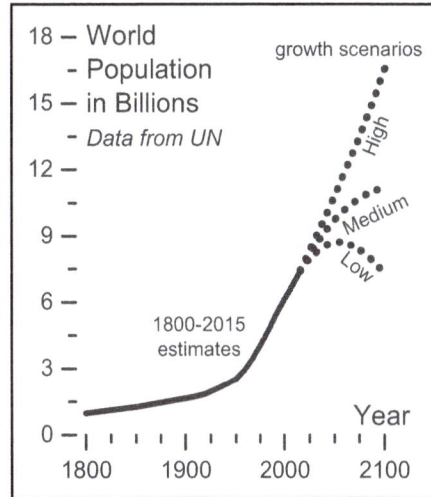

Figure 2.2 | Historical World Population
This graph depicts human population growth from the year and includes three possibilities for the future of human population.[1]
Author | User "Bdm25"
Source | Wikimedia Commons
License | CC BY-SA 4.0

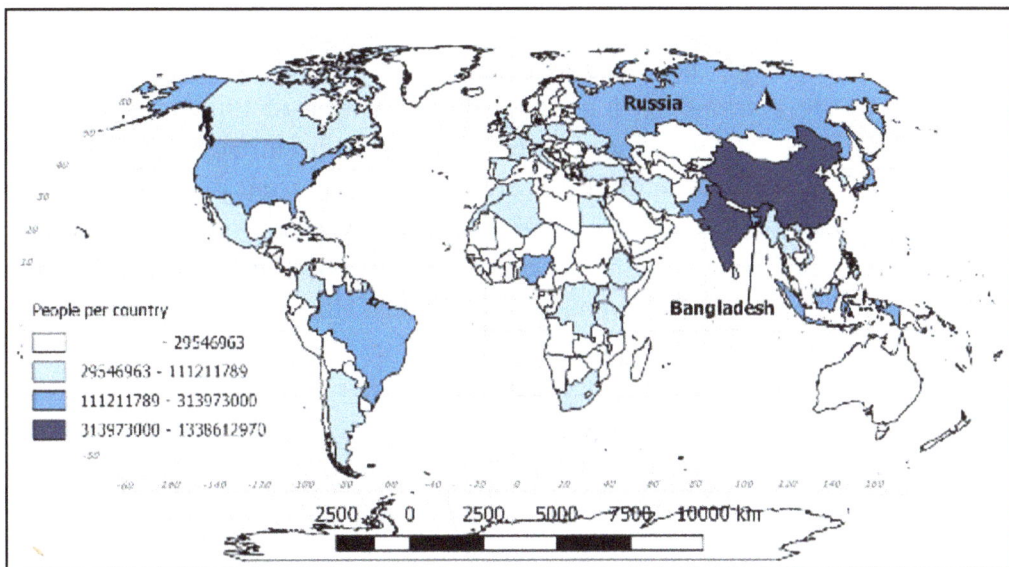

Figure 2.3 | Countries by Population 2015[1]
Author | David Dorrell
Source | Original Work
License | CC BY-SA 4.0

Hold that in your mind while you look at **Figure 2.4**. Although they are vastly different in size, the population of Bangladesh is almost 20 million larger than Russia. Not only that, but the population of Russia is shrinking and the population of Bangladesh is growing!

As **Figure 2.3** showed there are some spatial patterns that present themselves, but there is a great deal of noise in the signal. Many of the countries with large populations are physically large themselves. Places like China and India have had comparably large populations for a long time. Very often, explorations of population growth are short circuited by discussions of religion or levels of development. Although religion and development are not irrelevant, they are not as important as is often assumed. Individual characteristics have come to mean less than they have in the past.

Rank	Country	Population as of July 2016
1	China	1,382,323,332
2	India	1,326,801,576
3	Unites States	324,118,787
4	Indonesia	260,581,100
5	Brazil	209,567,920
6	Pakistan	192,826,502
7	Nigeria	186,987,563
8	Bangladesh	162,910,864
9	Russia	143,439,832
10	Mexico	128,632,004

Figure 2.4 | Top Ten Countries by Population[2]
Author | David Dorrell
Source | Original Work
License | CC BY-SA 4.0

The most obvious characteristic that often leads to higher population growth is poverty. There are many reasons for this, two of which were mentioned previously, but there are others. The effect of infant mortality drives some people to have a large number of children in the forlorn hope that some of them survive to adulthood. Another is the effect of migration, which can boost incomes by sending some population to other countries to work, but depopulate the places that are sending migrants.

In almost all countries, the rate of population growth has slowed. Two countries, China and India, account for 36 percent of the world's population. Any change in these two places will have a large impact on the values for the entire planet. According to the World Bank for 2013, the population of China is growing .5 percent per year, India is growing 1.2 percent per year, the United States is growing by .7 percent per year, and Indonesia is growing by 1.2 percent per year. The rates for all these countries have been falling for decades, even Indonesia and India.

The populations of the countries of Japan, Russia, Germany, Spain, and Ukraine are all shrinking whereas the populations of Nigeria, the Democratic Republic of Congo, Iraq, and Kenya are expanding rapidly (**Figure 2.5**). In developed countries, population decline has implications for social programs such as retirement, which is funded by a shrinking pool of workers. In very advanced societies, a worker shortage is driving rapid development of robotics. In poorer places rapid population growth can trigger large-scale migration and social disruption.

Why is it so difficult to find one characteristic that explains the population dynamics of a particular country? Because places matter. Each place is a unique combination of factors, and their interactions.

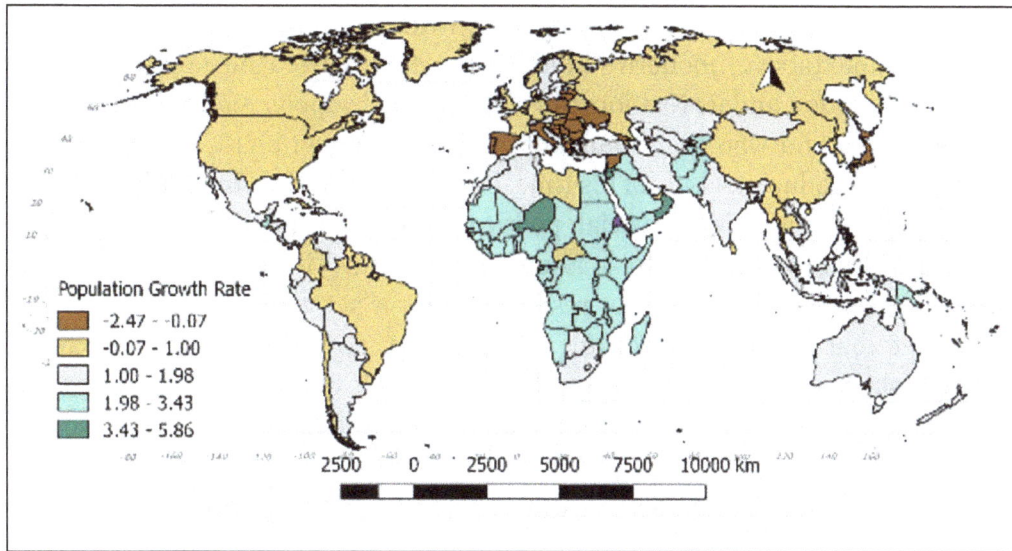

Figure 2.5 | Countries by Population Growth Rate 2015[3]
Author | David Dorrell
Source | Original Work
License | CC BY-SA 4.0

2.2 THINKING ABOUT POPULATION

2.2.1 The Greeks and *Ecumene*

No discussion of population is complete without a brief history of the philosophical understanding of population. This discussion starts as it often does, with the ancient Greeks. The Greeks considered that they lived in the best place on Earth. In fact, they believed in the exact center of the habitable part of the Earth. They called the habitable part of the Earth **ecumene**. To the Greeks, places north of them were too cold, and places to the south were too hot. Placing your own homeland in the center of goodness is common; many groups have done this. The Greeks decided that the environment explained the distribution of people. To an extent, their thinking persists, but only at the most extreme definitions. Many places that the Greeks would have found too cold (Moscow, Stockholm) too hot (Kuwait City, Las Vegas) too wet (Manaus, Singapore) or too dry (Timbuktu, Lima) have very large populations.

2.2.2 Modern Ideas About Population

Thomas Malthus (1766-1834) *"Population, when unchecked, increases in a geometrical ratio."* [4]

Ester Boserup (1910-1999) *"The power of ingenuity would always outmatch that of demand."* [5]

Modern discussions of population begin with food. From the time of Thomas Malthus (quoted above), modern humans have acknowledged the rapidly expanding human population and its relationship with the food supply. Malthus himself was a cleric in England who spent much of his time studying political economy. His views were a product not only of his time, but also of his place. In Malthus' case his time and place were a time of social, political, and economic change.

Karl Marx (1818–1883) took issue with Malthus' ideas. Marx wrote that population growth alone was not responsible for a population's inability to feed itself, but that imbalanced social, political, and economic structures created artificial shortages. He also believed that growing populations reinforced the power of capitalists, since large pools of underemployed laborers could more easily be exploited.

The post-World War II period saw a flurry of books warning of the dangers of population growth with books like Fairfield Osborn's *Our Plundered Planet* and William Vogt's *Road to Survival*. Perhaps most explicit was Paul and Anne Ehrlich's *The Population Bomb*.

These books are warnings of the dangers of unchecked population growth. Malthus wrote that populations tend to grow faster than the expansion of food production and that populations will grow until they outstrip their food supplies. This is to say that starvation, war, and disease were all predictions of Malthus and were revived in these Neo-Malthusian publications. Some part of the current conversation of environmentalism regards limiting the growth of the human population, echoing Malthus.

A common theme of these books is that they all attempt to predict the future. One of the advantages we have living centuries or decades after these books is the opportunity to see if these predictions were accurate or not. Ehrlich's book predicted that by the 1970s, starvation would be widespread because of food shortages and a collapse in food production. That did not happen. In fact, the global disasters predicted in all these books have yet to arrive decades later. What saved us?

Perhaps nothing has saved us. We have just managed to push the reckoning a bit further down the road.

If we have been saved, then the assumptions inherent in the predictions were wrong. What were they?

1. Humans would not voluntarily limit their reproduction.

2. Farming technology would suddenly stop advancing.

3. Food distribution systems would not improve.

4. Land would become unusable from overuse.

All of these assumptions have proven to be wrong, at least so far. Only the most negative interpretation of any particular factor in this equation could be accepted.

One the other hand, Ester Boserup, an agricultural economist in the twentieth century, drew nearly the opposite conclusion from her study of human population. Her reason for doing so were manifold. First, she was born one and one -half centuries later, which gave her considerably more data to interpret. Second, she didn't grow to adulthood in the center of a burgeoning empire. She was a functionary in the early days of the United Nations. Third, she was a trained as an economist, and finally, she was a woman. Each one of these factors was important.

2.2.3 Let's Investigate Each One of These Assumptions

In preindustrial societies, children are a workforce and a retirement plan. Families can try to use large numbers of children to improve their economic prospects. Children are literally an economic asset. Birth rates fall when societies industrialize. They fall dramatically when women enter the paid workforce. Children in industrialized societies are generally not working and are not economic assets. The focus in such societies tends to be preparing children through education for a technologically-skilled livelihood. Developed societies tend to care for their elderly population, decreasing the need for a large family. Developed societies also have lower rates of infant mortality, meaning that more children survive to adulthood.

The increasing social power of women factors into this. Women who control their own lives rarely choose to have large numbers of children. Related to this, the invention and distribution of birth control technologies has reduced human numbers in places where it is available.

Farming technology has increased tremendously. More food is now produced on less land than was farmed a century ago. Some of these increases are due to manipulations of the food itself—more productive seeds and pesticides, but some part of this is due to improvements in food processing and distribution. Just think of the advantages that refrigeration, freezing, canning and dehydrating have given us. Add to that the ability to move food tremendous distances at relatively low cost. Somehow, during the time that all these technologies were becoming available, Neo-Malthusians were discounting them.

Some marginal land has become unusable, either through desertification or erosion, but this land was not particularly productive anyway, hence the term marginal. The loss of this land has been more than compensated by improved production.

At this point, it looks like a win for Boserup, but maybe it isn't. Up to this point we have been mixing our discussions of scale. Malthus was largely writing about the British Isles, and Boserup was really writing about the developed countries of the world. The local realities can be much more complex.

At the global scale there is enough food, and that has been true for decades. In fact, many developed societies produce more food than they can either consume or sell. The local situation is completely different. There are developed countries that have been unable to grow food to feed themselves for over a century. The United Kingdom, Malthus' home, is one of them. However, no one ever calls the U.K. overpopulated. Why not? Because they can buy food on the world market.

Local-scale famines happen because poorer places cannot produce enough food for themselves and then cannot or will not buy food from other places. Places that are politically marginalized within a country can also experience famines when central governments choose not to mobilize resources toward the disfavored. Politically unstable places may not even have the necessary infrastructure to deliver free food from other parts of the world. This is assuming that food aid is even a good idea (a concept revisited in the agriculture chapter). These sorts of problems persist to this day and they have an impact of population, although often in unpredictable ways, such as triggering large-scale migration or armed conflict.

To recap, at the global level, population has not been limited by food production. However, people do not live at the "global level." They live locally with whatever circumstances they may have. In many places the realities of food insecurity are paramount.

Although discussions of population tend to start with food, they cannot end with it. People have more needs than their immediate nutrition. They need clothing and shelter as well. They also have desires for a high standard of living- heating, electricity, automobiles and technology. All of these needs and desires require energy and materials. The pressure put on the planet over the last two centuries has less to do with the burgeoning population and more to do with burgeoning expectations of quality of life.

2.2.4 Scale and the Ecological Fallacy

Numbers can be a little bit misleading. You may read that the "average woman" in the United Sates has 1.86 births and wonder, "What does this tell me about a particular woman?" And the answer is . . . it tells you nothing. Remember that number is an aggregate of the data for the entire country, which means it only works at that scale; it only tells you about the country as a whole. The **Ecological Fallacy** is the idea that statistics generated at one level of aggregation can be applied at other levels of aggregation.

Similarly, one of the biggest problems with maps is that they can make us think that a place is the same (homogenous) within a border. We see a country like the United States, and it has one color for the entire area on the map and we tell ourselves that the U.S. is just one place. And it is. But it's made of many smaller places. It's fifty states, and those states combined have 3144 counties (and county-like things). And each one of those things is a level of aggregation. It looks a bit like this (**Figure 2.6**).

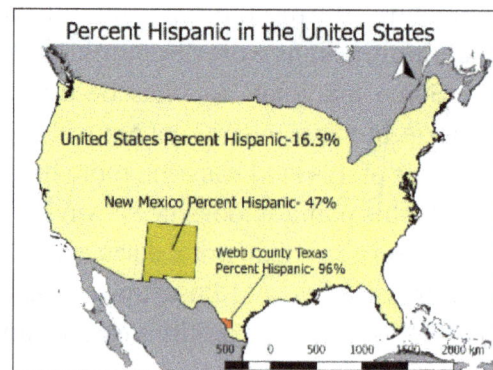

Figure 2.6 | Percent Hispanic aggregated to the County, State, and National level[6]
Author | David Dorrell
Source | Original Work
License | CC BY-SA 4.0

Webb County only uses the data from one county. New Mexico uses the data for its 33 counties, and the U.S. uses the data for all its counties. Each level of aggregation has its calculated value. They are all different. And they still tell you nothing about an individual person.

2.3 POPULATION AND DEVELOPMENT

We can return to the diverging ideas of Malthus and Boserup. Does population growth spur innovation or starvation? Is population growth good or bad?

Population growth has spurred innovation, but interestingly enough, not usually in the places that are experiencing that growth right now. If population growth is dramatically higher than economic growth, the result will not be new technologies and paradigm shifts; it will be emigration or civil unrest.

A country like Russia with a large land mass and a relatively small population may be underpopulated compared to its neighbor, China (**Figure 2.7**). Having large tracts of uninhabited land has historically invited the attention of outsiders. Underpopulation is not normally the problem that most people would associate with population. When we think of population problems, we generally think of overpopulation. What is overpopulation? Like so many other questions we have asked so far, the answer depends. **Overpopulation** means an inability to support a population with the resources available.

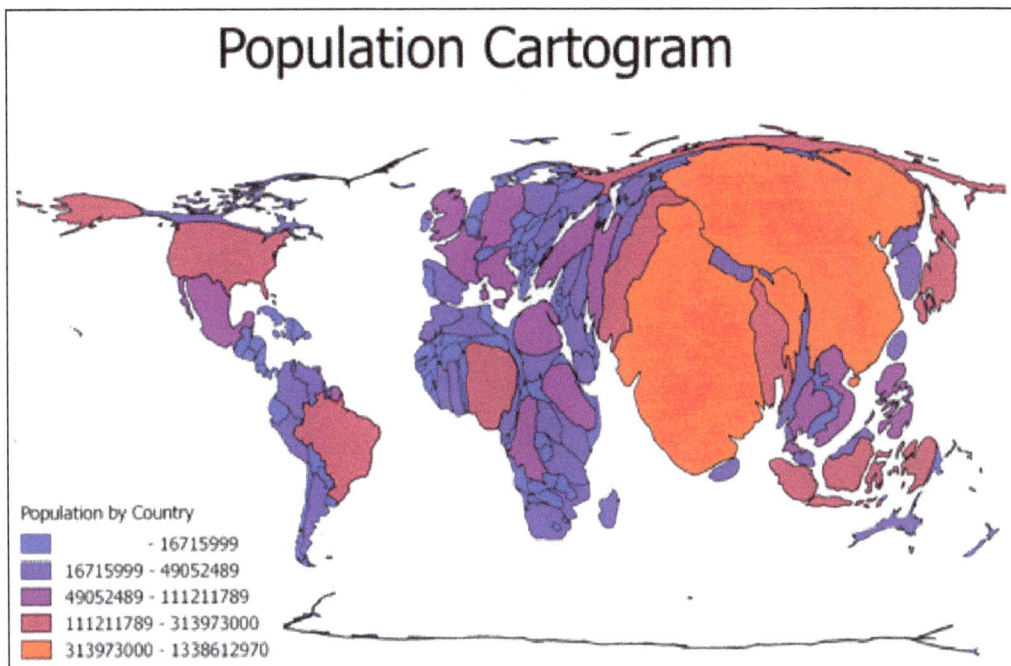

Population Cartogram

Population by Country
- - 16715999
- 16715999 - 49052489
- 49052489 - 111211789
- 111211789 - 313973000
- 313973000 - 1338612970

Figure 2.7 | Population Cartogram[7]
This population cartogram takes the borders of each country and adjusts the size of the country by the size of the population. What happens to China and India? What happens to Canada and Russia? What does this tell you?
Author | David Dorrell
Source | Original Work
License | CC BY-SA 4.0

2.4 POPULATION IS DYNAMIC

Although the world's population is still growing, the overall growth has slowed and the growth has become very uneven. Some places are still growing very rapidly. Others are growing much more slowly and some are shrinking in terms of population. We can compare differences between places using a series of different rates. Rates are ratios that divide the occurrence of a phenomenon with the population hosting the phenomenon. For example, **crude birth rate** is calculated as the number of births per 1000 people in a particular place in a particular year. The **crude death rate** is similar. It is calculated as deaths per 1000 people in a particular place in a particular year. These numbers can be used to compare places, but they have great limitations.

For one thing, they are aggregated variables describing the entire country; they tell you nothing about individuals within the country. Just because the average woman in a country has 2.3 children tells you nothing about a particular woman. The ecological fallacy is the idea that aggregated data tell you anything about individuals. It does not because it cannot. Once the data are lumped together, they lose their individual characteristics.

The rates do not necessarily relate to one another, either. The crude birth rate doesn't tell you anything about the average number of children born per woman, at what stage in their lives women tend to have children, etc. The crude death rates don't separate deaths of elderly people from deaths of infants. High rates of death are often found in developed societies; in many ways it's a sign of development, since developed countries tend to have older populations. A high rate of infant mortality (children under 1 year of age) is a near-universal sign of underdevelopment. The countries with the highest crude birth rates tend to be low income (**Figure 2.8**).

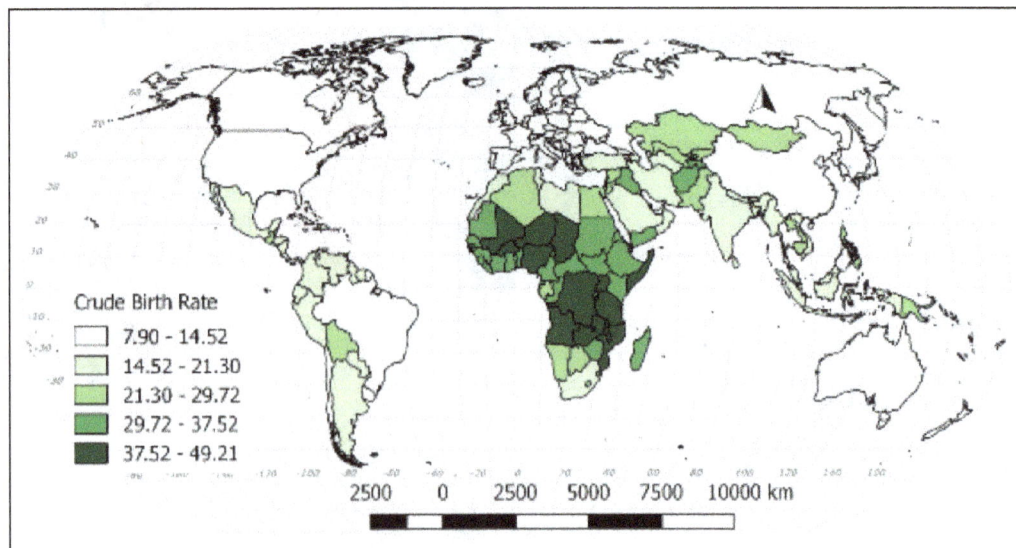

Figure 2.8 | Crude Birth Rate 2015[8]
Author | David Dorrell
Source | Original Work
License | CC BY-SA 4.0

The crude death rate is more nuanced (**Figure 2.9**). Some countries on the map, for example, Chad, have a high crude death rate due to a high rate of infant and child mortality. Russia, on the other hand, has a rapidly aging population and a partially collapsed social security network.

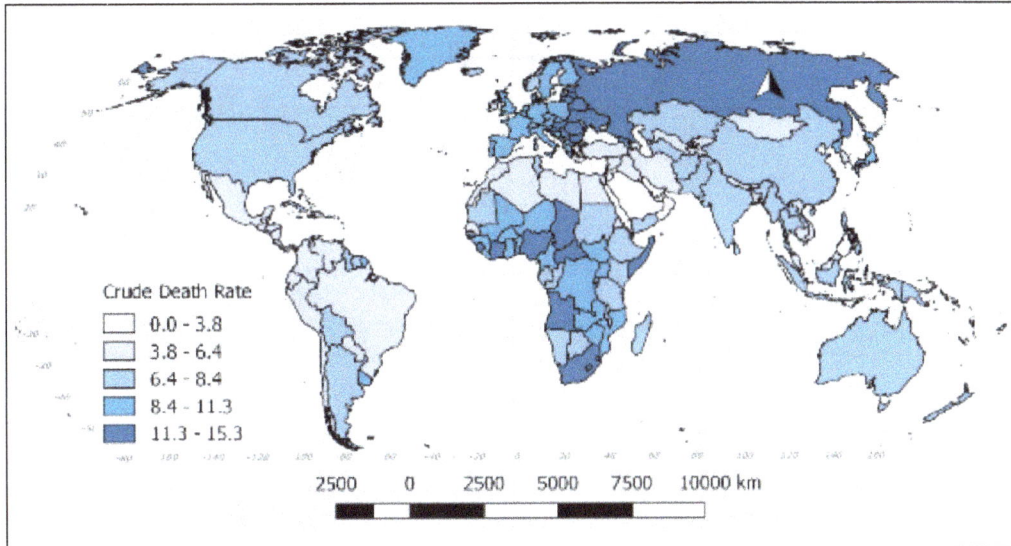

Figure 2.9 | Crude Death Rate 2015[9]
Author | David Dorrell
Source | Original Work
License | CC BY-SA 4.0

The **replacement level** of a population refers to the number of births that are necessary to offset deaths. This is often referred to in terms of average fertility of women. In modern societies, on average, women need to produce 2.1 children in a lifetime to keep a place demographically stable. This number is derived by counting a mother and her partner, and accounts for those who never reproduce.

Places with a fertility rate below 2.1 will shrink over time. Those places above that will grow, and those well above that will grow quickly (**Figure 2.10**). The preceding graphic demonstrates that in the same way that many places are growing very rapidly, many places are at or below replacement. The United States is below replacement. It is demographically buoyed by immigration.

What kinds of places are growing fastest? These are places that are poor or economically or politically unstable. This may seem counterintuitive. Why would people have children in places that are already so poor? Remember that individual families have children. Children can seem like a mechanism for surviving bad situations. What kinds of places are declining? This is more complicated, but in general the more educated and empowered the female population is, the lower the birth rate. This isn't a perfect, linear relationship, but it's useful as a start. Why is there so much variability? Because places matter. Uganda has a government advocating population growth, while Afghanistan and Somalia have little governance at all. Russia and Australia have similar fertility rates, and few other similarities.

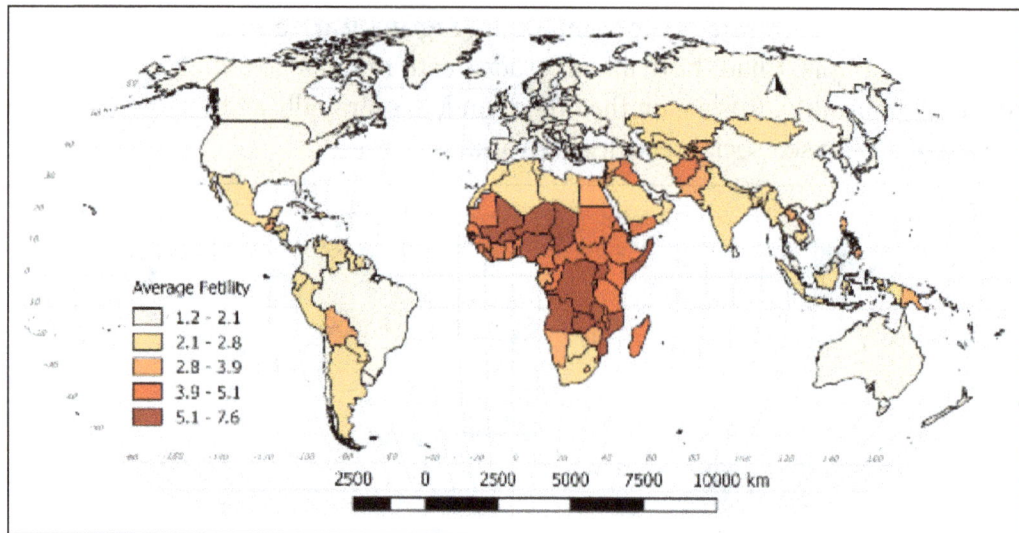

Figure 2.10 | Countries by Average Fertility 2015[10]
Author | David Dorrell
Source | Original Work
License | CC BY-SA 4.0

General measurements of population are useful, but often if is useful to know the age and gender structure of a population. This is shown using a population pyramid. A **population pyramid** breaks the population into groups sorted by age ranges, called **cohorts**, as well as by sex. The resulting shape tells you a great deal about the population dynamic of the country. If the shape is actually a pyramid, then the country has a high birth rate and a high death rate. Countries with stable populations look like a column. Some countries even have their greatest population in the older cohorts with comparably few young people. **Figure 2.11** shows some examples of current populations. Some examples of estimated future populations are shown in **Figure 2.12**.

First, it is important that you understand that all these numbers are estimates. Current population numbers are good enough for general comparison. As you can see, the differences between places becomes more pronounced as we look toward the future. The world population increases by over two billion people, but what is interesting are the shifting dynamics between countries. China shrinks by more than 30 million, the US grows by 64 million, and Niger- a poor Saharan country- grows by 52 million, more than double its current size! One of the most useful measures of population is **doubling time**, which is how much time it would take at current levels of population growth for a population to double. According to **Figure 2.2**, most of human history saw very slow growth with doubling times measured in centuries. During the 19th and 20th centuries, doubling time at the global level fell to as short as 35 years.

This has numerous implications. At the global level, those extra billions will need food and water, houses and clothing. That is to say that they will require resources. They will also have desires that require even more materials and energy expenditure. In places like Niger, this will be very difficult, if not impossible, to meet.

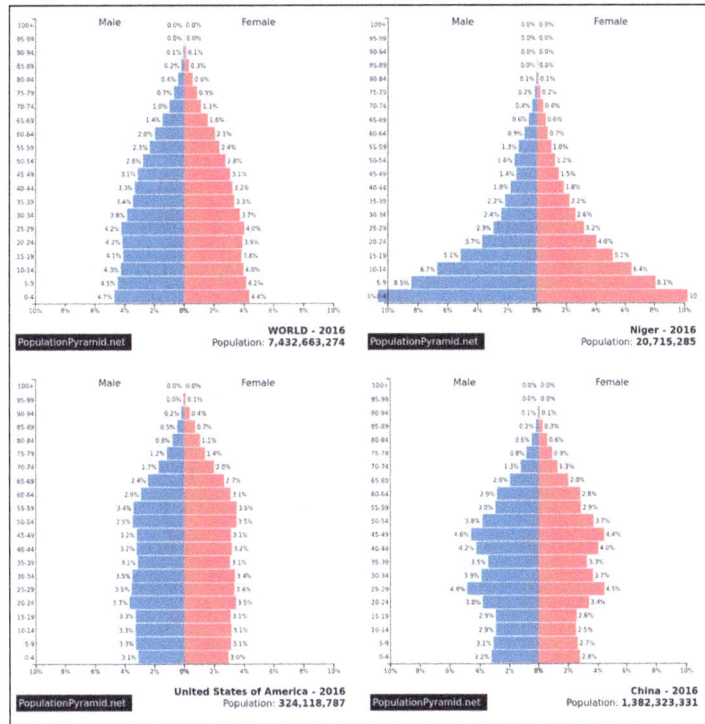

Figure 2.11 | Population Pyramids for select countries 2016
Authors | Martin de Wulf, David Dorrell
Source | PopulationPyramid.net
License | MIT License

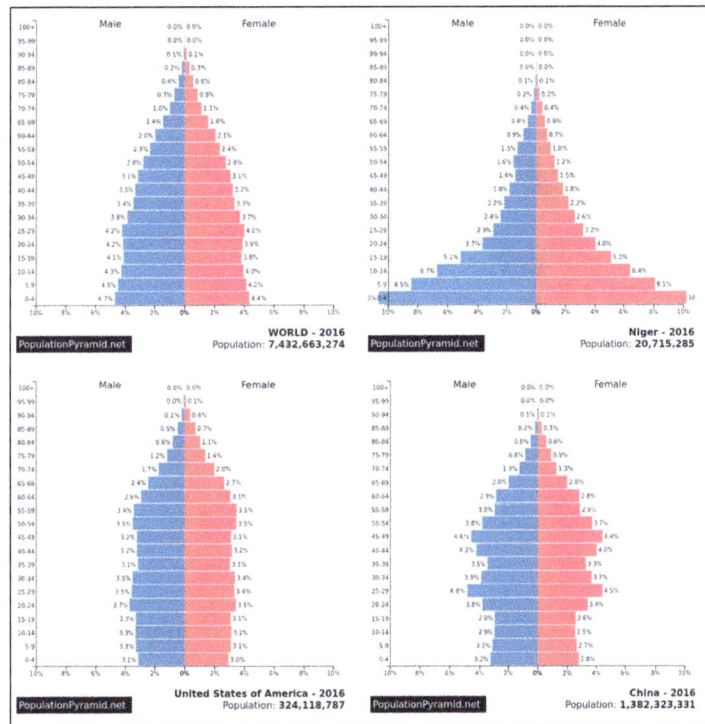

Figure 2.12 | Population Pyramids for select countries 2050
Authors | Martin de Wulf, David Dorrell
Source | PopulationPyramid.net
License | MIT License

In places like China, a completely different problem presents itself. These populations are both shrinking and aging. **Life expectancy**, the average lifespan in a country, has been increasing for decades in developed and some developing countries. At the same time, birth rates have fallen for a variety of reasons. This means that as time passes, the elderly portion of the population has grown. Many countries will see their populations age until large percentages will be unable to work. Societies for the past several centuries have prepared themselves for population growth, and much of modern society is predicated on it. Population growth is what has paid for social security for the elderly. Few places have prepared themselves for fewer workers in the future (although robotics may address this problem). This change will not happen at once, but the effects will be tremendous.

Another characteristic that must be acknowledged is **population momentum**. When much of your population is older than 45, it isn't reasonable to expect that population will continue to grow quickly. Countries with young populations should expect that their populations will grow when the large pool of young people have children of their own.

This map shows the tremendous differences in life expectancy from one country to another (**Figure 2.13**). In some places, people tend to live into their ninth decade. In others, they are unlikely to make it into their sixth.

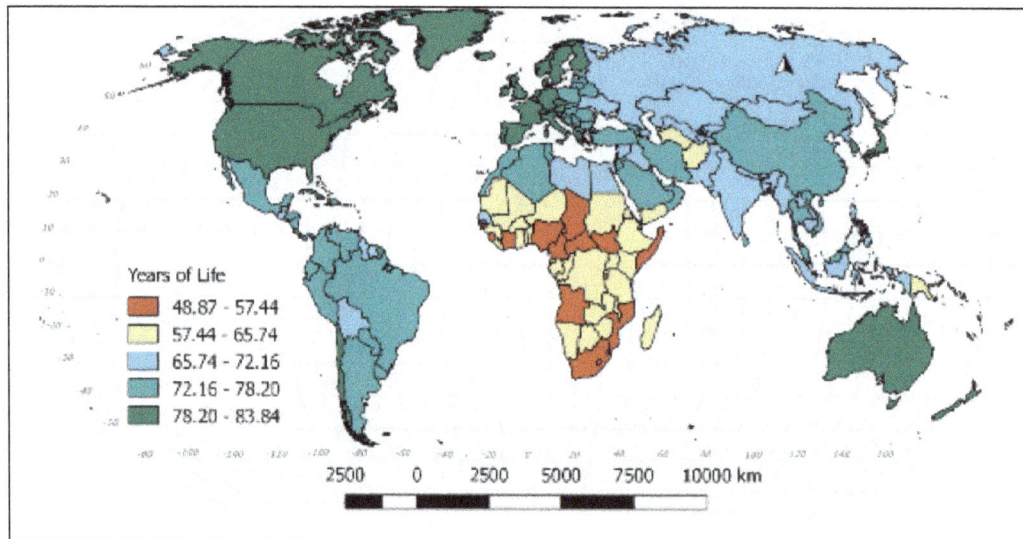

Figure 2.13 | Life Expectancy 2015[11]
Author | David Dorrell
Source | Original Work
License | CC BY SA 4.0

The **dependency ratio** is simply the number of people within a society who do not work compared to the number who do work. There are two main components of the dependency ratio- children under 15 years old and the elderly over 65 years old, although the degree to which either group is dependent is variable (**Figures 2.14** and **2.15**). Children need care and schooling, but generally produce little of economic value. Elderly populations can be too infirm to work and are the part of

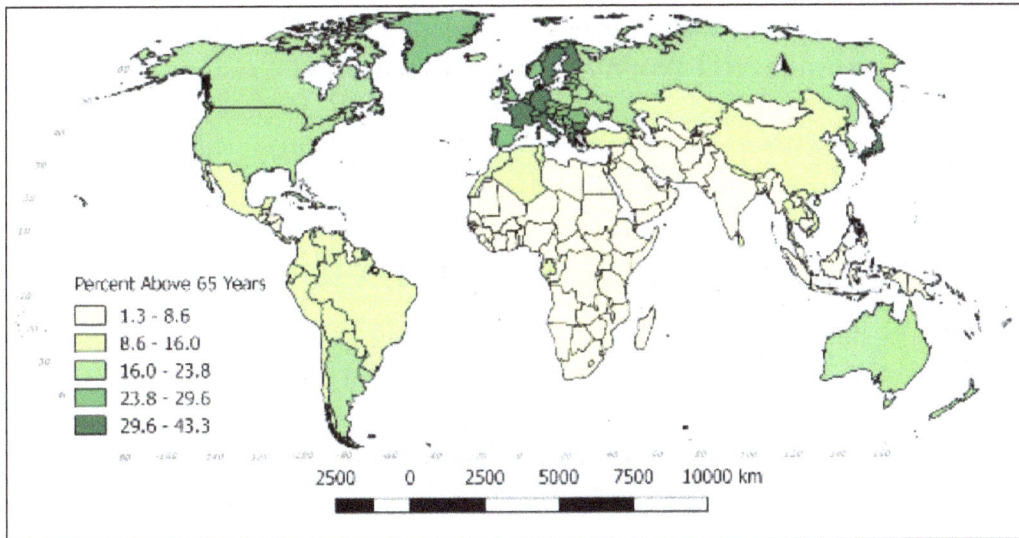

Figure 2.14 | Elderly Dependency Ratio 2015[12]
Author | David Dorrell
Source | Original Work
License | CC BY SA 4.0

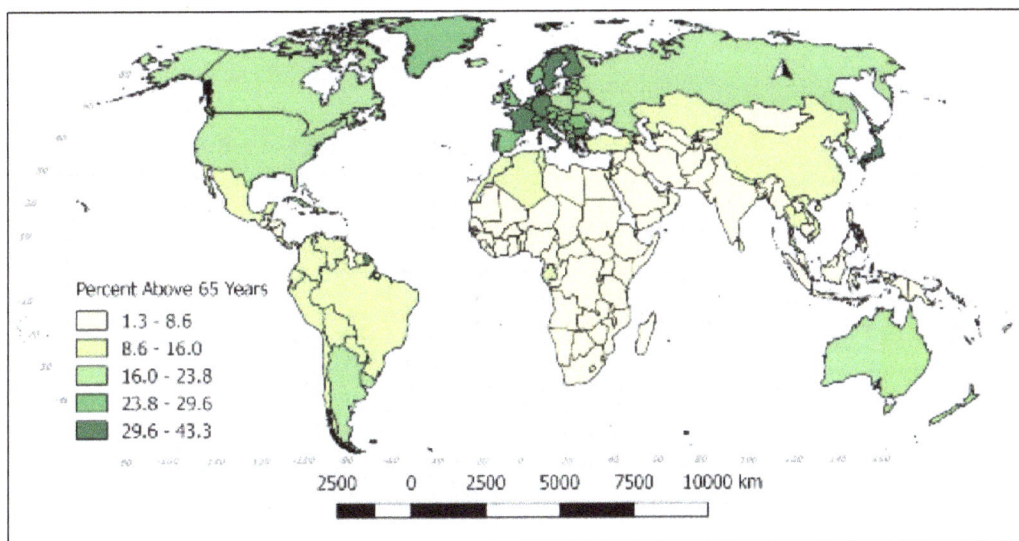

Figure 2.15 | Youth Dependency Ratio 2015[13]
Author | David Dorrell
Source | Original Work
License | CC BY SA 4.0

the population with the highest medical costs. Where does the wealth come from to take care of these two groups? It comes from the people working and producing wealth. If the dependency ratio is high, then each worker can be responsible for a large number of dependents and less wealth will be left for the workers.

Although the dependency ratio is used to compare places, this particular ratio can be somewhat misleading. In many less developed places, children are not dependent. They are not in school and they are employed. They are not consuming a family's resources, but are instead contributing to them. In other places, the elderly may still be in the workplace. The dependency ratio informs decisions regarding the future. Will a country need more schools or assisted living centers. What does this mean for retirement or pensions? Perhaps more importantly, will the supply of workers increase or decrease? Comparing China, Niger, and the U.S. shows you that different places have different options and challenges.

2.5 THE DEMOGRAPHIC TRANSITION

Geographers have modeled the population dynamics of places for decades. The result of these models is called the **Demographic Transition** model (**Figure 2.16**). It describes a series of stages that societies pass through as they develop and industrialize. These models represent the general demographic conditions that countries experience.

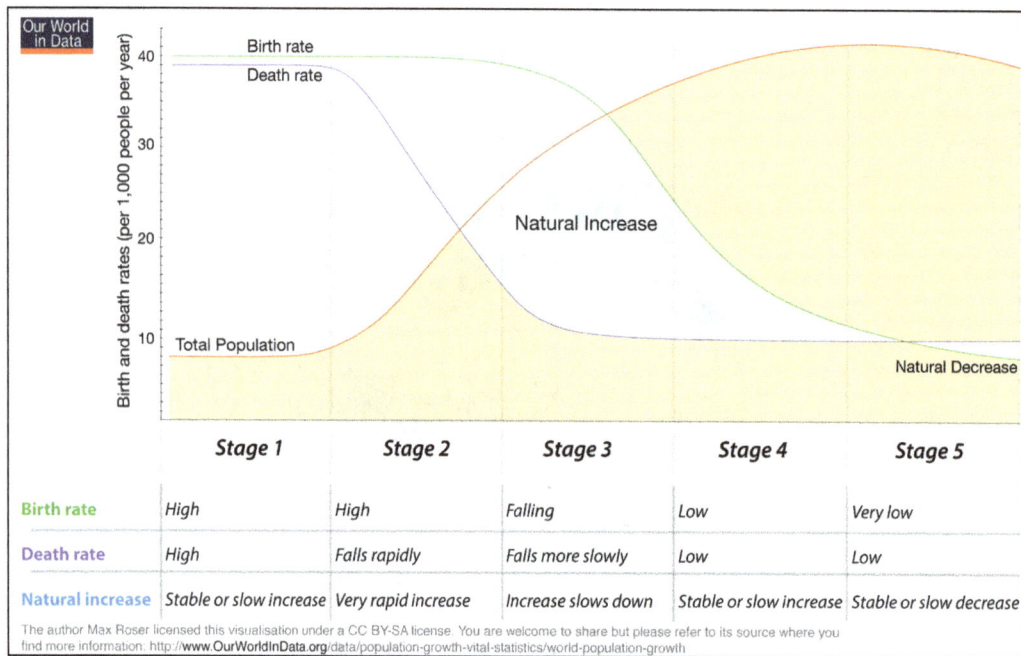

	Stage 1	Stage 2	Stage 3	Stage 4	Stage 5
Birth rate	High	High	Falling	Low	Very low
Death rate	High	Falls rapidly	Falls more slowly	Low	Low
Natural increase	Stable or slow increase	Very rapid increase	Increase slows down	Stable or slow increase	Stable or slow decrease

The author Max Roser licensed this visualisation under a CC BY-SA license. You are welcome to share but please refer to its source where you find more information: http://www.OurWorldInData.org/data/population-growth-vital-statistics/world-population-growth

Figure 2.16 | Demographic Transition Model
Author | Max Roser
Source | Wikimedia Commons
License | CC BY SA 4.0

In **Stage One**, which is pre-modern, birth rates are high, and death rates are high. As a result of these two factors, the population remains low, but stable.

By **Stage Two** as the society industrializes food becomes more steadily available, water supplies get cleaner and sanitation and medical care improves. Death rates fall, particularly infant mortality rates. However, there is a lag in the decline in birth rates. It takes time for people to adjust to the new reality of urban industrial life. Since the birth rate is still high and the death rate is low, population grows very rapidly.

Stage Three is characterized by a falling birth rate as the society begins to find an equilibrium between birth and deaths. The birth rates are still higher than death rates and population continues to increase. This continuation of population increase is known as population momentum. The end of Stage Three is characterized by the general balance between births and deaths.

Stage Four shows a return to population stability, but at a much greater number of people.

2.6 MEASURING THE IMPACT OF POPULATION

Remember the earlier comparison of Russia and Bangladesh? This is the section where we discuss the different ways of calculating the pressure that populations put onto the land that they inhabit. You'll recall that we began by looking simply at people per country. This is a good way to start, but the limitations are fairly obvious. Countries that are physically larger can hold more people. We need to use a method that changes from a measure of overall population to some kind of per capita measure. There are many of these and each has its merits.

Arithmetic density is the simplest one. It is simply the number of people divided by the area of the country. The area is usually measured in square kilometers, since most of the world uses the metric system (**Figure 2.17**).

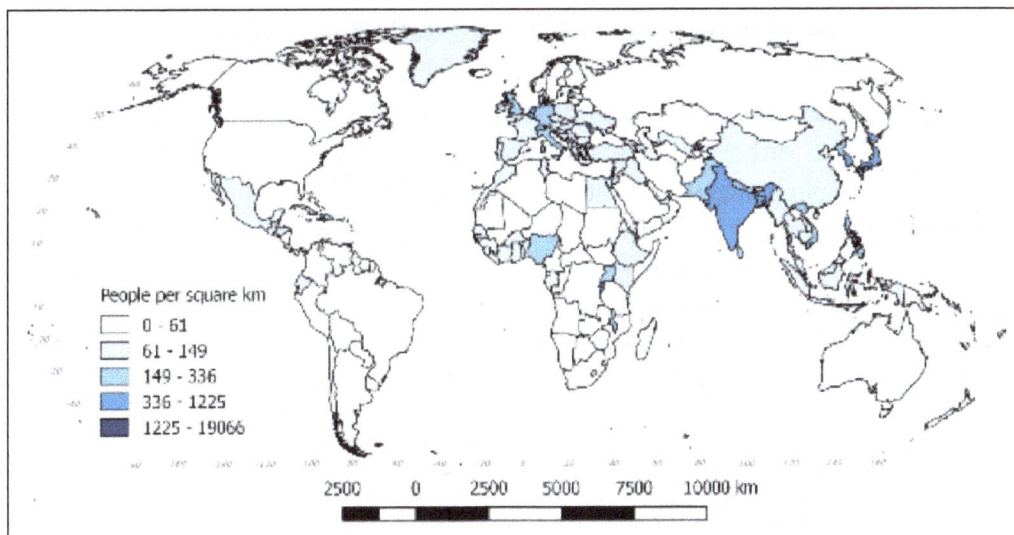

Figure 2.17 | Arithmetic Density 2015[14]
Author | David Dorrell
Source | Original Work
License | CC BY SA 4.0

Physiological density has the same numerator (population), but the denominator is different. Instead of using all the land in a country, it only accounts for arable (farmable) land (**Figure 2.18**). Places that are not used for agriculture- deserts, lakes, mountaintops and similar places - are subtracted from the land total. This is useful for demonstrating how much pressure is being put on the farmland that is available. Be aware that food that is gathered or hunted from nonagricultural land is not considered in this number.

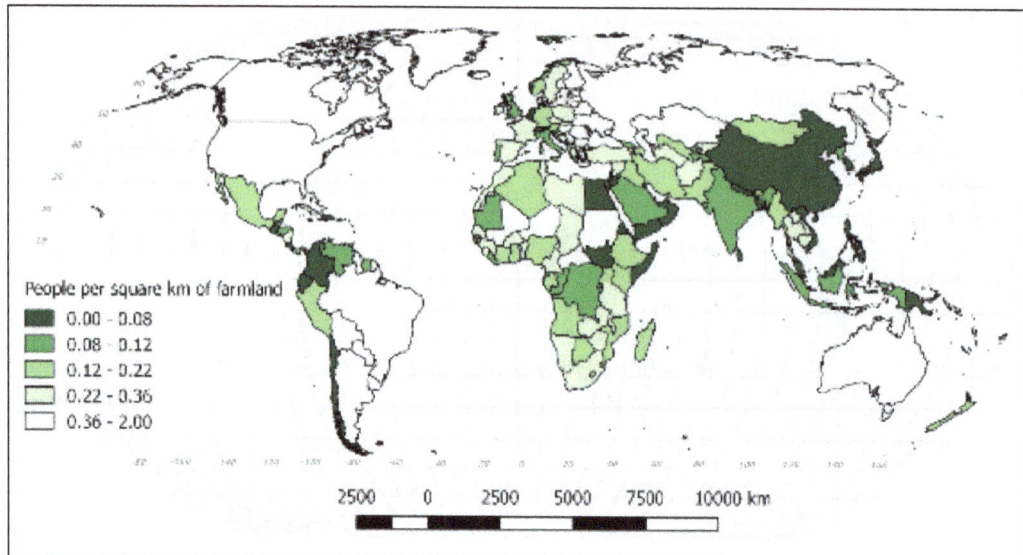

Figure 2.18 | Physiological Density 2015[15]
Author | David Dorrell
Source | Original Work
License | CC BY SA 4.0

Agricultural density has the same denominator as physiological density, but has a different numerator. Instead of using the entire population, it only uses farmers (**Figure 2.19**). This provides a number that is a good measure of development, or rather it's a good measure of underdevelopment. Developed countries have mechanized agriculture and few farmers per capita. Each farm tends

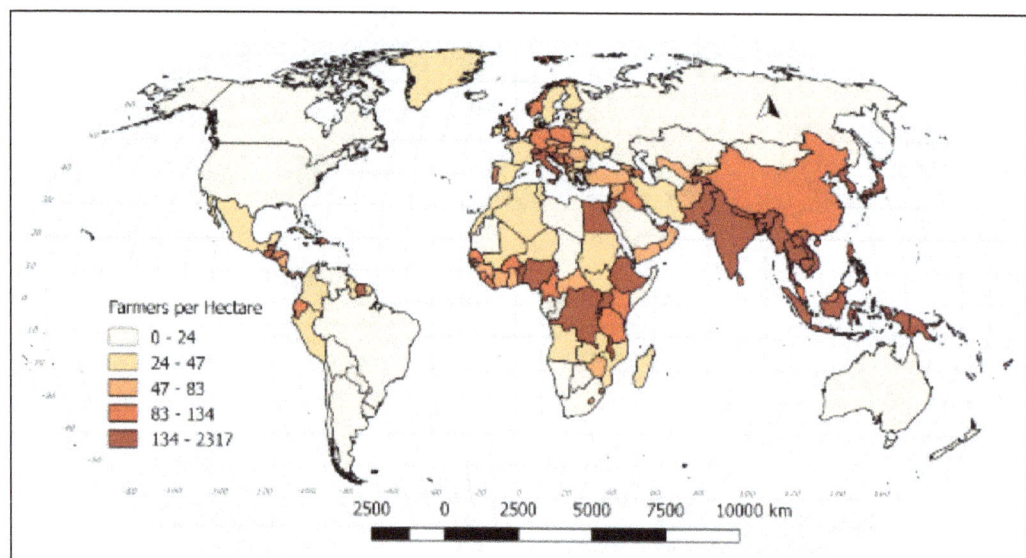

Figure 2.19 | Agricultural Density 2015[16]
Author | David Dorrell
Source | Original Work
License | CC BY SA 4.0

to be large in order to generate a sufficient income. Places with high agricultural densities have more farmers per hectare, meaning that farms will likely produce less revenue. Of course, an underlying assumption of this number is the idea that people are growing food to earn a living. If they are eating the produce directly, outside the cash economy, then the comparison is less valid.

Related to food production is the concept of **carrying capacity**. Carrying capacity is simply how many people can live from a given piece of land. However, it's not really that simple. Carrying capacity is not static throughout time. Not only do environmental characteristics change (due to desertification, for example) but technology changes as well. The carrying capacity of land in wealthy developed countries has expanded tremendously due to the application of technology. These technologies could be something as simple as irrigation ditches to something as complex as genetic modification of the plants and animals themselves. Carrying capacity is snapshot taken at a particular time.

2.7 FUTURE POPULATION

When we looked at the population pyramids we considered both the current population conditions and the projected population pyramids. When we looked at the first graph in the chapter, it didn't stop at the current year. It provided three different estimates of the near future. Estimating future populations is important, both at the global level and the national level, but also at the local level as well. Societies large and small attempt to plan their futures in terms of resource allocation and economic development. Population projections are difficult. They attempt to take current circumstances and use them to plot the likeliest future.

The problem is that the future may not be like the present. A disease that is suppressing fertility may be cured. The climate may change more than expected. We could experience another world war. Any number of unexpected large scale events could occur that completely invalidate the reasoning behind a projection. Nicholas Talib calls these Black Swan events. They are things that you don't know about until you do know about them. Irrespective projections will be made because they are necessary.

2.8 GEOGRAPHY OF HEALTH

Although the health of populations has been considered in most sections of this chapter, there are some aspects that can be addressed most appropriately on their own. The geography of health or medical geography is the study sickness or health across space. In the same way that it is possible to compare places by their population characteristics, it is possible to compare places by their health characteristics. Also in the same way that societies tend to pass through the stages of demographic transition, they also tend to pass through an epidemiological transition. The demographic transition and epidemiological transition are somewhat related. The improvements in food supply and better sanitation that lead to larger populations also lead to healthier populations. They are not exactly the same thing however. Epidemiology is the study of diseases themselves and the way that they function. The epidemiological transition is a way of representing the relationship between development and disease. In less developed places, infant mortality is high (**Figure 2.20**) and infectious disease is the greatest threat. In more developed societies, the health threats tend to be chronic afflictions- cancer, diabetes, or heart disease. One of the reasons for the difference between the developed and less developed places is the fact that in developed countries people tend to live longer. Countries in which people die in their forties will not have a problem with Alzheimer's disease.

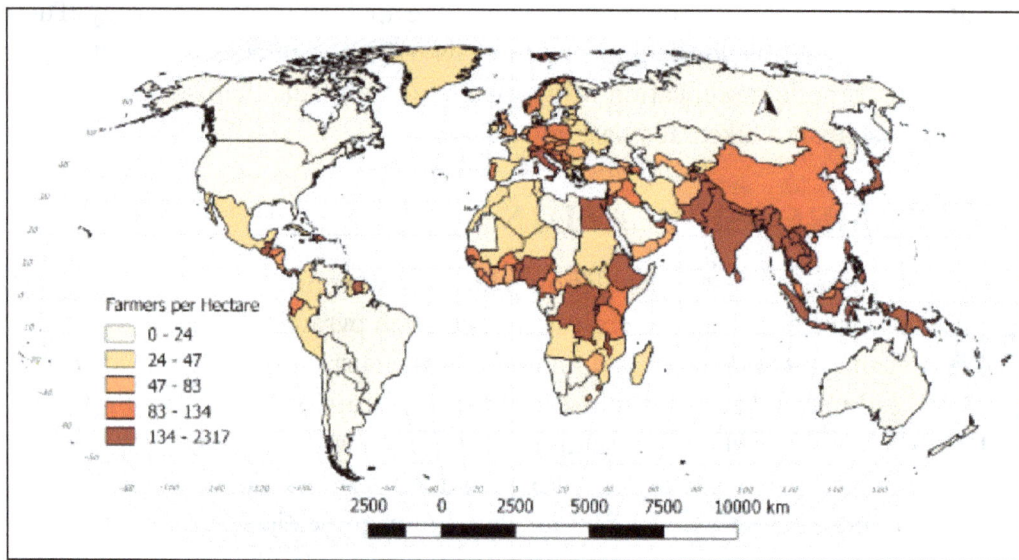

Figure 2.20 | Infant Mortality 2015[17]
Author | David Dorrell
Source | Original Work
License | CC BY SA 4.0

Mortality refers to death and **morbidity** refers to sickness. We can look at different rates of mortality and morbidity to gain some insight into the health of a population.

Diabetes was once related to development, but the relationship is weakening (**Figure 2.21**). As obesity becomes a greater problem in developing countries, so does diabetes.

On the other hand, malaria is a disease of poor countries, or at least it is a disease of poor places (**Figure 2.22**). The poorer parts of developing countries will likely have malaria problems, while the wealthier parts may not.

One of the changes that has occurred in medical geography is that the differences between the developed and less developed countries have been narrowing. Obesity was once considered fully within the sphere of developed world problems, but the obesity crisis has diffused into many parts of the developing world now. In the same way, infectious diseases such as HIV and Hepatitis C have become problems in developed countries.

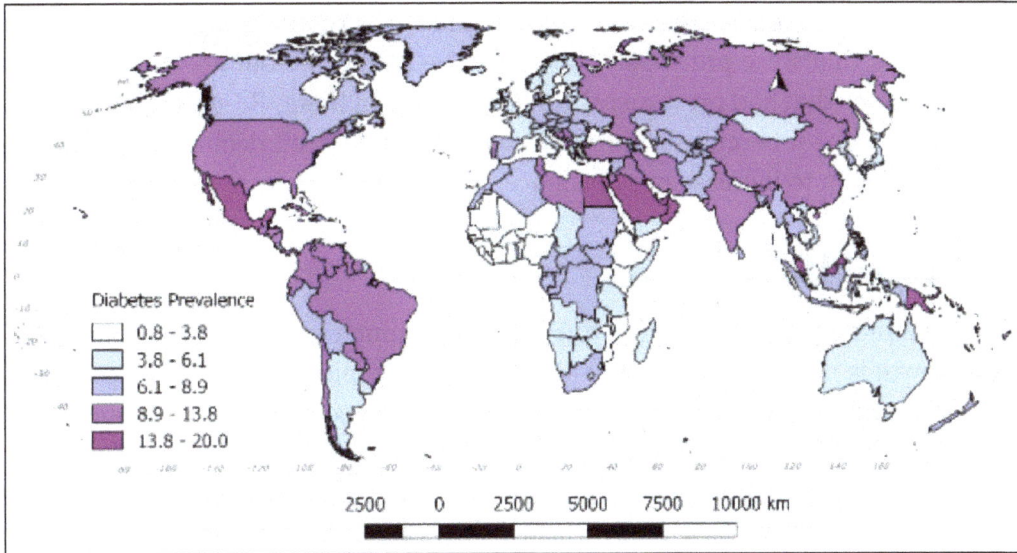

Figure 2.21 | Diabetes Morbidity 2015[18]
Author | David Dorrell
Source | Original Work
License | CC BY SA 4.0

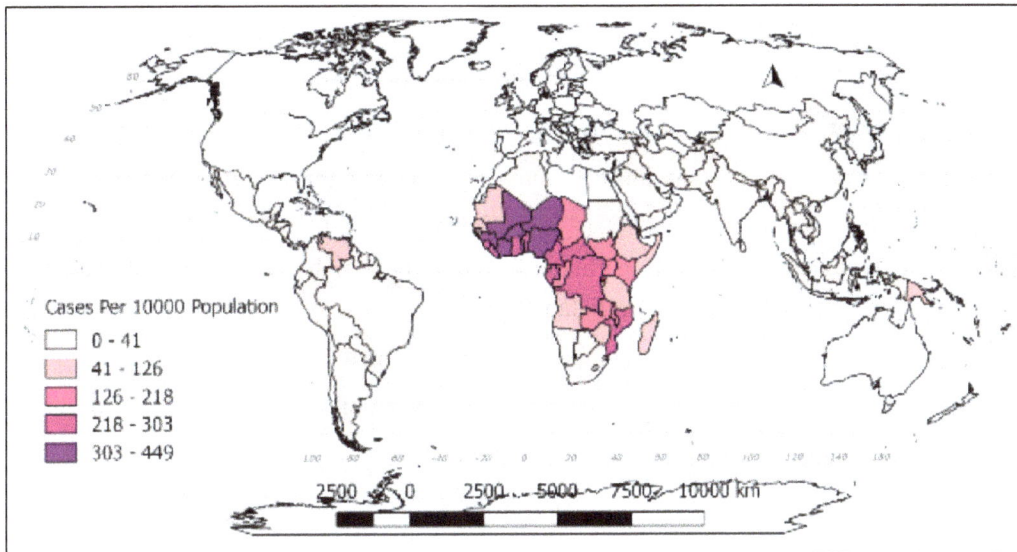

Figure 2.22 | Malaria Incidence 2015[19]
Author | David Dorrell
Source | Original Work
License | CC BY SA 4.0

Due to the work of numerous organizations, a number of diseases in many parts of the world have been reduced and some have been eradicated. Smallpox no longer exists in the wild and Dracunculiasis (Guinea worm) has gone from millions of cases thirty years ago to single digits in recent years. Other diseases have been dramatically reduced. Polio once infected millions of people per year across the world, but through vaccination programs, its range has been reduced to just two countries.

There is now even a category of diseases known as re-emerging diseases. These are diseases that were previously thought to be eradicable but are now returning to populations that had previously been largely free of them. Re-emerging diseases are diseases like tuberculosis and diphtheria that had been declining for decades that have begun to increase in prevalence.

2.9 SUMMARY

The human population is growing, but not as much as it has in the recent past. The growth is not even, with some places growing rapidly and other places losing population. Those places growing fastest tend to be poor. The effect of population growth is place-bound as well; highly developed places with low populations often use more resources than less developed places with more people. The impact of population can be measured in a number of ways, with each measurement providing a small insight into the dynamics of the human population. Uneven population growth and poverty are the two most important factors underlying the subject of our next chapter, Migration.

2.10 KEY TERMS DEFINED

Agricultural density: The number of farmers per unit area of arable land.

Arithmetic density: The population of a country divided by its total land area.

Carrying capacity: The maximum population size that the environment can sustain indefinitely.

Cartogram: map in which some thematic mapping variable—such as population—is substituted for land area or distance.

Cohort: A subset of a population, generally defined by an age range.

Crude birth rate: Total number of live births per 1,000 of a population in a year.

Crude death rate: Total number of deaths per 1,000 of a population in a year.

Demographic transition: The transition from high birth and death rates to lower birth and death rates as a country or region develops.

Dependency ratio: The ratio of those not in the labor force (generally ages 0 to 14 and 65+) and those in the labor force.

Doubling time: The period of time required for a population to double in size.

Ecological Fallacy: Characteristics about the nature of individuals are deduced from inference for the group to which those individuals belong.

Ecumene: The Greek concept of the habitable part of the Earth

Infant mortality rate: The number of infant deaths that occur for every 1,000 live births.

Life expectancy: The number of years that one is expected to live as determined by statistics.

Morbidity: The state of being diseased or unhealthy within a population.

Mortality: The number of people who have died within a population.

Overpopulation: A condition in which a place has outstripped its ability to provide for its own needs.

Physiological density: The number of people per unit area of arable land.

Population density: A measurement of population per areal unit, such as the world, a region, a country or other area.

Population momentum: The tendency for population growth to continue due to high concentrations of people in the childbearing years.

Population projection: An estimate of future population.

Population pyramid: Graphical illustration that shows the distribution of various age groups in a population.

Rate of natural increase: The crude birth rate minus the crude death rate. This rate excludes the effect of migration.

Replacement level: The average number of children a woman needs to have to ensure the population replaces itself. The number is roughly 2.1.

Total fertility rate: The average number of children that would be born to a woman over her lifetime.

2.11 WORKS CONSULTED AND FURTHER READING

Bank, World. 2017. "Metadata Glossary." DataBank. Accessed August 20. http://databank. worldbank.org/data/glossarymetadata/source/all/concepts/series.

Emch, Michael, Elisabeth D. Root, and Margaret Carrel. 2017. Health and Medical Geography. Fourth edition. New York: Guilford Press.

"Ester Boserup." 2017. Wikipedia. https://en.wikipedia.org/w/index.php?title=Ester_ Boserup&oldid=783397776.

Gould, W. T. S. 2009. Population and Development. Routledge Perspectives on Development. London ; New York: Routledge.

Gregory, Derek, ed. 2009. The Dictionary of Human Geography. 5th ed. Malden, MA: Blackwell.

Koch, Tom. 2017. Cartographies of Disease. Esri Press. http://www.myilibrary. com?id=965009.

Kurland, Kristen Seamens, and Wilpen L. Gorr. 2014. GIS Tutorial for Health. Fifth edition. Redlands, California: ESRI Press.

"List of Countries by Population (United Nations)." 2017. Wikipedia. https://en.wikipedia. org/w/index.php?title=List_of_countries_by_population_(United_ Nations)&oldid=796051350.

"Maps & More | GIS | CDC." 2017. Accessed August 20. https://www.cdc.gov/gis/mapgallery/index.html.

Newbold, K. Bruce. 2017. Population Geography: Tools and Issues. Third edition. Lanham: Rowman & Littlefield.

Taleb, Nassim Nicholas. 2010. The Black Swan: The Impact of the Highly Improbable. 2nd ed., Random trade pbk. ed. New York: Random House Trade Paperbacks.

"Thomas Robert Malthus." 2017. Wikipedia. https://en.wikipedia.org/w/index.php?title=Thomas_Robert_Malthus&oldid=795798343.

US Census Bureau, Demographic Internet Staff. 2017. "International Programs, World Population." Accessed August 20. https://www.census.gov/population/international/data/worldpop/table_history.php.

2.12 ENDNOTES

5. Data source: World Bank Health Nutrition and Population Statistics. http://databank.worldbank.org/data/glossarymetadata/source/all/concepts/series

6. Data source: World Bank Health Nutrition and Population Statistics. http://databank.worldbank.org/data/glossarymetadata/source/all/concepts/series

7. Malthus, Thomas R. 1798. An Essay on the Principle of Population.

8. Boserup, E. 1965. The Conditions of Agricultural Growth.

9. Data source: United States Census Bureau 2010 http://www2.census.gov/geo/tiger/TIGER_DP/2016ACS/ACS_2016_5YR_COUNTY.gdb.zip

10. Data source: Natural Earth Data. Image produced with the cartogram plugin in QGIS. https://www.naturalearthdata.com/downloads/ S

11. Data source: World Bank Health Nutrition and Population Statistics. http://databank.worldbank.org/data/glossarymetadata/source/all/concepts/series

12. Data source: World Bank Health Nutrition and Population Statistics. http://databank.worldbank.org/data/glossarymetadata/source/all/concepts/series

13. Data source: World Bank Health Nutrition and Population Statistics. http://databank.worldbank.org/data/glossarymetadata/source/all/concepts/series

14. Data source: World Bank Health Nutrition and Population Statistics. http://databank.worldbank.org/data/glossarymetadata/source/all/concepts/series

15. Data source: World Bank Health Nutrition and Population Statistics. http://databank.worldbank.org/data/glossarymetadata/source/all/concepts/series

16. Data source: World Bank Health Nutrition and Population Statistics. http://databank.worldbank.org/data/glossarymetadata/source/all/concepts/series

17. Data source: World Bank Health Nutrition and Population Statistics. http://databank.worldbank.org/data/glossarymetadata/source/all/concepts/series

18. Data source: World Bank Health Nutrition and Population Statistics. http://databank.

worldbank.org/data/glossarymetadata/source/all/concepts/series

19. Data source: World Bank Health Nutrition and Population Statistics. http://databank.
    worldbank.org/data/glossarymetadata/source/all/concepts/series   Data   source:
    World Bank Health Nutrition and Population Statistics. http://databank.worldbank.
    org/data/glossarymetadata/source/all/concepts/series

20. Data source: World Bank Health Nutrition and Population Statistics. http://databank.
    worldbank.org/data/glossarymetadata/source/all/concepts/series

21. Data source: World Bank Health Nutrition and Population Statistics. http://databank.
    worldbank.org/data/glossarymetadata/source/all/concepts/series

3 MIGRATION
Todd Lindley

STUDENT LEARNING OUTCOMES

By the end of this section, the student will be able to:

1. Understand: definitions of migration and associated significant terms
2. Explain: the geographic patterns of migration within and between countries as influenced by economic, socio-cultural, political, and environmental factors in the contemporary historical period
3. Describe: the general relationship between demographic factors and migration across time
4. Connect: factors of globalization to recent trends in migration

CHAPTER OUTLINE

3.1 MIGRATION AND GEOGRAPHY: A (VERY) BRIEF HISTORY

For most of human history, people did not move very far away from where they were born. **Migration** (a permanent move to a new location) over long distances was so dangerous, unpredictable, and risky that humans remained in a relatively small area of Eastern Africa until only about 65,000 years ago. At that time, some brave soul (or most likely many) set off on an adventure that would take tens of thousands of years to complete – the mass movement of humans to all corners of the **ecumene** (inhabited areas of earth). Scientists continue to disagree about the specific time periods (some studies suggest that the big move started 120,000 years ago!) and reasons that our ancestors finally decided to take flight, but significant evidence suggests that periodic climate change may have played a major role. The earliest evidence of human remains in the North America dates to approximately 13,000 years ago, when humans are hypothesized to have crossed an ice bridge from Eastern Russia into Alaska during the last Ice Age, before spending the next several thousand years spreading throughout North and South America and The Caribbean. Regardless of the time period, modern humans have been on the move for a very long time—a trend that has accelerated in recent years owing to cheap transportation and easier access to information by potential migrants.

Migration is also central to the formation of the world's largest religions. The spread of Christianity, for example, was facilitated by the massive movement of people within the Roman Empire along well-traveled transportation routes connecting modern Israel and Palestine with Turkey, Greece, Italy, and other parts of the empire where new ideas, cultures and beliefs were shared. Many of the most significant stories from Christianity and Judaism recount the experience of "foreigners" (sojourners) who are traveling to new lands. Similarly, Islam's prophet migrated from his city of birth, Mecca, to Medina in the seventh century, and religious ideas were spread as soldiers, merchants, and traders moved across North Africa and eventually into Europe and to Southeast Asia. Migration, in many ways, has been the most impactful of all human activities on the planet. Had humans never taken the journey, we all would still be in East Africa, just daydreaming about the rest of the earth! To learn more about humans' earliest journeys, you can click the link below: https://psmag.com/why-and-when-did-early-humans-leave-africa-c1f09be7bb70#.powpltf4t

3.2 DEFINITIONS AND DATA

Migration can be interregional (between regions), intraregional (within a region), or international (across national borders). Those moving in are **immigrants**, and those moving out are **emigrants**. **Net migration** is the difference between the number of immigrants and the number of emigrants in any given year. The United Nations provides data and analysis on immigration and emigration annually, but such figures depend largely upon government sources that are more reliable in some cases than in others. For example, the United States (US) maintains databases on immigrants of all types including **guest workers** (those given permission to enter the country legally for a specific job and for a specific period of time), students, tourists, **asylum seekers** (those seeking sanctuary from political, religious, gender, or ethnic persecution), and **undocumented migrants** (those inside of a country without proof of residency). However, the US does *not* maintain or report data on those that have emigrated, except in those notable cases where individuals have denounced citizenship. Meanwhile, countries like Mexico and the Philippines track their overseas citizens regularly, acknowledging the realities of dual citizenship or residency. **Return migration** (a permanent return to the country of origin) also represents a significant flow of people, but is often under-reported. For example, up to one quarter of Europeans that migrated to the US in the late nineteenth century eventually returned to Europe. In recent years, more people have migrated back to Mexico from the US than from Mexico to the US. The process of migration, then, is a complicated nexus of movement rather than a simple one-way, permanent, single-directional move from place A to place B.

In spite of the complex patterns and processes of migration, some general characteristics of migration and migrants were articulated by British demographer Ernst Ravenstein (1885), characteristics known as the **laws of migration**. Many of them still hold true 135 years later. Can you decide which are still true today?

1. Most move only a short distance.
2. Each migration flow produces a counter-flow of migrants.
3. Long-distance migrants tend to move to major cities.
4. Rural residents are more migratory than those in towns.
5. Females are more migratory than males.
6. Economic factors are the main reason for migration.

In short, each of the "laws" generally hold true in 2018 with the notable exception of number five. Slightly more men moved internationally than women, but the truth is much more complicated. In fact, Ravenstein's estimates of female migration proved incorrect as large-scale migration to North America increased in the early twentieth century, during which time most immigrants were men seeking land, wealth, and opportunity in the "New World."

A variety of non-government organizations, research groups, and humanitarian entities also track the movement of people across borders and within countries in order to provide a deeper understanding of the causes and effects of migration *locally*, *regionally*, *nationally*, and *globally*. For example, the Migration Information Source (https://www.migrationpolicy.org) offers a wealth of reports, analysis, and data visualization that dramatically enhance our geographic understanding of migration. The Pew Research Center (http://www.pewresearch.

org/topics/migration) conducts regular polls often focused on Latino populations in the US. Other non-profit organizations track the effects of immigration in the US and publish regular reports, but often they lack objectivity or editorial oversight, as the intent of such efforts is to achieve policy change to reduce immigration levels. For example, the Federation for American Immigration-FAIR (https://fairus. org) is an organization motivated by the explicit desire to reduce the number of immigrants and to secure the traditional cultural heritage of European Americans. Likewise, the Center for Immigration Studies (CIS) presents data to support its stated vision "of an America that admits fewer immigrants" and to reduce immigration of all kinds in the twentieth century (https://cis.org/About-Center-Immigration-Studies).

Geographers have identified general trends in global migration, also known as North-South migration, in which most emigrants originate in poorer, developing countries and most destinations have traditionally been wealthier, developed countries. For most Americans and Canadians, this pattern is very familiar, as recent decades have seen unprecedented numbers of Latinos immigrate to the US for the purpose of finding higher-paying jobs and better opportunities and escaping structural poverty in the developing world. Similarly, the recent patterns in Europe have seen record numbers of Eastern Europeans move west and north to earn higher wages than those available in the home country. However, such wage differentials do not tell the whole story. For example, wages in Chicago tend to be much higher than those in other parts of Illinois, but not everybody leaves rural Illinois just because they can earn a higher wage. Wages, though significant, only tell part of the story. Unless you are reading this text in Manhattan or Paris or Hong Kong, you could most likely move tomorrow and find a job elsewhere that pays more than what you earn now (if you are working). Geographers recognize that attachment to place, cultural factors, desire to stay close to family/friends, and other factors play a powerful role in the decision to move or stay.

Another pattern that has remained consistent over time is that of **highly skilled migrants**, who tend to enjoy a much greater freedom of movement than those with lower levels of education and fewer skills. For example, computer software engineers, database managers, and a host of other highly-demanded skills lead to efforts by countries and corporations to attract the best and the brightest minds to immigrate in order to bring those skillsets into a country where they are in short supply. Countries often offer travel **visas** (temporary permission to enter a country) to those with highly demanded skillsets. Countries like Australia, Canada, and New Zealand utilize a **points system** to determine which of the highly-skilled applicants will be granted permission to enter. The **brain drain** refers to the conceptual idea that when a wealthy country recruits the 'best brains' from a poorer country, it can be damaging to the sending country, as many of the most qualified and talented groups of people are poached away by higher-paying opportunities. As such, the term **brain gain** refers to the benefits received by a country that receives all those "brains" without having to produce them from scratch! Recently, countries have also acknowledged the concept of **brain waste**, in which receiving countries fall short in utilizing the full range of **human capital** inherent in many immigrant populations. For example, nearly half of all immigrants into the US from 2011-2015 held at least a bachelor's degree, but more than 2 million immigrants with college degrees continue to work low-skilled jobs because employers or governments do not recognize foreign-held degrees. Similarly, a brain-drain/brain-gain phenomenon occurs *within* some countries, such as the US. California and New York, for example, continue to draw the most highly-trained and qualified people away from other states. Governments that wish to keep the highly skilled at home take such transfers of educated and highly-skilled people very seriously.

Framing migration as a loss or gain, however, also does not tell the whole story. Most countries that send migrants also receive them. The US receives large numbers of immigrants, but it also is a country of emigration, whereby retirees choose to live outside of the US, or long-term migrants (who usually are US citizens) choose to return to their home country in retirement. Migration is neither inherently good nor bad; rather, it is complicated. This chapter seeks to tease out some (but not all) of the important characteristics of migration in the twenty-first century to help you gain a better understanding of a topic that too often is used by politicians to gain votes or credibility. Let us put those simplistic debates aside for a few minutes to consider the basic elements of migration around the world.

3.3 GLOBAL, NATIONAL, REGIONAL, AND LOCAL PATTERNS

3.3.1 Global Patterns

Though geopolitical and global economic forces change over time, it is useful to understand contemporary global, national, and regional patterns of migration as processes that vary by geography. The vast majority of people do not migrate internationally, yet migration makes a powerful impact globally. Just imagine what our world would look like if nobody ever moved! Latin America would have no coconut trees, the American Midwest would still be dominated by buffalos, and most humans would look very much alike—how boring! Instead, migrants across the globe diffuse new ideas, new genetic footprints, new diseases, new cooking styles, and new sports. While a small number of countries receive large numbers of newcomers each year, only 3.4 percent of the world's population live outside of their birth country, so most places in the world are *not* significantly impacted directly by international migration. However, the dynamics of migration are undergoing a significant transformation, and the future is very difficult to predict. Humans have been on the move for over 60,000 years, so it is unlikely to come to a halt anytime soon.

The number of international migrants worldwide reached 244 million in 2015, representing a 44 percent increase since the year 2000 (**Figure 3.1**). During the same period, global population grew by just 20 percent, so cross-border migration seems to be accelerating in many parts of the world as more people have access to information, infrastructure, and communication—all elements that facilitate the large-scale movement of people. In 2015, nearly 70 percent of all migrants originated in either Europe or Asia, with the largest numbers coming from India, Mexico, Russia, and China. Meanwhile, those countries hosting the most immigrants were the US (47

Europe	76 million
Asia	75 million
North America	54 million
Africa	21 million
Latin America/Carribean	9 million
Oceana	8 million

Figure 3.1 | Immigrant populations around the world by region
Authors | Dilip Ratha, Christian Eigen-Zucchi, and Sonia Plaza
Source | World Bank eLibrary
License | CC BY 3.0 IGO

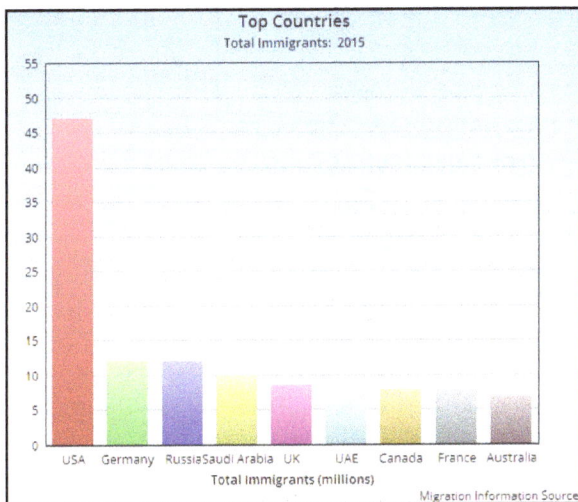

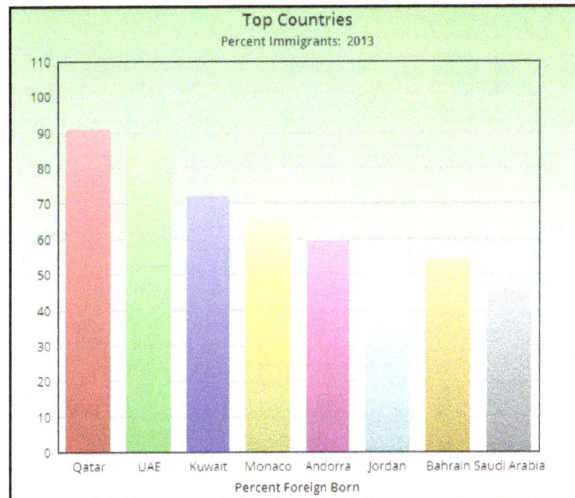

million), followed by Germany (12 million), Russia (12 million), and Saudi Arabia (10 million) (**Figure 3.2**). Generally speaking, migrants tend to move away from low/middle income countries into high income countries because the most common driving force is economic opportunity. Most migrants move for better jobs, higher incomes, and better opportunities overall. It is worth noting, however, that about one third of international migration takes place *between* lower income countries (e.g. from Bangladesh to India or from Afghanistan to Iran).

Figure 3.2 | Top countries of immigration by number and percent
Authors | Dilip Ratha et al.
Source | World Bank eLibrary
License | CC BY 3.0 IGO

It is also important to note that patterns can change very quickly, depending upon economic or political conditions. Spain, for example, was a major recipient of immigrants from 2000-2008, but when its economy dipped and job opportunities decreased, people began to leave *en masse*. In every year since 2011, more people have left Spain than have arrived. Ireland, on the other hand, experienced mostly net emigration from the mid 1800's until the late 1990's, when large numbers of Irish and descendants returned "home" and new immigrants began to choose

Ireland as a popular destination due to its improved economic opportunities under the European Union.

3.3.2 Regional Migration Flows: Europe

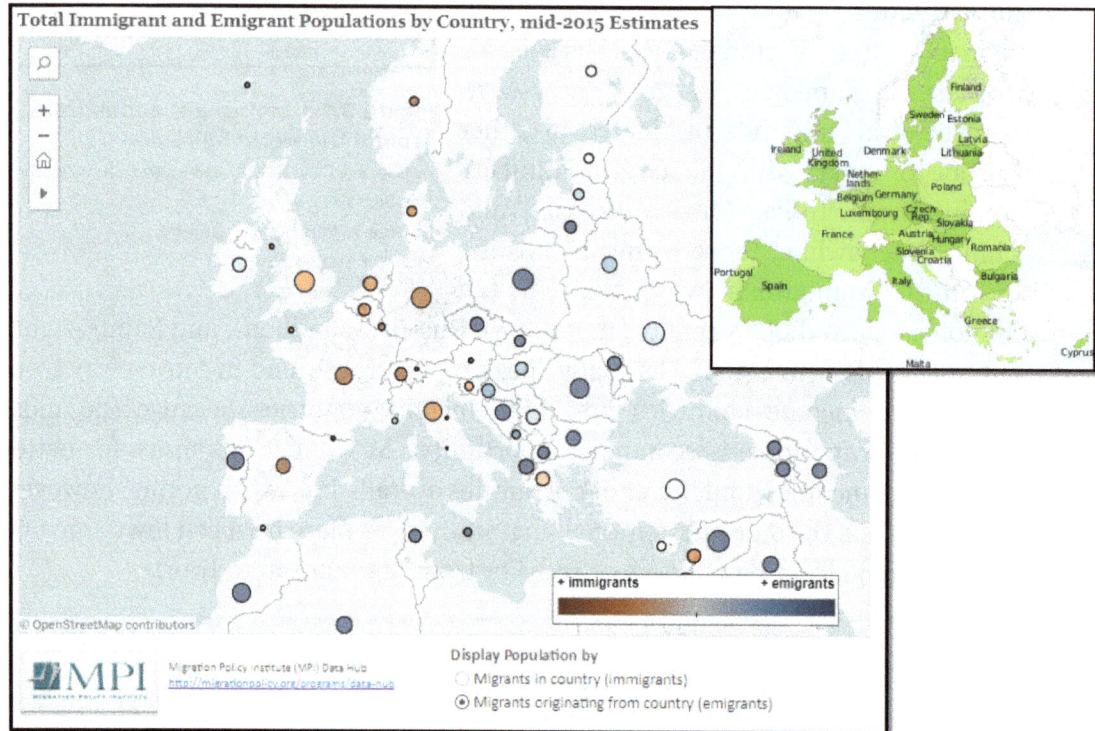

Figure 3.3 | Total European immigrant and emigrant populations by country, 2015

Figure 3.3a (left)
Authors | OpenStreetMaps and Migration Policy Institute
Source | Migration Policy Institute
License | ODbL

Figure 3.3b (right)
Author | User "Ssolbergj"
Source | Wikimedia Commons
License | CC BY SA 3.0

Consistent with Ravenstein's "Laws of Migration," most international moves continue to take place across relatively short distances. In no place is this more evident than in Europe, where it also quite common to find Spaniards in France, Germans in Switzerland, Romanians in Germany, etc. As portrayed in the map above, movement within Europe tends to be from East (blue dots) to West (red/orange dots) (**Figure 3.3**). Given the ease of travel within, the small size of, and short distances between many European countries, it is not surprising that many people move across borders for a variety of reasons. Of the 508 million residents of the European Union (EU), about 10 percent (54 million) are foreign born, of which 35 million originate outside of the EU and 19 million moved from one EU country to another. Movement within and to the EU has accelerated in recent decades for two reasons. First, Europe has intentionally worked towards greater economic integration and cooperation since WWII by removing barriers to movement (See more about the EU in Chapter 8). Citizens and legal residents of participating EU member states may travel, live, study, and work seamlessly across national borders under the *Schengen Agreement*

(Figure 3.4). While the EU facilitates the free movement of capital, products, and services in the region, the agreement permits the free flow of people across international borders without significant delay, hassle, or regular security checks. Millions of working-aged residents of Poland, Romania, and Bosnia, for example, have moved to countries like Germany, France, and Spain for higher wages. In some ways, the European region has become similar to the US in that Spaniards may travel to France just as easily as Californians may travel to Oregon.

The second reason for increased movement and mobility in Europe, however, is that more people from nearby Africa and the Middle East have attempted to enter Europe to seek work, escape conflict, or find better educational opportunities. Although economic and political life has been relatively more attractive in Europe than in neighboring regions for some time, the forces of globalization have accelerated such flows. Human smuggling organizations transport hopeful migrants in exchange for large amounts of money. As such, the EU has increased maritime military patrols in the region, which has forced human smugglers and would-be immigrants into riskier and more perilous routes. Thousands die each year trying to gain entry to another country (**Figure 3.5**). In 2016, approximately 7,400 died during a migration route, and nowhere is the trek riskier than in the Mediterranean, where 4,800 migrants perished in 2016, most of whom were from African countries. Additionally, conflicts in the Middle East, most notably Syria, have driven hundreds of thousands to seek refuge in Europe.

The free flow of people across the European region has come into question in recent years as northern member states mistrust the

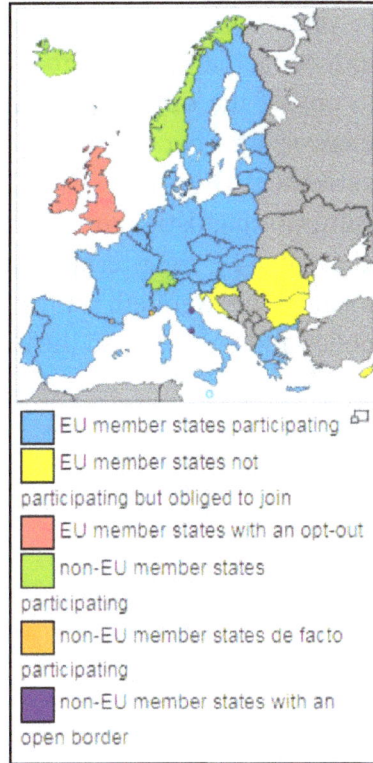

Figure 3.4 | Schengen Countries (2017)
Author | User "Ssolbergj"
Source | Wikimedia Commons
License | Public Domain

Figure 3.5 | Migrant Fatalities in 2017
Author | International Organization for Migration Data Analysis Center
Source | International Organization for Migration Data Analysis Center
License | CC BY SA 4.0

vetting process put into place by southern and eastern European states. An individual that manages to enter any of the Schengen states can travel quite easily to any of the other European states. Denmark and Sweden, for example, have increased border security and border checks in spite of being signatories to the agreement **(Figure 3.6)**. The bold experiment in Europe to integrate so many cultural, linguistic, and political systems faces constant challenges and has no guarantee for long-term success. Twice in the last century, the region was war-torn. As the twentieth century unfolds, Europe remains a focal point for migration policy and practice into the twenty-second century.

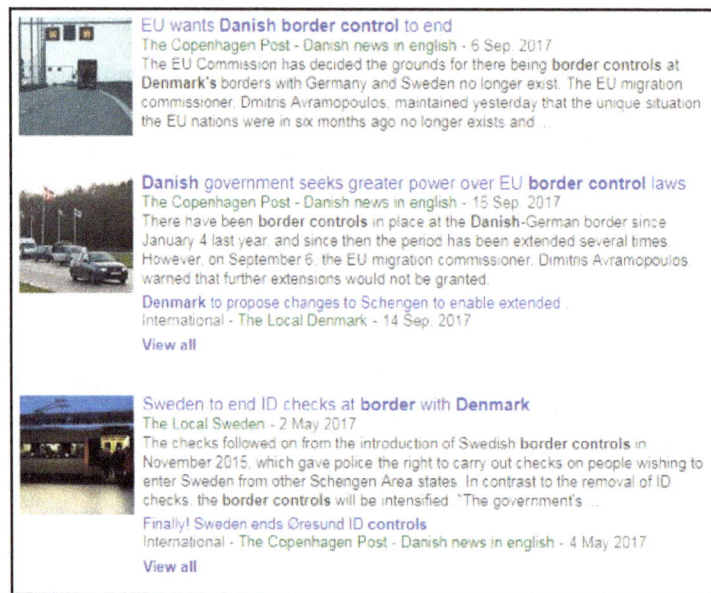

Figure 3.6 | Google News headlines on Danish border control
Search Results Screen Capture for terms "Danish border control."
6 September 2017.
Author | Google News
Source | Google News
License | © Google. Used with permission.

3.3.3 Regional Migration Flows: North America

North America represents a special case in that the US routinely receives significantly more immigrants than any other country in the world. Roughly 14.5 percent (46 million) of all residents in the US were foreign born in 2015, representing seventeen percent of the labor force. In Canada, the numbers were roughly twenty-one percent for both categories. The **International Organization for Migration (IOM),** sponsored by the United Nations, offers a great tool for visualizing migration around the world **(https://www.iom.int/ world-migration)**. Click on the tool to find out how many in the US came from China or how many from the US lived in Mexico last year (hint: toggle the In/Out button). Do you notice any patterns? Click on a few other Caribbean countries to view the flows of migration into North America. You will notice that the largest source areas to the US were Latin America and Asia, but sending countries and volumes, in fact, vary significantly over history. The figure below shows changes over time in the US immigrant population, and you can see that immigration from Latin America and Asia has only occurred in the most recent period. Do you know or think you know when your ancestors arrived? Can you visualize the time period during which they first came to North America? Which population groups are not included in the chart? Why do you think this might be?

There four distinct phases of immigration, but let's examine each one briefly. First, during the era of frontier expansion (1820-1880), a rapid westward movement occurred. Settlers were given the rare opportunity to acquire large amounts of good farmland in the "New World," which pushed people out of the large cities like New York and Boston and towards the American Midwest into the future states of Ohio, Illinois, Iowa, Nebraska, etc. Many immigrants came to the US specifically in search of land, because population pressure in Europe made land acquisition very difficult. The *Homestead Act* of 1862 offered up to 160 acres of farmland to citizens or those planning to become citizens as long as they continuously lived on and farmed the land for five years. More than 1.5 million claims for land were made in subsequent years, resulting in a massive movement of people. It was also during this era that the transcontinental railroad was completed, linking the East Coast and West Coast for the first time in a meaningful way. Most immigrants during this time came from England, Scotland, Ireland, Germany, and Scandinavia. More than 1.5 million also came from Ireland in just a ten-year period (1845-1855) when the potato crop failed due to blight (disease), a period known as the Great Hunger. The only significant non-European immigrants arrived as a result of the *Burlingame-Seward Treaty* of 1868, which welcomed Chinese workers, as China also opened its borders to American businesses, students, and missionaries. Migration from China was short-lived, however, during the subsequent era.

Phase 2 (1880-1910) ushered in a dramatic shift as the number of new arrivals soared to unprecedented levels. Moreover, the source countries also changed as immigrants began to arrive from southern and eastern European countries such as Italy, Greece, Russia, and Poland. American imagery, stories, legends, books, and films tend to focus heavily on this period, because it was the most intense era of movement from Europe, and most were arriving via Ellis Island by ship before settling into distinctly ethnic neighborhoods in cities across the US It was also an era of massive industrialization that demanded a large source of low-cost workers, which immigration provided. Coal mines, steel mills, and factories were growing at unprecedented levels as the US raced to "catch up" with the great cities of the world, building bridges, skyscrapers, museums, sports stadiums, and other large structures at unprecedented levels. The period also produced a new era of exclusion as the Chinese Exclusion Act of 1882 prohibited immigration, naturalization, and citizenship for anyone of Chinese descent, including those that had immigrated previously under legal means.

During Phase 3 (1910-1965), the US dramatically curtailed the number of immigrants permitted to enter the country legally. The great western settlement had ended, and high birth rates in the US created enough new workers to meet the demand of the continued industrialization of the country. A series of new exclusionary laws prevented all "Asiatic People" from entering the country, as American nationalism fanned by white power and eugenic movements sought to exclude and divide the people of the world into categories of "civilized," "savage," and "semi-civilized." Also banned were "polygamists, anarchists, beggars, and importers of prostitutes." (For an excellent graphical overview of US immigration legislation, visit the Pew Research Center website: http://www.pewhispanic.org/2015/09/28/selected-u-s-immigration-legislation-and-executive-actions-1790-2014). During this period, immigration levels declined precipitously as the country grappled with questions of heritage, identity, and cohesion. National legislation was crafted to maintain the racial and ethnic balance of the country as mistrust between nations erupted in two world wars. Quotas were established to allow immigration of only a limited number of people from specific European countries, while all others were kept out. Ellis Island closed its doors in 1954 after having processed more than 12 million immigrants. However, a dramatic social shift was on the horizon, as the 1960's would dramatically alter US immigration in the coming decades.

Immigration into the US in the current era, Phase 4 (1965-Present), is characterized by (1) enormous growth and (2) a new diversity of source countries, dominated by the regions of Asia and Latin America. While the percent of foreign-born residents is similar to that of the early 1900's, the 47 million (in 2016) living in the US represents an unprecedented historical high with one out of every four children having at least one parent who is an immigrant. The Immigration and Naturalization Act of 1965 dramatically shifted the nature of immigration into the US from one with low numbers from a handful of pre-determined countries to a system that favors those with particular skills (e.g. scientists, engineers, and doctors) and for family reunification. As more Americans entered college, demand for manual labor also grew. Jobs that were once done by children and teenagers, such as harvesting crops, cutting grass, or washing dishes in restaurants were increasingly being filled by immigrants willing to work long hours for little pay. Asian exclusion laws and quotas were abolished, as progressive civil rights era provisions took a "colorblind" approach to welcome those from all regions of the world as long as they had the skills and education that would benefit the American economy. Laws also made it much easier for existing citizens to legally bring immediate relatives from abroad, so that families would not be divided by restrictive immigration policies. Speaking to a sparse crowd in 1965 about the new immigration bill, President Lyndon B. Johnson said, this "is not a revolutionary bill. It does not affect the lives of millions . . . It will not reshape the structure of our daily lives or add importantly to either our wealth or our power."[1] Yet in the three decades after its passage, more than 18 million people arrived legally, tripling the number of immigrants from the previous three decades. The "face" of immigration changed dramatically. In 2016, the largest number of foreign born were those from Mexico, China, India, the Philippines, and Vietnam. Moreover, immigrants are responsible for 25 percent of all new businesses and more than half of all start-up companies valued at more than $1 billion. Sometimes, Presidents are wrong.

3.3.4 Regional Migration Flows: Asia

Migration within Asia is difficult to summarize given the massive size and scope of a region with 4.4 billion people! However, if you revisit the world migration map (https://www.iom.int/world-migration), you will notice that even the largest countries in the world (China and India) contain relatively-few foreign born people compared to those in Europe and North America. Can you think of reasons why this might be the case? The first explanation is simple. Large-scale human settlement and political development occurred much earlier than in Europe or in North America. Early Chinese civilizations had already emerged more than 5,000 years ago, and many in Asia can trace ancestral lineages back thousands of years as well. As such, Asia (for the most part) wasn't "discovered," conquered, or colonized during the time of European imperialism. Apart from a small number of missionaries, entrepreneurs, and adventurers, Asia did not become a major destination for those from foreign lands (there are several notable exceptions, particularly in the case of the British Empire). Second, most Asian countries simply do not allow for permanent immigration, except in the case of marriage (which also can be a very slow and tedious process). Nonetheless, in the age of globalization and migration, more people are on the move in Asia in the twenty-first century than ever before.

With a population of 1.3 billion and a fast-growing economy, China tends to dominate the East Asia region. The Chinese **Diaspora** is a term that refers to the 46 million people that identify themselves as Chinese but live outside of China, with the largest number making their homes in Southeast Asia and others in Australia, North America, or Europe. Many fled China during times of political instability before and after WWII and under early communist rule, but today, many of those who leave China are wealthy and educated and do so because their skills are in demand in other parts of the world.

The most significant movement of people anywhere in the history of the world has been taking place recently inside both China and India as rural farmers have been leaving the countryside and moving into cities. Nearly 200 million people have left the interior of China to re-settle and seek work in China's dynamic cities, mostly located along the eastern part of the country. While they do not cross international borders, the distance and socioeconomic differences between rural and urban areas in China are very similar to an international move for many. The motivation is largely economic in that farmers can barely earn a few dollars a day in rural areas; new manufacturing jobs in coastal cities pay several times that.

3.4 DEMOGRAPHIC TRANSITION, MIGRATION, AND POLITICAL POLICY

3.4.1 What is the relationship between population structure and migration?

Even though people generally migrate to find/make a better life for themselves and for their families, the benefits and pitfalls of migration affect different countries and regions in very different ways. Demography (age structure) has a lot to do with understanding how such a differentiation occurs. (Hint: You may need to go back to Chapter 2 to review your understanding of the demographic transition model or click here): http://www.bbc.co.uk/schools/gcsebitesize/geography/population/population_change_structure_rev4.shtml.

The demographic transition model explains how countries experience different stages of population growth and family sizes, but the model also works well to understand sources and destinations for migrants. Before the explanation continues, take a look at the model to see if you can predict the stages during which you would expect large-scale emigration versus immigration. Do you feel confident in your guesses? Why or why not?

Geographers note that countries of emigration tend to be late in stage 2 or early stage 3, while countries of immigration tend to be late stage 4 or stage 5 countries. Let's think about this for just a moment. At what age are people most likely to take the risk to move to a completely different country to seek their fortune? Would you move to Slovenia next year, if you could earn triple your current salary? Most international migrants tend to be relatively young (18-35) during the prime of their working years. Countries that tend to have an abundance of working-age people also tend to be early in stage 3 of the model. Countries like Mexico, Guatemala, the Philippines, and India had large families a generation ago. As those children enter their working years, there are not enough jobs created due to an "oversupply" of laborers. Meanwhile, countries in late stage 4 find that because of low fertility rates a generation earlier, the economy now faces a shortage of working-age residents to do all kinds of jobs. Birth rates were high during the 1950's in the US, but as women had fewer children in subsequent generations, fewer workers were entering the workforce every year. Unsurprisingly, young workers from Mexico, El Salvador, Jamaica, etc., began to immigrate as the demand for their labor increased. In Europe, the same pattern occurred as those from higher fertility countries like Turkey migrated to lower fertility countries like Germany.

3.4.2 Why do some countries benefit from migration while others do not?

Demographic realities can be **push factors** or **pull factors** that serve to push people away from a place or pull them towards a place as explained in the previous paragraph. Push and pull factors can be cultural, economic, or ecological. Baptists might be "pulled" towards the American South, and Mormons might be "pulled" to Utah for cultural reasons. Meanwhile, the devastation of Hurricane Maria (2017) "pushed" hundreds of thousands of Puerto Ricans off the island. The most common destination for Puerto Ricans was Florida for economic (jobs), ecological (warm weather), and cultural (existing Spanish speakers) reasons. When the push factors and pull factors fit together nicely, migration can benefit both the sending and receiving regions.

Very often in the case of **forced migration**, however, migration does not benefit both parties. Those migrants who flee their country based upon claims of danger based upon race, religion, nationality, or other pertinent identifiers are known as **asylum seekers.** They seek a country willing to take them in permanently for fear of imprisonment, retribution, or death in their country of origin. When asylum seekers have satisfactorily demonstrated a claim in court, their status changes to **refugee.** Under international agreement, refugees cannot be forced to return to any country where they are deemed to be in danger, so refugee status provides displaced people with a legal protection against deportation. The number of refugees worldwide at the end of 2016 was 22.5 million, the highest total of refugees since the end of WWII. There were also 40.3 million people uprooted within their own country, known as **internally displaced people (IDP),** bringing the total number of displaced people to an all-time high of nearly 63 million. In recent years, large numbers of asylum seekers have left Syria, Iraq, Afghanistan, Palestine, Somalia, Sudan, Cuba, Venezuela, and Myanmar (formerly Burma) as wars and political conflict have endangered millions. During previous wars, wealthier countries in Europe, North America, and Australia accepted large numbers of asylum seekers, but anti-immigrant sentiments have risen in many parts of the world, leaving the majority of would-be refugees without refuge. Opponents of resettlement argue that the cost is just too great, and they fear that accepting refugees would encourage more unwelcome immigration in the future, so this has significantly reduced the number of migrants that countries are willing to accept.

Though twenty-first century anti-immigrant sentiment remains high in countries like the US, Australia, and the U.K., the benefits for receiving countries are well documented. Apart from the demographic advantages already described, employers and consumers tend to benefit markedly from the low-cost, readily-available supply of labor provided by immigrants. Everything from the cost of fruit, construction, fast food, and lawn care tends to be cheaper owing to an immigrant labor force. Evidence also indicates that wages in low-skilled jobs may be suppressed by immigrants, but the overall economic benefit is widely reported by economists to be favorable when unemployment rates are low. Immigrants have higher rates of employment, are more likely to start businesses and are less likely to commit crimes than the native-born population in the US Sending countries also tend to benefit from emigration in two ways. First, emigration provides an opportunity for young workers who cannot find employment at home. Second, emigrants tend to send home the majority of their overseas earnings. Money sent home by overseas workers is called **remittances.** Countries that send large numbers of workers overseas benefit from the large infusion of foreign capital into the local economy, which tends to spur new investment opportunities. When migration is working smoothly, both sending and receiving countries can benefit.

Besides low-skilled workers and refugees, a third category of immigrant has increased dramatically under globalization. Highly-skilled immigrants represent a unique contradiction. On the one hand, countries seek to increase border security, limit asylum seekers, and build walls. On the other, those same countries actively recruit and seek to attract immigrants with specific skills, training, and educational levels. Most commonly, wealthy countries like the U.K. regularly recruit nurses, scientists, and engineers from poorer countries to meet the needs of an aging population. As British residents have fewer children and society gets older, there is more demand specifically for healthcare workers of all kinds. As such, the best and brightest minds from poorer countries become attracted to the much higher wages outside of their countries and they leave—resulting in a **brain drain.** Nearly sixty percent of all doctors born in Ghana and eighty-five percent of nurses born in the Philippines have left the country to work elsewhere! Wealthy countries reap the benefits of this **brain gain.** Although highly-skilled and educated individuals and families that emigrate for higher pay undoubtedly benefit from emigration, the countries that experience brain drain persistently lose a very valuable resource. Besides losing most of its doctors, Ghana also now faces a major shortage of those qualified to teach the next generation of medical professionals, so the negative effect crosses multiple generations.

3.5 CULTURE, GLOBALIZATION, AND ECONOMICS OF MIGRATION IN THE TWENTY-FIRST CENTURY

3.5.1 What is the connection between globalization and increased international migration?

Globalization is defined as the set of forces and processes that involve the entire planet, making something worldwide in scope. Although 246 million people live outside of their birth country, more than 96 percent of the world's people do not ever move outside of their birth country, so in the case of migration, globalization might not be as powerful as once believed. While geographers try to understand where people move and why, a more significant question might be why so many people *do not* move. If you live in one of the places on Earth where you interact with people from all over the globe on a regular basis, then it might seem to you that globalization is operating at full speed. Yet, the vast majority of humans don't leave their home country and have limited interactions with people from other places, even if the products they consume and produce might be worldwide in scope.

As we entered the new millennium in 2000, scholars were convinced that a new age was upon us. Some went so far as to say that geography was "dead" and that place didn't matter. A best-selling book written by Thomas Friedman, *The World Is Flat* (2005), decreed that humanity was becoming more connected all the time to the point that it didn't matter if you were in the streets of Bombay or in a classroom at Harvard. The best minds and the best ideas would always rise to the top, regardless of their origin. Inherent in this argument was the assumption that more people would be on the move, and international migration would accelerate as people and products would just zoom across the "flat earth" at lightning speed.

In 2017, however, geography has re-emerged to re-stake its claim. While more people than ever are, in fact, living outside of their birth countries, there is a growing resistance across the planet to "outsiders." In response to the attacks of September 11, 2001, President George W. Bush firmly asserted that "every nation in every region now has a decision to make. Either you are with us, or you are with the terrorists." Perhaps that is the moment in which the earlier hope of a fully-integrated world with open borders and full mobility was deemed too optimistic. In the years since that speech, western nations have taken a collective stance against immigration often based upon religious or ideological grounds. In 2017, for example, Donald Trump called for a "total and complete shutdown of Muslims entering the United States" in response to a shooting in San Bernadino, California. Meanwhile, Britain moved to cut immigration levels dramatically as it exited the European Union and took a more isolationist position, and Australia moved to prevent refugees from arriving on its shores.

3.5.2 How is migration in the twenty-first century different from that of the twentieth century?

While the percent of those who migrate has not increased significantly in the twenty-first century, the destinations and origins of immigrants have shifted dramatically, so that more people are migrating to and from more places than ever before. Such a statement may sound confusing at first, but one need look no further than college campuses throughout the world to understand this dynamic. Georgia Gwinnett College in suburban Atlanta, for example, is the most ethnically-diverse college in the American South. In a typical geography class at that school, thirty-three percent speak a language other than English at home. A recent survey of the author's students (n=115) from just one semester found that students (or their families) came from Mexico, El Salvador, Peru, the Philippines, Laos, China, Colombia, Korea, Haiti, Jamaica, Vietnam, Ukraine, Dominican Republic, Liberia, Latvia, Romania, India, Pakistan, Scotland, and Paraguay. While past waves of migration were dominated by a small number of sending countries and a small number of receiving countries, migration in the twenty-first century is much less predictable.

Australia offers another example, as thousands of prospective immigrants have been traveling thousands of miles by land and sea to seek refuge in the small continent. The government eventually made the difficult decision to intercept would-be immigrants at sea, but it still faces a complex dilemma of where to redirect those who have risked their lives to make the perilous journey. For now, they are being taken to remote islands of Papua New Guinea, Christmas Island, and Nauru for resettlement, resulting in an odd mix of refugees and South Pacific Islanders living side by side. Unpredictability is now the rule rather than the exception as twenty-first century technology allows for a rapid flow of information. Those wishing to move can now find out much more quickly about opportunities, transportation options, and routes to take than has ever been the case before. Geography does not represent nearly as formidable of a barrier to travel than it did in the twentieth century. Other than asylum seekers, most immigrants are far more likely to travel by plane than by boat in the twenty-firs century, and they may or may not seek permanent residence.

The final way in which migration is different today relates to the concept of **transnationalism** (exchanges and interactions across borders that are a regular and sustained part of migrants' realities and activities that transcend a purely "national" space). Those traveling to the US in 1900 were leaving everything behind to seek a new homeland, learn a new culture, and often to speak a new language. Migrants of today, however, are not forced to fully disconnect from "home." Even after a long-distance migration, people can stay connected to friends, family, news stories, and relationships across the world. A journey that took several months in 1900 today takes less than half a day. Meanwhile, apps like Facebook, WhatsApp, Snapchat, and Skype allow newcomers to stay intimately involved in the lives of those left behind—for *free*! **Transnational** families contain members that are living in multiple countries simultaneously, speak multiple languages, and are ready to move in a moment's notice, based upon market conditions in any given place at a particular time. Critics argue that immigrants of today do not assimilate as readily as those from the nineteenth and twentieth centuries, but in that era, people simply had no other choice. **Transnationalism** is a feature of twenty-first century immigrants that is here to stay as immigrants to new places have the option to live their lives in more than one world, carry more than one passport, and move more seamlessly from place to place than at any time in human history. Take note that the term appears again in this textbook (section 11.4, Political Geography) in that corporations also operate within and between multiple countries, depending upon various factors. An important distinction exists, however, from the concept of *supranationalism* (Section 8.3, Political Geography) in that supranationalist organizations do not officially reside under the direction of any single state (country), while transnationalism involves moving between countries – not operating outside of them.

3.6 THE FUTURE OF HUMAN MOVEMENT AND CONCLUSION

The future dynamics of migration are very difficult to predict, but certain geographic realities provide clues for where patterns are likely to change. In those parts of the world where societies are quickly aging and fertility rates continue to decline, we can expect rates of immigration to increase. Certain countries have experienced and will continue to experience dramatic population decline (most notably Germany, Italy, Russia, and Japan), and the demand for young, working-aged immigrants will certainly continue to draw more people to those places. Japan and China represent unique cases because both are aging and yet have been resistant to allowing outsiders to become citizens or permanent residents. In spite of cultural preferences for ethnic homogeneity, it would seem likely that the culture will shift and become more accepting of outsiders, as the country needs them to take care of the elderly, pay taxes, and provide an infusion of energy into the respective countries.

Meanwhile, people living in places without sufficient opportunities will continue to move away in search of jobs and better circumstances, regardless of the attempts made by wealthy nations to keep them out. Source countries and destinations will continue to shift, as they always have. More Mexicans have returned than have left the US in recent years, and it is no longer the leading sending source country of migration to the US. Rather, more people are now coming from Central America. The demographic pressures in many African countries will absolutely drive more working-aged people out of the continent in search of better opportunities even as the journey becomes increasingly dangerous. As the Internet becomes more pervasive on that continent, more people will find the information that they need in order to plan their emigration. Finally, the highly-skilled people of the world will continue to be increasingly mobile and largely unaffected by borders or increased security. Computer programmers, nurses, doctors, engineers, and high-tech workers of all sorts will use the globe to their advantage and seek out places that best fit their desires. Television shows like "House Hunters International" demonstrate how the vast numbers of people who work online or in highly-skilled careers can virtually live anywhere. Wealthy people from across the globe are leaving their passport countries by the millions, as Wi-Fi networks are now available across the world. The poor and wealthy alike will continue to move about the planet, reconstituting the human geography of our world well into the twenty-first century but for very different reasons and with very different experiences.

3.7 KEY TERMS DEFINED

Asylum seeker – those who leave the sovereign territory of one country in order to achieve refugee status in another, based upon claims of danger because of race, religion, nationality, or other pertinent identifiers.

Brain drain – the collective loss of skills, education, training, and wealth that occurs when highly-skilled and educated people move away from a country (usually away from a relatively poor country).

Brain gain – the collective gain of skills, education, training, and wealth that occurs when highly-skilled and educated people move into another country (usually to a relatively wealthier country).

Brain waste – a phenomenon in which international migrants with high levels of education and/or training often are not eligible to work in their area of training due to regulations or certification requirements, resulting in a "wasted" potential in certain groups.

Diaspora – a group of people sharing a common historical and ethnic connection to a territory, but who no longer live in that territory or country. Some members of a diaspora may have been removed from the traditional homeland for multiple generations but still identify with it as a "homeland."

Ecumene – human inhabited areas of Earth

Emigrant – an individual who moves *away* from one country into another for a prolonged period. The definition of "prolonged" varies by country and is defined by the World Bank as minimum of one year.

Forced migration – a type of movement in which individuals or groups are coerced into moving by an external set of forces, most notably environmental, economic, social, or political factors.

Globalization – all those processes, technologies, and systems that result in greater connections, communication, and movement among increasingly distant people and places on Earth.

Guest worker – someone without legal permanent status who has been granted permission to reside in a country's territory in order to work for a specific set of time on a particular kind of work.

Highly skilled migration – patterns movement by those with skills that are in high demand on the global market. Examples include nurses, doctors, IT specialists, actors/artists, and athletes who tend to enjoy greater levels of movement across borders than others.

Highly skilled migration – patterns movement by those with skills that are in high demand on the global market. Examples include nurses, doctors, IT specialists, actors/artists, and athletes who tend to enjoy greater levels of movement across borders than others.

Immigrant – an individual who moves for a prolonged period *to* another country. The definition of "prolonged" varies by country. In 2016 there were 246 million immigrants in the world.

Internally Displaced People (IDP) – those who have moved or been forced to move from a homeland for the same reasons as refugees but have not crossed an international boundary and do not have refugee status.

Laws of migration – generalizations about international migration as detailed by nineteenth-century demographers

Migration – a permanent move to a new location

Net migration – the difference between the number of immigrants and the number of emigrants in any given year

Points system – a national immigration policy that seeks to attract people with a specific set of skills, experience, and job training to satisfy unmet demand among those currently in the country. Regardless of origin country, anyone with the prescribed set of skills, linguistic ability, and education may apply to migrate to that country if they have acquired enough points to do so. Canada, Australia, New Zealand, and England all have a points system.

Pull Factor – those forces that encourage people to move *into* a particular place

Push Factor – those forces that encourage people to move *away* from a particular place

Refugee – an individual who, owing to a well-founded fear of being persecuted for reasons of race, religion, nationality, membership of a particular social group or political opinion, is outside the country of his nationality, and is unable, or unwilling, to avail themselves of the protection of that country. An individual who has been granted "refugee" status is afforded a certain set of rights and privileges, most notably, the right not to be forcibly returned to the country of origin.

Remittances – money sent "home" by international migrants. Remittances represent the largest single source of external funding in many developing countries. The global figure for 2016 was US$600 billion.

Return migration – a return of a migrant to the country or place of origin

Transnationalism – exchanges and interactions across borders that are a regular and sustained part of migrants' realities and activities that transcend a purely "national" space.

Undocumented migrants – those inside of a country without proper authorization or proof of residence.

Visa – the legal permission granted by a receiving country to those seeking to enter. Examples include tourist, temporary work, and student visas. A visa is different from a passport.

3.8 WORKS CONSULTED AND FURTHER READING

Boyle, Paul, and Keith Halfacree. *Exploring contemporary migration*. Routledge, 2014.

Castles, Stephen, Hein de Haas, and Mark J. Miller. *The age of migration: international population movements in the modern world*. New York: Guilford Press, 2014.

Collins, Nathan. "Why — and When Did Early Humans Leave Africa?" *Pacific Standard*, September 21, 2016. https://psmag.com/why-and-when-did-early-humans-leave-africa-c1f09be7bb70#.powpltf4t.

Cornish, Audie. "This Simple Puzzle Test Sealed The Fate Of Immigrants At Ellis Island." *National Public Radio*, May 17, 2017. http://www.npr.org/2017/05/17/528813842/this-simple-puzzle-test-sealed-the-fate-of-immigrants-at-ellis-island.

Czaika, Mathias and Hein de Haas. "The Globalization of Migration: Has the World Become More Migratory?" *International Migration Review* 48, no. 2 (Summer 2014): 283–323. https://doi.org/10.1111/imre.12095.

Donald, Adam. "Immigration Points-Based Systems Compared." *BBC News,* June 1, 2016. http://www.bbc.com/news/uk-politics-29594642

Eurostat. "Migration and migrant population statistics." http://ec.europa.eu/eurostat/statistics-explained/index.php/Migration_and_migrant_population_statistics.

Global Migration Group: http://www.globalmigrationgroup.org.

International Organization for Migration. "Missing Migrants Project." https://missingmigrants.iom.int.

Jennings, Ralph. "Taiwan Gambles on Visa-Free Entry For Citizens of the Poorer Philippines." *Forbes*, May 15, 2017. https://www.forbes.com/sites/ralphjennings/2017/05/15/modern-taiwan-gambles-on-visa-waiver-for-citizens-of-the-poorer-philippines/#5a8489656ab7

Koh, Yoree. "Study: Immigrants Founded 51 percent of US Billion-Dollar Startups. The Wall Street Journal." *The Wall Street Journal*, March 17, 2016. https://blogs.wsj.com/digits/2016/03/17/study-immigrants-founded-51-of-u-s-billion-dollar-startups.

Migration Policy Institute. https://www.migrationpolicy.org.

NBC News. "Trump Calls for 'Complete Shutdown of Muslims Entering the US'" *Meet the Press*, December 7, 2016. https://www.nbcnews.com/meet-the-press/video/trump-calls-for-complete-shut-down-of-muslims-entering-US-581645891511.

Oxford Bibliographies. (2013). "The Chinese Diaspora." http://www.oxfordbibliographies.com/view/document/obo-9780199920082/obo-9780199920082-0070.xml.

Ratha, Dilip. "Understanding the Importance of Remittances." *Migration Information Source*, October 1, 2004. https://www.migrationpolicy.org/article/understanding-importance-remittances.

Russell, Sharon Stanton. "Refugees: Risks and Challenges Worldwide." *Migration Information Source*, November 1, 2002. https://www.migrationpolicy.org/article/refugees-risks-and-challenges-worldwide.

United Nations Department of Economic and Social Affairs/Population Division. *International Migration Report 2017.* http://www.un.org/en/development/desa/population/migration/publications/migrationreport/docs/MigrationReport2017.pdf.

United Nations Department of Economic and Social Affairs/Population Division. *International Migration.* http://www.un.org/en/development/desa/population/migration.

Voice of America. "Ghana Faces Worrying Brain Drain." *Voice of America News*, January 9, 2010. https://www.voanews.com/a/ghana-faces-worrying-brain-drain-81090582/111371.html.

World Bank Group. "Migration and Remittances Factbook 2016." https://openknowledge.worldbank.org/bitstream/handle/10986/23743/9781464803192.pdf?sequence=3&isAllowed=y.

Zong, Jie and Jeanne Batalova. "Annual Refugee Resettlement Ceiling and Number of Refugees Admitted, 1980-2017." *Migration Information Source*, June 7, 2017.   https://www.migrationpolicy.org/article/refugees-and-asylees-united-states#Admission_Ceiling

3.9 ENDNOTES

1. Public Papers of the Presidents of the United States: Lyndon B. Johnson, 1965. Volume II, entry 546, pp. 1037-1040. Washington, D. C.: Government Printing Office, 1966

4 Folk Culture and Popular Culture

Dominica Ramírez and David Dorrell

Figure 4.1 | Double Decker Bus
Double decker bus referencing the Beatles in Ferrol, Spain near the historic Camino de Santiago. This is the intersection of two iconic cultural symbols. Both the bus and the pilgrimage route invite the public to take journeys, whether sonic or geographic.
Author | Dominica Ramírez
Source | Original Work
License | CC BY SA 4.0

STUDENT LEARNING OUTCOMES

By the end of this section, the student will be able to:

1. Understand: the origins and diffusion of culture and globalization

2. Explain: how culture changes across space and time

3. Describe: popular and folk culture, diffusion and the changing pace of globalization

4. Connect: globalization and cultural conflict

CHAPTER OUTLINE

4.1 INTRODUCTION

What is **culture**? When some people speak of culture, they are thinking of high culture (e.g. ballet or opera). Others may think of current, prominent topics (i.e. pop culture). Academic settings, though, are referring to something else. Culture is a learned behavior and a human construct. Culture exists to answer questions. Some of the questions that are answered are philosophical or ideological, for example, "Where did we come from?" or "What is acceptable behavior?" Other questions revolve around daily life. "How do we secure shelter, clothe ourselves, produce food, and transmit information?" Culture provides us with guidance for our lives. It both asks and answers questions. Children, from an early age, start asking "What is my place in the world?" Culture helps to answer that.

The word culture itself comes the Latin word *cultura*, meaning cultivation or growing. This is precisely what humans do with both their material and abstract cultural components. Humans, since early childhood, learn to shape, create, share, and change their culture. Culture is the very vehicle we use to navigate through our environments. Culture is a form of communication and it evolves. And as a type of compass, it leads.

4.1.1 Components of Culture

At the most simplistic levels, culture can be either concrete and tangible, or abstract. Either way, culture is used to express identity for both individuals and groups. And whether it is concrete or abstract, again it is a human construct used as a way to create a sense of belonging. People convey culture through various outlets such as festivals, food, and architecture. They are able to meet their worldwide fundamental needs while maintaining individual group qualities. Culture can be classified into three different categories: mentifacts (ideas or beliefs), artifacts (goods or technology), and sociofacts (forms of social organization).

All cultures have underlying beliefs and thought processes. These beliefs include religion (Chapter 6) and language (Chapter 5) but go far beyond that. Other important beliefs include things like nationalism (see Chapter 8), customs or prejudices. These beliefs can be expressed through various avenues, using a variety of auditory, visual, and tactile means. For example, nationalism can be expressed through song, cuisine, dress, and public events. You can hear nationalism in the form of an anthem, see it in the form of a flag, taste it when you eat a dish that represents a group of people, and feel it in the form of a piece of jewelry.

Technology is a human construct. From our earliest inventions (fire, weapons) to a supercomputer, the things that people build are products of their perceived needs, their technical abilities and their available resources. Technology includes clothing, foods and housing. Another name for technology is **material culture**. Think of material culture as the material that archaeologists study. The materials that we use are often left behind for later people to study, like the clay tablets the Sumerians used to record their writing or the remnants of an Iroquois longhouse used for shared living. Other components of culture (like ideas) leave fewer traces. We can find material culture related to burial practices that date back millennia but may not always have the material evidence to show how people grieve.

Lifestyle is a component of culture that can be overlooked, but it is vitally important. In many cultures, a family is a very large unit and people can tell in great detail their exact relationship to everyone else in a place. In the modern context, a family could consist of a single parent and a child and it is possible to live in a neighborhood filled with unrelated people and not know the name of a single neighbor.

Culture can be seen either through the lens of a microscope or through that of a telescope. Folk culture is local, small and tightly bound to the immediate landscape. Popular culture is large, dispersed, and globalizing. These two forms of culture are not totally separated. They are related and both currently exist in the world. Prior to about 2008, most people on Earth lived in rural communities, often practicing a folk culture. The world as a whole is moving toward popular culture.

4.1.2 Cultural Reproduction

As human beings, we reproduce in two ways: biologically and socially. Physically we reproduce ourselves through having children. However, culture consists solely of learned behavior. In order for culture to reproduce itself, it has to be taught. This is what makes culture a human creation. How is culture transmitted? Human beings are natural mimics. This is the way we learn to speak, and it is the way we learn the rest of our culture as well. We learn through observation and, subsequently, through practice. At another scale, mimicry is the mechanism that drives cultural diffusion. Human beings copy the things we like. The old line "Imitation is the sincerest form of flattery," perfectly describes the human desire to incorporate successful adaptations.

Figure 4.2 | Ribs restaurant in Madrid, Spain
This restaurant is near one of the busiest tourist areas in the city, but it is on a side-street off the iconic Gran Vía. Could you imagine this restaurant on a street corner in the United States?
Author | Dominica Ramírez
Source | Original Work
License | CC BY SA 4.0

Figure 4.3 | Spanish restaurant in Tivoli Gardens, Copenhagen, Denmark
Notice the use of yellow and red (the colors of the Spanish flag). Also, the bull and matador image is prominent. Is that the best way to promote churros, a typical fried food of Spain?
Author | Dominica Ramírez
Source | Original Work
License | CC BY SA 4.0

How people have shared culture has changed drastically over time. This can partly be attributed to the channels in which we share culture. In the past, culture was shared orally and in person. Words eventually became written, the written became electronic, and we now have access to things from anywhere at any time.

The debate over authenticity has also become a topic of debate. If culture is created, recreated, and is fluid, how can we define authenticity? Has culture become **placeless**? Placelessness or the irrelevance of place has become a central topic in the contemporary philosophy of geography. Can people have an All-American dining experience in the heart of Madrid, Spain? Can they experience Spanish street-food at a historic theme park in Copenhagen, Denmark? As the following pictures show, icons representing other places are now common in the landscape.

4.1.3 Culture Hearths

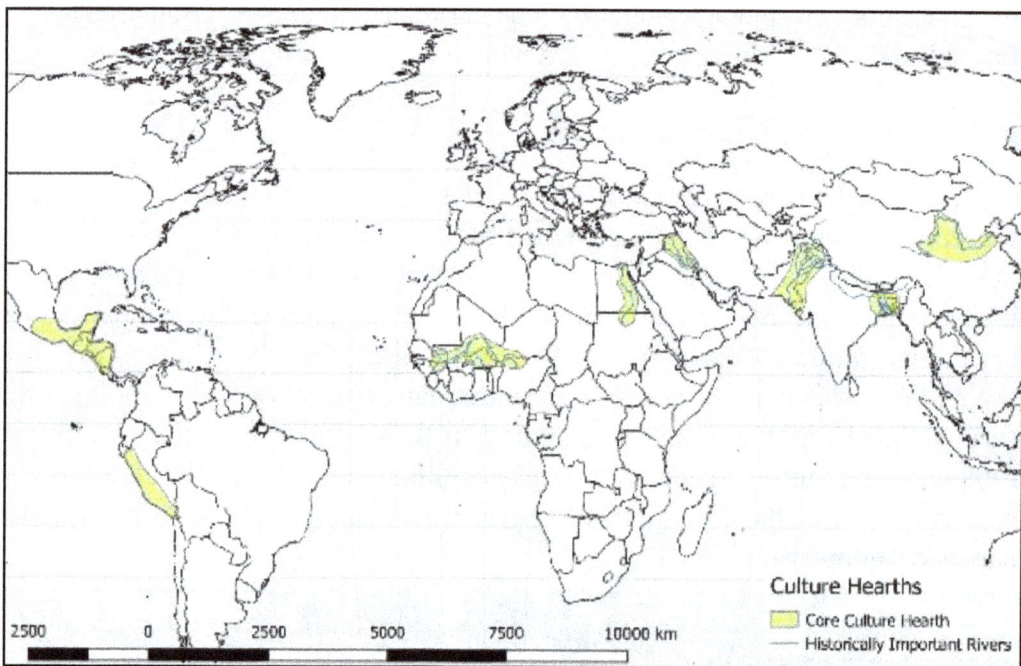

Figure 4.4 | Depicts culture hearths and their associated rivers, where applicable[1]
Author | David Dorrell
Source | Original Work
License | CC BY SA 4.0

Human beings have always had learned behaviors, it's one of the defining characteristics of human beings. Cultural evolution describes the increasing complexity of human societies over time. Our earliest cultures were simple. We lived in small groups, ranged across fairly large areas, and lived off the natural landscape. Human impact on the environment is less than it is now, but there was an impact. Earlier peoples burned forests to clear land and flush game, and in some places hunted the megafauna to extinction.

Recognizable cultures have places of origin. The word culture refers to cultivation or growing. We care for something, we nurture an idea as it grows. All cultural elements have a place of origin. Some places have been responsible for a great deal of cultural development. We call these places **culture hearths**. Culture hearths provided many of the cultural elements (technologies, organizational structures, and ideologies) that would diffuse to other places and later times. Cultural hearths provide operational scripts for societies.

Culture hearths are closely associated with the foods that they domesticated. food is an important cultural element due to the fact that it is both a technology as well as form of expression. Although the preceding map shows areas of ancient civilization, these are not the only places that have contributed to contemporary cultures. Ideas can arise anywhere, but ancient ideas collected in these places.

Cultures incorporate pieces of other cultures. Some recognizable themes in creation myths are stories of great floods or other cataclysmic events, and these types of stories recycled through history. Languages without writing will often reuse another language's writing system. Once again, desirable characteristics get copied.

4.2 THE CULTURAL LANDSCAPE

Cultures' beings rely on natural resources to survive. In the case of rural cultures, those resources tend to be local. For urban cultures, those resources can either be local, or they can be products brought from great distances. Either way, cultures influence landscapes and in turn landscapes influence cultures.

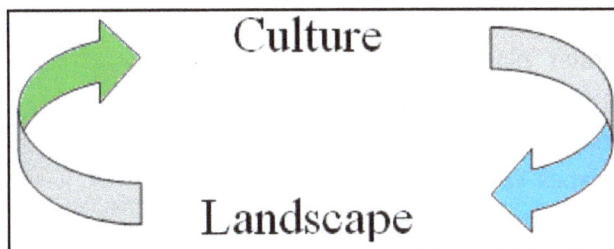

Figure 4.5 | Culture-Landscape Relationship
Author | David Dorrell
Source | Original Work
License | CC BY SA 4.0

The physical landscape consists of places like the Appalachian Mountains that stretch across a large portion of North America, the Mongolian-Manchurian grasslands, the Amazon river basin, or any other environment. These are landscapes that have been formed over thousands, if not millions of years by forces of nature. In order to live in places a different as these, humans have needed to adapt their lifestyles. The relationship between people, their culture, and the physical landscape is known as human-environment interaction. This relationship is reciprocal; culture adapts to a particular place, and that place is changed by people. **Cultural ecology** refers to the types of landscapes created by the interaction of people and their physical environment.

Humans have been thinking about the relationship between people and their environments for a considerable amount of recorded history. In the book *On*

Airs, Waters, and Places, the Greek philosopher Hippocrates wrote that different climates produced different kinds of people. He believed that cold places produced emotionally distant people, and hot climates produced lazy, lethargic people. The ideal place (which coincidentally was his own place) was in the middle of the known world and produced the best kind of people. These ideas would now be considered environmentally deterministic. Remember that Environmental determinism is the idea that a particular landscape necessarily produces a certain kind of people. Ideas like this were still fashionable into the twentieth century. The problem with the idea is that it's simplistic and reductionist. A cold environment doesn't force people to be aloof. It forces them to invent warmer clothes. Technology is the difference, not behavior. Instead of determinism, the more common term to use is now possibilism. Physical landscapes set limits on a group of people that may or may not require a large adaptation, or a large modification of the environment itself. Humans can now survive in very inhospitable environments, most notably, the International Space Station.

Landscapes are cultural byproducts. The way that we use the local resources generates the visible landscape. Architecture, economic activities, clothing and entertainment are all visible to anyone interested in looking at a place. Because the physical landscape varies across space, and because culture varies across space, then the cultural landscape is variable as well. Different people can have different adaptations to similar places. Conversely, places far from one another may have similar adaptations to climate or other factors.

Cultural landscapes can be considered as both history and narrative. Power is written into the landscape. We make statues to commemorate the wealthy and the politically connected in rich places. We place garbage dumps and airports in poor places. Looking at the landscape as a record of history, power, and representation is known as landscape-as-text. The landscape can be read in the same that a book can be read.

The largest differences between landscapes that we see now are the differences between the rural and the industrial and between places that are less integrated with the rest of the world and those that are heavily integrated (globalized). Global places are becoming homogenized.

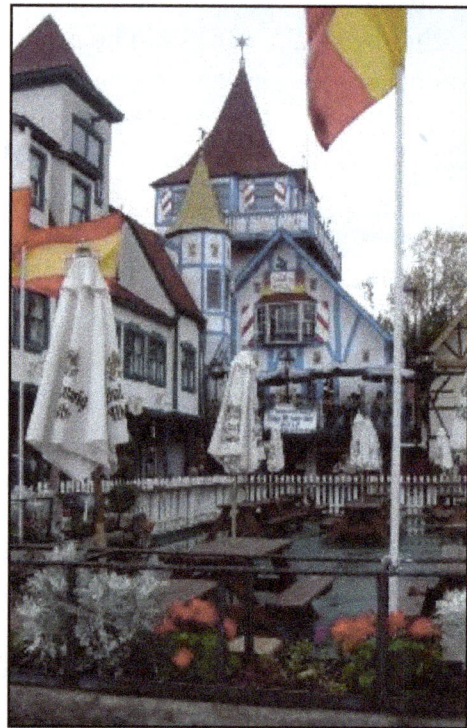

Figure 4.6 | Helen, Georgia
What is the narrative here? How have people changed/adapted the mountainous landscape of this region of Georgia, USA, to look like a town in mountainous Bavaria, Germany?
Author | Dominica Ramírez
Source | Original Work
License | CC BY SA 4.0

Cultural Change

A sensible question to ask might be "Where did all the cultures come from?" As people moved into new places, they adapted and changed, and the new places were changed in turn. People change over time as well. Circumstances change in a place. Groups who move into a forest will need a to adapt if they cut down all the trees. Groups that adopt a new crop will see their lives change. Divergence could be as simple as borrowing a word to describe an invention. All cultures change.

Culture Regions

We can sort the world into regions based on cultural characteristics. A region is an area characterized by similarity or a cohesiveness that sets it apart from other regions. Regions are mental constructs, the lines between places are imaginary. When someone talks about the English-speaking world or Latin America, they are referring to culture regions.

4.2.1 Cultural Case Study: The Diffusion of Dancehall

A cultural attribute *could* diffuse just about anywhere, but that isn't how diffusion usually works. Some places are interested in the innovation (new thing) and some are not. The following example takes one narrowly defined cultural attribute and traces a path to other places. Dancehall, a form of reggae, developed in Jamaica in the late 1970s and grew over time to prominence on the island. On of Jamaica's main exports is music and Dancehall became an exported commodity like many other musical forms. In the case of Dancehall, the music is sufficiently defined to allow us to see where it has diffused to. Large markets like North America are visible to us and we see Dancehall in some of the music of Drake or Rihanna, but other places may seem less obvious. In Brazil, artists such as Lai Di Dai has taken the genre and adapted it to local tastes (which may or may not include changing the language of delivery).

Figure 4.7 | Lai Di Dai
from the video for "Chega na Dança"
Author | Lai Di Dai
Source | YouTube
License | Fair Use

In Germany, the band Seeed has found great success with Dancehall.
And in Denmark we find artists like Raske Penge.

Figure 4.8 | Seeed
from the live video "Live 2013 (Berlin + Mönchengladbach)"
Author | Seeed
Source | YouTube
License | Fair Use

Figure 4.9 | Raske Penge
from the video "Original Bang Ding"
Author | Raske Penge
Source | YouTube
License | Fair Use

This musical style has moved far beyond its origin. Similar diffusion is found with other musical genres, from the earliest forms of pop music through to the present day.

4.3 FOLK CULTURE

The term *folk* tends to evoke images of what we perceive to me traditional costumes, dances, and music. It seems that anything with the prefix *folk* refers to something that somehow belongs in the past and that is relegated to festivals and museums. The word *folk* can be traced back to Old Norse/English/Germanic and was used to refer to an army, a clan, or a group of people. Using this historic information, folk culture (folktales, folklore, etc.) can be understood as something that is shared first among a group of people and then with the more general population. It is a form of identification. Folk is ultimately tied to an original landscape/geographic location as well. Folk cultures are found in small, homogeneous groups. Because of this, folk culture is stable through time, but highly variable across space.

Folk customs originate in the distant past and change slowly over time. Folk cultures move across space by relocation diffusion, as groups move they bring their cultural items, as well as their ideas with them.

Folk culture is transmitted or diffused in person. Knowledge is transmitted either by speaking to others, or through participating in an activity until it has been mastered. Cooking food is taught by helping others until an individual is ready to start cooking. Building a house is learned through participating in the construction of houses. In all cases, folk cultures must learn to use the resources that are locally available. Over time folk cultures learn functional ways to meet daily needs as well as satisfy desires for meaning and entertainment. Folk cultures produce distinctive ways to address problems.

Houses tend to be similar within a culture area, since once a functional house type is developed, there is little incentive to experiment with something that may not work. Food must be grown or gathered locally. People prefer variety, so they produce many crops, plus relying on only a few foods is dangerous. Clothing is made from local wool, flax, hides, or other materials immediately available. Local plants serve as the basis of folk medicinal systems. People are entertained by music that reinforces folk beliefs and mythologies, as well as reflects daily life. Folktales or folklore exists as foundational myths, origin stories or cautionary tales.

Holidays provide another form of entertainment. Special days break the monotony of daily life. A holiday such as Mardi Gras, which has its roots in the Catholic calendar provides an occasion to flout cultural norms and relieve tension. Another way of providing escape from monotony is provided by intoxicants. Although often not considered when discussing culture, human beings have been altering their own mental states for millennia. The production of alcohol, cannabis, tobacco or coca demonstrates that folk cultures understood the properties of psychoactive substances. Later these substances would be commercialized into modern products.

As folk cultures have receded there has been a return to valuing the folk. The Slow-food movement and the growth of **cultural tourism** has largely been driven for the desire experience elements of folk culture. As early as the German Grimm Brothers (19th Century Germany) people have wanted to preserve and promote folk culture. John Lomax (1867-1948) traveled the United States trying to record as many folk songs and folk tales (including slave narratives) as they faded from human memory.

Folk culture can also be expressed as craftsmanship versus factory work. Hand production of goods requires a great amount of knowledge to select materials, fabricate components, assemble and finish a product. Contrast this with industrial production in workers need to know very little about the final product, and have little relationship with it. This difference in modes of production was first discussed by Ferdinand Tönnies and *Gemeinschaft* and *Gesellschaft*. These two words denote the relationship between people and their communities, and by extension, their landscapes. *Gesellschaft* is the way life is lived in a small community. *Gemeinschaft* is the way that life is lived in a larger society.

4.4 THE CHANGING CULTURAL LANDSCAPE

It is understood that folk culture has been declining in the face of popular culture for some time. What is driving this decline? There are many things, with different underlying processes. Politically, in the last few centuries many places have been incorporated into states. These states have often pursued nationalistic policies that made life difficult for minorities of nearly every variety. The growth of a state-sanctioned national culture is the beginning of popular culture. Something as innocuous as public schooling or an official language can serve as a vehicle for promoting national values. Even if there is some overlap between the old culture and the new, the old has been prised loose from its central position in communal life.

Economics also plays a role. Small, rural communities have been shrinking globally for centuries, since the beginning of the Industrial revolution. Leaving the spatial confines of a folk culture makes reproducing that culture very difficult, due to its close connection to place. When people migrate to places practicing popular cultures, the pressure to acculturate and assimilate are tremendous.

Changes in infrastructure has also aided the diffusion of popular culture. Roads bring in outside people, as well as reduce the friction of leaving a place. The internet has dramatically reduced the friction of distance regarding the diffusion of popular culture. It took decades for tomatoes to diffuse from the Americas to Italy, but we know about a new iPhone months before it even gets released.

The United States is huge laboratory of cultural interchange. Innumerable distinctive folk cultures were already in the Americas when the Europeans arrived. Waves of people from folk cultures arrived for decades, and they changed the larger culture of the United States.

Places of folk culture aren't the only places that are changing. Many places with an established popular culture are subject to interaction between different pop culture spheres. At one time immigrants to the United States came from folk cultures. Now they are often from areas of popular culture. The Spanish-speaking world has its own pop music stars, and chart-topping musicians from different countries will often collaborate together garnering airplay and sales around the world. World regional cuisines are subject to becoming fads as well.

4.5 POPULAR CULTURE

Popular culture is culture that is bought. Think about your daily life. You work to buy food and clothing, pay your rent, and entertain yourself. The origin of each ingredient in your food could be hundreds or even thousands of miles in either direction. Your clothing almost certainly wasn't made locally, or even in this country. Your house might look just like any house in any subdivision in North America, placeless and with little connection to local resources.

Popular culture is driven by marketing. Entire industries exist to convince us that our desires and needs will be best met through shopping. Why is this? Because without sales, the companies that produce pop culture will go bankrupt.

Popular culture industries must continuously reinvent themselves. Being popular today is not a guarantee of longevity. In order to convince consumers that last year's t-shirt is now unacceptable, it is necessary to promote **fashion**. Fashion is not just a concept related to clothing. It is the reason that automobile companies make cosmetic changes to their products every year. It is why fast food restaurants continually change some parts of their menus. Without the cachet of fashion, consumers may feel socially disadvantaged. This explains why some people with very limited incomes will spend money on expensive luxuries.

In terms of popular culture holidays are simply reasons to sell merchandise. The commercialization of Christmas has been increasing for over a century in western countries. Now it is possible to see Christmas displays in Japan or China, places with few Christians, but many available consumers. The same sort of marketing can be seen in the expansion of Halloween globally, and in the growth of Cinco de Mayo in the United States.

Hierarchical diffusion plays prominently in popular culture diffusion. Larger places tend to generate many of pop culture's hit songs, clothing styles, and food trends. Diffusion in popular culture is highly related to technology. Although it wasn't invented as such, the internet has become a venue for advertisement. Clickbait headlines and ad revenue have created an atmosphere where every conversation is a sales pitch.

This hierarchical diffusion means that innovations tend to diffuse from large, well-connected places to other large-well-connected places first, then trickle-down to smaller and smaller places. The gap in time that it takes for a new idea or product is known as cultural lag. In some places, there is almost no cultural lag. To very remote places, some innovations take a very long time to arrive. Bear in mind that there are places in the United states that still have no internet service.

Popular culture covers large populations with access to similar goods and services, but the pressing need to sell drives almost incessant modification, generally at a superficial level. Because of this we usually describe pop culture as stable across space, but highly variable across time.

Figure 4.10 | Fast Food Restaurant
Where is this fast food restaurant? It could be almost anywhere on Earth. In this case it is in Malmö, Sweden.
Author | Dominica Ramírez
Source | Original Work
License | CC BY SA 4.0

The commodification of folk intoxicants mentioned earlier has had a decided effect on the modern world. Low alcohol beers have a minimal effect on the human body compared to commercial distilled spirits. The opium poppy, dangerous enough in its raw form, can be processed into heroin. The same can be said of coca (now used to produce cocaine) or tobacco. In folk form, these substances tended to be used for ritual purposes. In the context of popular culture, they are used in great quantity. Nevertheless, old folk patterns are still visible in the pop culture landscape. Italians still drink wine, a product they have produced for centuries, only now it might be bought from somewhere else. The Russian climate was good for producing grain and potatoes, which eventually was distilled into vodka.

Selling culture goes well beyond just food and clothing. Popular music and other forms of entertainment (video games, movies, etc.) are huge commercial entities marketing products well outside their places of origin. Movies made in the United States are often being made with the understanding that their international box office sales will be larger than their domestic sales. This is also true of other mass media products. These products are often related to other pop culture products. Companies rely on the familiarity consumers will have with a movie character in order to sell clothing, toys, video games, conventions and more movies in that series. When Marshall McLuhan wrote that "The medium is the message" he meant that television would be able to sell itself. The same expression could be expanded to pop culture in general.

The following graphic shows the top ten music charts for the week of September 10-16, 2017. Look closely at them.

US Top Ten
1. **Attention** by Charlie Puth
2. **There's Nothing Holdin' Me Back** by Shawn Mendes
3. **Slow Hands** by Niall Horan
4. **Believer** by Imagine Dragons
5. **Strip That Down** by Liam Payne featuring Quavo
6. **Wild Thoughts** by BJ Khaled featuring Rihanna & Bryson Tiller
7. **Look What You Made Me Do** by Taylor Swift
8. **Despacito** by Luis Fonsi & Daddy Yankee featuring Justin Bieber
9. **No Promises** by Cheat Codes featuring Demi Lovato
10. **Feels** by Calvin Harris featuring Pharrell Williams, Katy Perry, & Big Sean

Croatia Top Ten
1. **No Roots** by Alice Merton
2. **More Thank You Know** by Axwell & Ingrosso
3. **Feels** by Calvin Harris featuring Pharrell Williams, Katy Perry, & Big Sean
4. **New Rules** by DuaLipa
5. **Ok** by Robin Schulz featuring James Blunt
6. **Mi Gente** by J Balvin & Willy William
7. **Look What You Made Me Do** by Taylor Swift
8. **Frka** by Nipplepeople
9. **Sign Of The Times** by Harry Styles
10. **On My Mind** by Disciples

Colombia Top Ten
1. **Olha A Explosao** by MC Kevinho
2. **Feels** by Calvin Harris featuring Pharrell Williams, Katy Perry, & Big Sean
3. **Walk on Water** by Thirty Seconds to Mars
4. **La Estrategia** by Cali Y El Dandee
5. **Vivo Pensando En Ti** by Felipe Peláez & Maluma
6. **Besame** by Valentino, Manuel Turizo
7. **Bonita** by J Balvin & Jowell & Randy
8. **Robarte Un Beso** by Carlos Vives & Sebastian Yatra
9. **Shape of You** by Ed Sheeran
10. **Bailame** by Nacho

India Top Ten
1. **Mi Gente** by J Balvin & Willy William
2. **Walk on Water** by Thirty Seconds to Mars
3. **Makeba** by Jain
4. **Despacito** by Luis Fonsi & Daddy Yankee featuring Justin Bieber
5. **Hawayein** by Pritam & Arijit Singh
6. **Attention** by Charlie Puth
7. **Qismat** by Ammy Virk
8. **Feels** by Calvin Harris featuring Pharrell Williams, Katy Perry, & Big Sean
9. **Mere Rashke Qamar** by Nusrat Fateh Ali Khan
10. **Dusk Till Dawn** by ZAYN & Sia

Mexico Top Ten
1. **El Cido** by Timbiriche
2. **Ready For It?** by Taylor Swift
3. **Una Lady Como Tu** by Manuel Turizo
4. **Mi Gente** by J Balvin & Willy William
5. **Look What You Made Me Do** by Taylor Swift
6. **Robarte Un Beso** by Carlos Vives & Sebastian Yatra
7. **Felices los 4** by Maluma
8. **Me Rehúso** by Danny Ocean
9. **Bonita** by J Balvin & Jowell & Randy
10. **What Lovers Do** by Maroon 5 featuring SZA

China Top Ten
1. **A Million On My Soul** by Alexiane
2. **Despacito** (Remix) by Luis Fonsi & Daddy Yankee
3. **Look What You Made Me Do** by Taylor Swift
4. **Feels** by Calvin Harris featuring Pharrell Williams, Katy Perry, & Big Sean
5. **Attention** by Charlie Puth
6. **Go Away and Fly** by Jin Zhiwen
7. **Shape of You** by Ed Sheeran
8. **You Be Love** by Avicii
9. **Mi Gente** by J Balvin & Willy William
10. **Faded** by Alan Walker

Figure 4.11 | Top Ten Charts for Select Countries

Notice the amount of repetition from one chart to the next? Not every chart is a copy of the others, but the similarities certainly bear out the idea of an international music industry.

Compiled By | David Dorrell
Source | Original Work

4.6 THE INTERFACE BETWEEN THE LOCAL AND THE GLOBAL

The basis of popular culture is commerce. As long as a product can be sold, it can survive in the marketplace. This brings up an interesting process. Commodification is the process in which a cultural attribute is changed into a mass-market product. Bear in mind that the mass-market product may not resemble the original product very much. Using a similar fast-food model as McDonald's, Taco Bell sells ostensibly Mexican products across the world with the notable exception of Mexico. Panda Express is similar for Chinese food. These chains are not in the business of making hamburgers, tacos, or orange chicken. They are in the business of making a profit. Authenticity is irrelevant, and probably harmful in the drive to sell more.

The reverse of this process also applies. One aspect of marketing is the incorporation of global products into the local market. Companies will change their products or their entire product lines, if doing so is cost-effective and generates a good return on investment. There is even a Central America-based fried chicken chain in the United States. Brands like Sony and Hyundai are highly visible in the American landscape. K-Pop bands tour suburban arenas that also host Celine Dion and Paul McCartney.

4.7 GLOBAL CULTURE

Globalization is the integration of the entire world into a single economic unit. This is associated with frictionless movement of money, ideas, and (to a lesser extent) people. This growing reality has created a newer type of popular culture, global culture.

Historically, popular culture was restricted to areas the size of States, or at the very most areas within culturally related spheres (e.g. the English-speaking world). United States culture was defined by a set of characteristics (language, law, settler colonial history, etc.) that translated to a few other places, such as Canada or Australia, but mostly remained place bound. This is no longer the case. As was mentioned previously, video games are designed in one country to appeal to a global market. The same is true of music, movies, clothes, smartphones and office productivity software.

At a superficial level at least, the components of life are becoming more homogeneous across large parts of the world. National popular culture producers are merging into international producers, and these international producers have global ambitions. Any sizable popular culture content distributor (EMI Records, Sony, Vivendi Universal, AOL Time Warner and BMG) is a transnational corporation. In fact, the five listed global record labels account for 90% of global music sales.

Starbucks, Toyota, Wyndham and others have helped reduce the friction of distance by reducing spatial variation. They aren't doing this to help people, or to hurt them. Although they will cater to local needs to some degree, they are not in the business of promoting local flavors.

William Gibson wrote "The future is already here, it's just not evenly distributed." In terms of globalization, he was correct. There are still people living in remote areas practicing something similar to a paleolithic lifestyle. On the other end of the scale there are people with great wealth who have access to powerful technologies and are able to live anywhere they desire.

Sometimes globalization even has an effect on folk culture. In many places, economic realities have forced people to perform religious activities of relive special events for tourist dollars. Attending luaus in Hawaii or watching voladores in Veracruz in a quest for authenticity is in itself changing the folk culture that is the center of attention.

This assessment may seem particularly bleak for folk culture, but it isn't necessarily completely bad. People survive, and they try to keep the practices that are most valuable to their lives. Folk cultures have a much longer timeline than pop cultures, and have proven to be resilient.

4.8 RESISTANCE TO POPULAR CULTURE

Although popular culture has been expanding rapidly at the expense of folk culture, it is not without resistance. Although it is not strictly-speaking true, global culture is perceived as largely corporate, secular, and western. Each of these aspects have their own critics.

Anti-globalists fall into two main groups. The first are leftists who oppose the power that has accumulated to corporations and the authoritarian state. The other group are rightists who prefer that power be centered at the state level, and who fear that globalization naturally undermines state sovereignty.

In some places, globalization is the same thing as modernization is the same thing as westernization which is perceived as secular (or even atheistic), materialistic and corrupt. Movements such as Al-Qaeda or Islamic State are violently opposed to popular culture, although they would not have a problem if their idea of the ideal culture were to become fully global. Rejecting modern popular culture often also involves elevating a nostalgic, often imaginary golden age as the only acceptable model of society.

Many people feel that symbols and representations of popular culture are erasing the very personality of regional cultures. Resistance to popular culture can come in many forms. Let us revisit the concept of fundamental needs, which are universal, and how they vary geographically. Locally sourced food products, customs, and recipes are pitted against global fast-food giants that provide inexpensive and easy access to ready-made food products. There has been an uprising of farm-to-table businesses, the Slow Food movement, and an emphasis on fair-trade products.

4.9 SUMMARY

Culture, the learned portion of human behavior is a very broad and very deep topic of human study. Historically humans have lived in small groups practicing folk culture. This was particularly true of the cultures the sprang from the diffusion of agriculture. Many of the folk culture attributes date to this time of human development. The industrial age also ushered in the era of popular culture. Pop culture provides the cues that people use to live, work, and interact. Relatively recently has been the rise of global culture, a phenomenon in which large numbers of people in diffuse places are committing the same or similar culture practices.

4.10 KEY TERMS DEFINED

Commodification: The process of transforming a cultural activity into a saleable product.

Cultural ecology: Study of human adaptations to physical environments.

Cultural Landscape: Landscapes produced by the interaction of physical and human inputs.

Cultural reproduction: The process of inculcating cultural values into successive generations.

Cultural tourism: A variety of tourism concerned with exploring the culture of a place.

Culture: Learned human behavior associated with groups.

Culture Hearth: Historic location of cultural formation.

Fashion: The latest and most socially esteemed style of clothing or other products and behaviors.

Folk Culture: Culture practiced by a small, homogeneous, usually rural group. AKA Traditional culture.

Formal region: A region that has defined boundaries, often a governmental unit such as a country, province, or county.

Functional region: A region defined by a relationship, such as the market area of a product, a commuter zone or an employment market.

Globalization: The global movement of money, technology, and culture.

Heterogeneous: A population composed of dissimilar people.

Homogeneous: A population composed of similar people.

Material culture: The objects and materials related to a particular culture.

Perceptual region: Internally defined region that exists as the expression of a cultural type.

Placelessness: The state of having no place. In the modern context, a place exactly like any other place.

Popular Culture: Culture created for consumption by the mass of population.

Resistance: Actively pursuing a policy of obstruction of a particular process or undertaking.

4.11 WORKS CONSULTED AND FURTHER READING

Black, Jeremy. 2000. *Maps and History: Constructing Images of the Past*. Yale University Press.

"Catholic Pilgrimages, Catholic Group Travel & Tours By Unitours." n.d. Accessed March 16, 2013. http://www.unitours.com/catholic/catholic_pilgrimages.aspx.

Dicken, Peter. 2014. *Global Shift: Mapping the Changing Contours of the World Economy*. SAGE.

Dorrell, David. 2018. "Using International Content in an Introductory Human Geography Course." In *Curriculum Internationalization and the Future of Education*.

Gillespie, Marie. 1995. *Television, Ethnicity and Cultural Change*. Psychology Press.

Gregory, Derek, ed. 2009. *The Dictionary of Human Geography*. 5th ed. Malden, MA: Blackwell.

Jarosz, Lucy. 1996. "Working in the Global Food System: A Focus for International Comparative Analysis." *Progress in Human Geography* 20 (1):41–55. https://doi.org/10.1177/030913259602000103 .

Knowles, Anne Kelly, and Amy Hillier. 2008. *Placing History: How Maps, Spatial Data, and GIS Are Changing Historical Scholarship*. ESRI, Inc.

Massey, Doreen B. 1995. *Spatial Divisions of Labor: Social Structures and the Geography of Production*. Psychology Press.

Mintz, Sidney Wilfred. 1985. *Sweetness and Power: The Place of Sugar in Modern History*. Viking.

Sorkin, Michael, ed. 1992. *Variations on a Theme Park | Michael Sorkin*. 1st ed. Macmillan. http://us.macmillan.com/variationsonathemepark/michaelsorkin.

"The Global Music Machine." n.d. Accessed December 13, 2017. http://www.bbc.co.uk/worldservice/specials/1042_globalmusic/page3.shtml .

"The Internet Classics Archive | On Airs, Waters, and Places by Hippocrates." n.d. Accessed December 15, 2017. http://classics.mit.edu/Hippocrates/airwatpl.html.

"The Medium Is the Message." 2017. *Wikipedia*. https://en.wikipedia.org/w/index.php?title=The_medium_is_the_message&oldid=81254

4.12 ENDNOTES

1. Data Source: World Borders Dataset http://thematicmapping.org/downloads/world_borders.php

5 The Geography of Language

Arnulfo G. Ramírez

STUDENT LEARNING OUTCOMES

By the end of this section, the student will be able to:

1. Understand: the differing bases of ethnic identity
2. Explain: the relationship between ethnicity and personal identity
3. Describe: the degrees of relevance of ethnicity in a society
4. Connect: ethnicity, race and class as they relate to political power

CHAPTER OUTLINE

5.1 INTRODUCTION

Language is central to daily human existence. It is the principal means by which we conduct our social lives at home, neighborhood, school, work place and recreation area. It is the tool we use to plan our lives, remember our past, and express our cultural identity. We create meaning when we talk on the cell phone, send an e-mail message, read a newspaper and interpret a graph or chart. Many persons conduct their social lives using only one language. Many others, however, rely on two languages in order to participate effectively in the community, get a job, obtain a college degree and enjoy loving relation- ships. We live in a discourse world that incorporates ways of speaking, reading and writing, but also integrates ways of behaving, interacting, thinking and valuing. Language is embedded in cultural practices and, at the same time, symbolizes cultural reality itself.

5.2 LANGUAGE AND ITS RELATIONSHIP TO CULTURE

This first part of the chapter will enable you to understand three major questions regarding the nature of human language:

1. What knowledge of language is available to every speaker?
2. What communicative uses of language do speakers utilize in interactive situations?
3. How does language reflect cultural beliefs and practices?

5.2.1 Language as a Mental Capacity

To understand the nature of human language, one needs to approach the concept as a complex system of communication. An important distinction should be considered when using the term *language*. It can be viewed as an internal mental capacity (***langue***) as well as an external manifestation through speech (***parole***). As human beings, we are able to produce and understand countless number of utterances which are characterized by the use of grammatical elements such as *words, phrases* and *sentences.*

With a limited number language forms, we can produce numerous utterances that can be easily understood by other members of the **speech community** who share a similar cultural background and language knowledge. This underlying mental capacity is embodied in the concept that language is rule-governed creativity, operating at different grammatical levels in the formation of utterances or sentences.

To illustrate, many examples of utterances or sentences can be derived using a limited set lexical and grammatical words as listed below.

Lexical Words	*Grammatical Words*
Nouns (book, class, pencil, student/s, teacher)	Prepositions (on, of, for, from, with)
Verbs (forms of "to be", have, want, write)	Conjunctions (and, but, which)
Adjectives (big, good, red, green)	Determinants (a/an/the, her/their, my, this/that)
Adverbs (far, near, where, very, no/not)	Pronouns (I, he/she/it, you, mine/yours)

-1-

Some possible grammatical sentences based on the list of lexical and grammatical words are noted here:

The pencil is near the book. / The student is near the teacher. / The teacher writes with a pencil. /

The green pencil is not far from the book. / I want to be a teacher. / The red pencil is mine. /

She wants to write a good book for her students. / This is my book, but I want you to have it. /

No, the red pencil is not mine. / The teacher wants her students to have a good class./

Where is the teacher? / The book is where? / My class has many students. /

5.2.2 Language as a Means of Communication

A salient aspect of language involves the use of the communication system to perform a broad range of conversational acts/functions in "face-to-face" situations. Four major types of conversational acts have been proposed:

- *assertives* (speaker informs/answers/agrees/confirms/rejects/suggests)
- *directives* (speaker directs/ invites/ questions/orders someone else to do something),
- *commissives* (speaker makes an offer/promise involving some future action),
- *expressives* (speaker apologizes/evaluates/greets/thanks/expresses opinion/reacts).

Minor *secondary acts* consist of language use that serves to emphasize (repetition of words/phrases), expand (add additional information) and comment on on-going talk. *Complementary acts* can function as conversational fillers ("you know"), starters ("well"), stallers ("uh"), and hedges ("I mean").

Participants in a conversation tend to follow culturally specific norms. Speaker A (greets, gives an order, asks a question, apologizes, bids farewell) and Listener B (responds accordingly, and uses appropriate conversational language, necessary to maintain the dialogue). Cultural norms specify "What to say /not say in a particular conversational situation?" "How to initiate/end the conversation?" "With whom to talk/not talk during a conversational encounter?" "What locations are appropriate/not appropriate for the use of certain language forms?

Language use is joint action carried out usually by two people. Its use may vary due to such factors as the personal characteristics of the participants (friends, strangers, native/non-native speakers, family members, age/sex differences). The conversation may also be influenced by the location (home, school, work, shopping center, political meeting) and the topic of conversation (advice, complaint, news about the family, plans for the weekend).

5.2.3 Language as Cultural Practice

Speakers view language as a symbol of their social identity. As the sayings go: "You are what you speak" and "you are what you eat." The words that people use have cultural reality. They serve to express information, beliefs and attitudes that are shared by the cultural group. Stereotype perceptions come into play when we think about race (Asian, African, European, Native American), religion (Christian, Muslim, Jewish, atheist), social status (working class, middle class, wealthy, upper class) and citizenship status (US born, visa holder foreigner, undocumented worker). *Cultural stereotypes* are formed by extending the characteristics of a person or group of persons to all others, as in the belief that "all Americans are individualists and all Chinese are group-followers, collectivists."

Along with cultural beliefs about groups of people, individuals manifest specific views regarding languages themselves. Some make judgments about Language X as being "difficult to learn", "not useful in society" and "too boring". Others might view Language Y as the means "to get ahead","to make friends", "to complete a college requirement" or "to participate in the global marketplace".

According to royal court gossip in the 16th century, King Charles V of Europe had definite opinions about the languages he spoke: French was the language of love; Italian was the best language to talk to children; German was the appropriate language to give commands to dogs; Spanish was the language to talk to God.

Cultural meanings are assigned to language elements by members of the speech community who, in turn, impose them to others who want to belong to the group. Expressions such as "bug off", "you know", "you don't say" and "crack house" have a common meaning to members of a cultural group. Members in a speech group tend to share a common social space and history and have a similar system of standards for perceiving, believing, evaluating and acting. Based on one's experience of the world in a given cultural group, one uses this knowledge (**cultural schemata**) to predict interactions and relation- ships regarding new information, events and experiences.

Schemata function as knowledge structures that allow for the organization of information needed to perform daily cultural routines (eating breakfast, going shopping, planning a party, visiting friends). We can examine cultural patterns of behavior in relation to **cultural scripts**. The concept of cultural scripts is a metaphor from the language of theater. They are the "scripts" that guide social behavior and language use in everyday speaking situations.

"Attending a wedding", for example, calls for a variety of **speech situations** (locations and occasions requiring the use of different styles of language). First, there are a series of initial activities (dressing with proper attire, driving to the ceremony, greeting other persons attending the ceremony), then the actual wedding ceremony (participating in the diverse wedding rituals), and finally the post-wedding activities (attending the wedding banquet, engaging in the different activities—eating, dancing, toasting the wedding couple, interacting with other attendees, and taking leave at the end of the festive celebration.

Each speech situation may consist of a range of *speech events,* different ways of speaking involving various genres/styles: colloquial/informal language, reading of a text, song, prayer, farewell speech.

At the same time, each speech event might encompass a broad range of *conversational acts* such as greetings, questions, suggestions, advise, promises and expressions of gratitude. For individuals who live in a bilingual or multilingual world, verbal behavior is even more dynamic since questions such as *Who* speaks *What* language to *Whom, When* and *Where* come into play during most conversational situations.

5.3 CLASSIFICATION AND DISTRIBUTION OF LANGUAGES

This second section will facilitate your understanding of the dimensions of language across geographic areas and cultural landscapes. Three main questions are addressed in this section:

1. How are languages classified with respect to issues of national identity and genealogical considerations?

2. What are the major language families of the world and how many speakers make use of the respective languages?

3. How does language use vary in the United States with respect to dialects of English and multilingualism?

5.3.1 Diffusion of Languages

Language, like any other cultural phenomenon, has an inherent spatiality, and all languages have a history of diffusion. As our ancestors moved from place to place, they brought their languages with them. As people have conquered other places, expanded demographically, or converted others to new religions, languages have moved across space. Writing systems that were developed by one people were adapted and used by others. Indo-European, the largest language family, spread across a large expanse of Europe and Asia through a mechanism that is still being debated. Later, European expansion produced much of the current linguistic map by spreading English, French, Spanish, Portuguese, and Russian far from their native European homelands.

Language is disseminated through diffusion, but in complex ways. Relocation diffusion is associated with settler colonies and conquest, but in many places, hierarchical diffusion is the form that best explains the predominant languages. People may be compelled to adopt a dominant language for social, political or economic mobility. Contagious diffusion is also seen in languages, particularly in the adoption of new expressions in a language. One of the most obvious examples has been in the current convergence of British and American English. The British press has published books[1] and articles[2] decrying the Americanization of British English, while the American press has done the same thing in reverse[3]. In reality, languages borrow bits and pieces from other languages continuously.

The establishment of official languages is often related to the linguistic power differential within countries. Russification and Arabization are just two implementations of processes that use political power to favor one language over another.

5.3.2 Classification of Languages

There is no precise figure as to the total number of languages spoken in the world today. Estimates vary between 5,000 and 7,000, and the accurate number depends partly on the arbitrary distinction between languages and dialects. *Dialects* (variants of the same language) reflect differences along regional and ethnic lines. In the case of English, most native speakers will agree that they are speakers of English even though differences in pronunciation, vocabulary and sentence structure clearly exist. English speakers from England, Canada, Australia, New Zealand and United States of America will generally agree that they speak English, and this is also confirmed with the use of a standard written form of the language and a common literary heritage. However, there are many other cases in which speakers will not agree when the question of national identity and mutual intelligibility do not coincide.

The most common situation is when similar spoken language varieties are mutually understandable, but for political and historical reasons, they are regarded as different languages as in the case of Scandinavian languages. While Swedes, Danes and Norwegians can communicate with each other in most instances, each national group admits speaking a different language: Swedish, Danish, Norwegian and Icelandic. There are other cases in which political, ethnic, religious, literary and other factors force a distinction between similar language varieties: Hindi vs. Urdu, Flemish vs. Dutch, Serbian vs. Croatian, Gallego vs. Portuguese, Xhosa vs Zulu. An opposite situation occurs when spoken language varieties are not mutually understood, but for political, historical or cultural motives, they are regarded as the same language as in the case of Lapp and Chinese dialects.

Languages are usually classified according to membership in a **language family** (a group of related languages) which share common linguistic features (pronunciation, vocabulary, grammar) and have evolved from a common ancestor (**proto-language**). This type of linguistic classification is known as the *genetic* or *genealogical* approach. Languages can also be classified according to sentence structure (S)ubject+(V)erb+(O)bject, S+O+V, V+S+O). This type of classification is known as **typological classification**, and is based on a comparison of the formal similarities (pronunciation, grammar or vocabulary) which exist among languages.

Language families around the world reflect centuries of geographic movement and interaction among different groups of people. The Indo-European family of languages, for example, represents nearly half of the world's population. The language family dominates nearly all of Europe, significant areas of Asia, including Russia and India, North and South America, Caribbean islands, Australia, New Zealand, and parts of South Africa. The Indo-European family of languages consists of various **language branches** (a collection of languages within a family with a common ancestral language) and numerous *language subgroups* (a collection of languages within a branch that share a common origin in the relative recent past and exhibit many similarities in vocabulary and grammar.

Indo-European Language Branches and Language Subgroups

Germanic Branch
Western Germanic Group (Dutch, German, Frisian, English)
Northern Germanic Group (Danish, Swedish, Norwegian, Icelandic, Faeroese)

Romance Branch
French, Portuguese, Spanish, Catalan, Provençal, Romansh, Italian, Romanian)

Slavic Branch
West Slavic Group (Polish, Slovak, Czech, Sorbian)
Eastern Slavic Group (Russian, Ukrainian, Belorussian)
Southern Slavic Group (Slovene, Serbo-Croatian, Macedonian, Bulgarian)

Celtic Branch
Britannic Group (Breton, Welsh)
Gaulish Group (Irish Gaelic, Scots Gaelic)

Baltic-Slavonic Branch
Latvian, Lithuanian

Hellenic Branch
Greek

Thracian-Illyrian Branch
Albanian

Armenian Branch
Armenian

Iranian Branch
Kurdish, Persian, Baluchi, Pashto, Tadzhik

Indo-Iranian (Indic) Branch
Northwestern Group (Panjabi, Sindhi, Pahari, Dardic)
Eastern Group (Assamese, Bengali, Oriya)
Midland Group (Rajasthani, Hindi/Urdu, Bihari)
West and Southwestern Group (Gujarati, Marathi, Konda, Maldivian, Sinhalese)

Other languages spoken in Europe, but not belonging to the Indo-European family are subsumed in these other families: Finno-Ugric (Estonian, Hungarian, Karelian, Saami, Altaic (Turkish, Azerbaijani, Uzbek) and Basque. Some of the language branches listed above are represented by only one principal language (Albanian, Armenian, Basque, Greek), while others are spoken by diverse groups in some geographic regions (Northern and Western Germanic languages, Western and Eastern Slavic languages, Midland and Southwestern Indian languages).

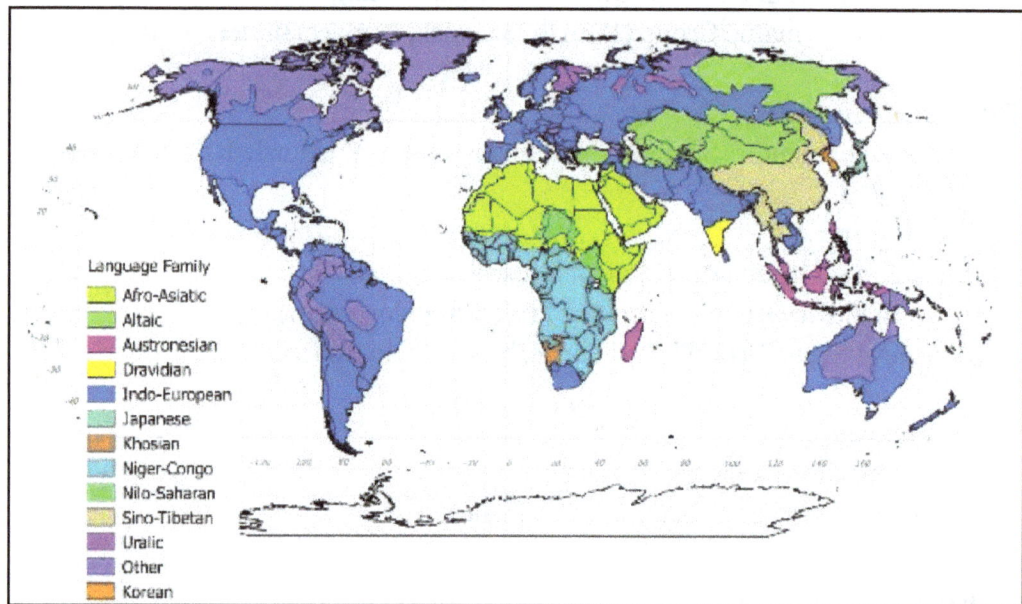

Figure 5.1 | Language Families of the World[4]
Author | David Dorrell
Source | Original Work
License | CC BY SA 4.0

Major Language Families of the World by Geographic Region

Europe
Caucasian Family
Abkhaz-Adyghe Group (Circassian, Adyghe, Abkhaz)
Nakho-Dagestanian Group (Avar, Kuri, Dargwa)
Kartvelian Group (Kartvelian, Georgian, Zan, Mingrelian)

Africa
Afro-Asiatic Family (Arabic, Hebrew, Tigrinya, Amharic)
Niger-Congo Family (Benue-Congo, Adamawa, Kwa)
Nilo-Saharan Family (Chari-Nile, Nilo-Hamitic, Nara)
Khoisan Family (Sandawe, Hatsa)

Asia
Sino-Tibetan Family (Chinese, Tibetan, Burmese)
Tai Family (Laotian, Shan, Yuan)
Austro-Asiatic Family (Vietnamese, Indonesian, Dayak, Malayo-Polynesian)
Japanese (an example of an isolated language)

Pacific
Austronesian Family (Malagasy, Malay, Javanese, Palauan, Fijian)
Indo-Pacific Family (Tagalog, Maori, Tongan, Samoan)

Americas
Eskimo-Aleut Family (Eskimo-Aleut, Greenlandic Eskimo)
Athabaskan Family (Navaho. Apache)
Algonquian Family (Arapaho, Blackfoot, Cheyenne, Cree, Mohican, Choctaw)
Macro-Siouan Family (Cherokee, Dakota, Mohawk, Pawnee)
Aztec-Tanoan Family (Comanche, Hopi, Pima-Papago, Nahuatl, Tarahumara)
Mayan Family (Maya, Mam, Quekchi, Quiche)
Oto-Manguean Family (Otomi, Mixtec, Zapotec)
Macro-Chibchan Family (Guaymi, Cuna, Waica, Epera)
Andean-Equatorial Family (Guahibo, Aymara, Quechua, Guarani)

The number of language families distributed around the world is sizable. The linguistic situation of specific member groups of the language family might be influenced by diverse, interacting factors: settlement history (migration, conquest, colonialism, territorial agreements), ways of living (farming, fishing, hunting, trading) and demographic strength and vitality of the speaker groups. Some languages might converge (many local varieties becoming one main language), while others might diverge (one principal language evolves into many other speech varieties). When different linguistic groups come into contact, a **pidgin** type of language may be the result. A pidgin is a composite language with a simplified grammatical system and a limited vocabulary, typically borrowed from the linguistic groups involved in trade and commerce activities.

Tok Pisin is an example of a pidgin spoken in Papua New Guinea and derived mainly from English. A pidgin may become a *creole* language when the size of the vocabulary increases, grammatical structures become more complex and children learn it as their native language or mother tongue. There are cases in which one existing language gains the status of a **lingua franca**. A lingua franca may not necessarily be the mother tongue of any one speaker group, but it serves as the medium of communication and commerce among diverse language groups. Swahili, for instance, serves as a lingua franca for much of East Africa, where individuals speak other local and regional languages.

With increased globalization and interdependence among nations, English is rapidly acquiring the status of lingua franca for much of the world. In Europe, Africa and India and other geographic regions, English serves as a lingua franca across many national-state boundaries. The linguistic con- sequence results in countless numbers of speaker groups who must become *bilingual* (the ability to use two languages with varying degrees of fluency) to participate more fully in society.

Some continents have more spoken languages than others. Asia leads with an estimated 2,300 languages, followed by Africa with 2,138. In the Pacific area, there are about 1,300 languages spoken and in North and South America about 1,064 languages have been identified. Europe, even with its many nation-states, is at the bottom of the list with about 286 languages.

Language	Family	Speakers in Millions	Main Areas Where Spoken
Chinese	Sino-Tibetan	1197	China, Taiwan, Singapore
Spanish	Indo-European	406	Spain, Latin America, Southwestern United States
English	Indo-European	335	British Isles, United States, Canada, Caribbean, Australia, New Zealand, South Africa, Philippines, former British colonies in Asia and Africa
Hindi	Indo-European	260	Northern India, Pakistan
Arabic	Afro-Asiatic	223	Middle East, North Africa
Portuguese	Indo-European	202	Portugal, Brazil, southern Africa
Bengali	Indo-European	193	Bangladesh, eastern India
Russian	Indo-European	162	Russia, Kazakhstan, part of Ukraine, other former Soviet Republics
Japanese	Japanese	122	Japan
Javanese	Austronesian	84.3	Indonesia

Ten Major Languages of the World in the Number of Native Speakers[5]

Other important languages and related dialects, whose total number includes both native speakers and second language users, consist of following: Korean (78 million), Wu/Chinese (71 million), Telugu (75 million), Tamil (74 million), Yue/ Chinese (71 million), Marathi (71 million), Vietnamese (68 million) and Turkish (61 million).

Language Spread and Language Loss

Of the top 20 languages of the world, all these languages have their origin in south or east Asia or in Europe. There is not one from the Americas, Oceania or Africa. The absence of a major world language in these regions seems to be precisely where most of the linguistic diversity is concentrated.

- English, French and Spanish are among the world's most widespread languages due to the imperial history of the home countries from where they originated.

- Two-thirds (66%) of the world's population speak 12 of the major languages around the globe

- About 3 percent of the world's population accounts for 96 percent of all the languages spoken today. Of the current living languages in the world, about 2,000 have less than 1,000 native speakers.

- Nearly half of the world's spoken languages will disappear by the end of this century. Linguistic extinction (*language death*) will affect some countries and regions more than others.

- In the United States many endangered languages are spoken by Native American groups who reside in reservations. Many languages will be lost in Amazon rain forest, sub-Saharan Africa, Oceania, aboriginal Australia and Southeast Asia.

- English is used as an official language in at least 35 countries, including a number of countries in Africa (Botswana, Kenya, Namibia, Sudan, Tanzania, Uganda among others), Asia (India, Pakistan, Philippines), Pacific Region (Fiji, Solomon Islands, Vanuatu, New Zealand), Caribbean (Puerto Rico, Belize, Guyana, Jamaica), Ireland and Canada.

- English is not by law (*de jure*) the **official language** in the United Kingdom, United States and Australia. English does enjoy the status of "national language" in these countries due to its power and prestige in institutions and society.

- English does not have the highest number of native speakers, but it is the world's most commonly studied language. More people learn English than French, Spanish, Italian, Japanese, German and Chinese combined.[6]

Dialects of English in the United States

At the time of the American Revolution, three principal dialects of English were spoken. These varieties of English corresponded to differences among the original setters who populated the East Coast.

Northern English

These settlements in this area were established and populated almost entirely by English settlers. Nearly two-thirds of the colonists in New England were Puritans from East Anglia in southwestern England. The region consists of the following states: Massachusetts, New Hampshire, Maine, Connecticut, Rhode Island, Vermont, New York and New Jersey.

Southern English

About half of the speakers came from southeast England. Some of them came from diverse social- class backgrounds, including deported prisoners, indentured servants, political and religious persecuted groups. The following states comprise the region: Virginia, Delaware, North Carolina, South Carolina and Georgia.

Midlands English

The settlers of this region included immigrants from diverse backgrounds. Those who settled in Pennsylvania were predominantly Quakers from northern England. Some individuals from Scotland and Ireland also settled in Pennsylvania as well as in New Jersey and Delaware. Immigrants from Germany, Holland and Sweden also migrated to this region and learned their English from local English-speaking settlers. This region is formed by the following areas/states: Upper Ohio Valley, Pennsylvania, Maryland, West Virginia, western areas North and South Carolina.

Dialects of American English have continued to evolve over time and place. Regional differences in pronunciation, vocabulary and grammar do not suggest that a type of linguistic convergence is under- way, resulting in some type of "national dialect" of American English. Even with the homogenizing influences of radio, television, internet, and social media, many distinctive varieties of English can be identified. Robert Delaney (2000) has outlined a dialect map for the United States which features at least 24 distinctive dialects of English. Dialect boundaries are established using diverse criteria: language features (differences in pronunciation, vocabulary and grammar) settlement history, ethnic diversity, educational levels and languages in contact (Spanish/English in the American Southwest). The dialect map does not represent the English varieties spoken in Alaska or Hawaii. However, it does include some urban and social (ethnolinguistic) dialects.

General Northern English, spoken by nearly two-thirds of the country.

New England Varieties
1. Eastern New England
2. Boston Urban
3. Western New England
4. Hudson Valley
5. New York City
6. Bonac (Long Island)

7. Inland Northern English Varieties
>8. San Francisco Urban
>9. Upper Midwestern
>10. Chicago Urban

Midland English Varieties
>11. North Midland (Pennsylvania)
>12. Pennsylvania German-English

Western English Varieties
>13. Rocky Mountain
>14. Pacific Northwest
>15. Pacific Southwest

16. Southwest English

17. South Midland Varieties
>18. Ozark
>19. Southern Appalachian (Smoky Mountain English)

General Southern English Varieties
Southern
>20. Virginia Piedmont
>21. Coastal South
>22. Gullah (coastal Georgia and South Carolina)
>23. Gulf Southern
>24. Louisiana (Cajun French and Cajun English) [7]

Multilingualism in the United States

Language diversity existed in what is now the United States long before the arrival of the Europeans.

It is estimated that there were between 500 and 1,000 Native American languages spoken around the fifteenth century and that there was widespread language contact and bilingualism among the Indian nations. With the arrival of the Europeans, seven colonial languages established themselves in different regions of the territory:

- English along the Eastern seaboard, Atlantic coast
- Spanish in the South from Florida to California
- French in Louisiana and northern Maine
- German in Pennsylvania
- Dutch in New York (New Amsterdam)

- Swedish in Delaware
- Russian in Alaska

Dutch, Swedish and Russian survived only for a short period, but the other four languages continue to be spoken to the present day. In the 1920's, six major minority languages were spoken in significant numbers partly to due to massive immigration and territorial histories. The "big six" minority languages of the 1940's include German, Italian, Polish, Yiddish, Spanish and French. Of the six minority languages, only Spanish and French have shown any gains over time, Spanish because of continued immigration and French because of increased "language consciousness" among individuals from Louisiana and Franco-Americans in the Northeast.

The 2015 Census data for the United States reveals valuable geographic information regarding the top 10 states with the extensive language diversity.

- California: 45 percent of the inhabitants speak a language other than English at home; the major languages include Spanish, Chinese, Korean, Vietnamese, Arabic, Armenian and Tagalog.

- Texas: 35 percent of the residents speak a language other than English at home; Spanish

- is widely used among bilinguals; Chinese, German and Vietnamese are also spoken.

- New Mexico: 34 percent of the state's population speak another language; most speak Spanish but a fair number speak Navajo and other Native American languages.

- New York: 31 percent of the residents speak a second language; Chinese, Italian, Russian, Spanish and Yiddish; some of these languages can be found within the same city block.

- New Jersey: 31 percent of the state's residents speak a second language in addition to English; some of the languages spoken include Chinese, Gujarati, Portuguese, Spanish and Italian.

- Nevada: 30 percent of the population is bilingual; Chinese, German and Tagalog are used along with Spanish, the predominant second language of the Southwest.

- Florida: 29 percent of the residents speak a second language, including French (Haitian Creole), German and Italian

- Arizona: 27 percent of the residents claim to be bilingual; most speak Spanish as in New Mexico while others use Native American languages.

- Hawaii: 26 percent of the population claims to be bilingual; Japanese, Chinese, Korean and Tagalog are spoken along with Hawaiian, the state's second official language.

- Illinois and Massachusetts: 23 percent of their respective populations speak a second language at home; residents of Illinois speak Chinese, German, Spanish and Polish, especially in Chicago; residents of Massachusetts speak Spanish, Haitian Creole, Chinese, Portuguese, Vietnamese and French. [8]

Top Ten Languages Spoken in U.S. Homes Other Than English

Data from the 2015 American Community Survey ranks the top ten languages spoken in U.S. homes other than English. The data highlight the size of the speaker population, bilingual proficiency (fluency in the home language and English) and degree of English proficiency (LEP, limited English proficiency).

Rank	Language Spoken at Home	Total	Bilingualism %	Limited English %
1.	Spanish	64,716,000	60.0	40.0
2.	Chinese	40,046,000	59.0	41.0
3.	Tagalog	3,334.000	44.3	55.7
4.	Vietnamese	1,737,000	67.6	32.4
5.	French	1,266,000	79.9	20.1
6.	Arabic	1,157,000	62.8	37.2
7.	Korean	1,109,000	46.8	53.2
8.	German	933,000	85.1	14.9
9.	Russian	905,000	56.0	44.0
10.	French Creole	863,000	58.8	41.2

Chinese includes Mandarin and Cantonese. French also comprises Haitian and Cajun varieties. German encompasses Pennsylvania Dutch. [9]

While a record number of persons speak a language at home other than English, a substantial figure within each immigrant group claimed an elevated command of English. Overall, some 60 percent of the speaker groups using a second language at home were also highly fluent in English. Limited fluency in English among young children ranged from a high of 55.7 percent in the Tagalog speaker group to a low of 14.9 percent in the German group which included Pennsylvania Dutch users.[10]

Most immigrant language groups have tended to follow an **intergenerational language shift** in the United States. This first generation is basically monolingual, speaking the native language of the group. The second generation is bilingual, speaking both the home language and English. By the third generation, the cultural group is essentially monolingual, speaking only English in most communicative situations.

More recently, some immigrant groups, particularly those with advanced training and degrees in professional fields (technology, health sciences and business), come to the United States with a high degree of fluency in English. At the same time, the variety of English these persons speak is usually marked by the country of origin (India, Philippines, Singapore among others). With globalization "new Englishes" have emerged (Indian English, Filipino English, Nigerian English) which challenge the notion of a Standard English variety (British or American) for use around the world.

5.4 LANGUAGE IN THE PHYSICAL, BUSINESS AND DIGITAL WORLDS

This third part of the chapter will enable you to comprehend the uses of language across different environments. Three major questions are addressed in this section:

1. How is language used to indicate place in the physical landscape?

2. How is language exploited for the purpose of advertising products and services?

3. How is language employed in the digital world to connect multiple senders and recipients in diverse techno formats?

5.4.1 Language and Place Names

The names people give to their physical environment provide a unique source of information about cultural character of various social groups. Place names often reveal the history, beliefs and values of a society. *Toponymy* is the study of place names, and the names people assign to specific geographic sites offer us the opportunity to recognize a country's settlement history, important features of the landscape, famous personalities, and local allusion to distant places and times. Place names can change overnight, often depending of political factors and social considerations. The change of Burma to

Myanmar and Zaire to Congo are two recent example of changes due to political developments. In the United States, interest in changing the names of places associated with Civil War heroes from the South is an on-going effort, at times with dramatic confrontations between different social groups.

Toponyms provide us with valuable geographic insights about such matters as where did the settlers come from, who settled and populated the area, and what language did the early settlers speak. Many of the place names found in the United States can be classified in terms of the following major categories:

- natural landscape features (hills/mountains, rivers, valleys, deserts, coastline)

    Hollywood Hills, Blue Ridge Mountains, Chattahoochee River, Rio Grande, San Fernando Valley, Monument Valley, Mohave Desert, Biscayne Bay,

- urbanized areas (cities, towns, and streets)

  Williamsburg, VA; Lawrenceville, GA; Chattanooga, TN; New Iberia, LA; Buford Highway, GA; Martin Luther King, Jr. Drive, Ponce de Leon Avenue

- directional place names (East, North, South, West)

  North & South Dakota, West Virginia, North & South Carolina, South Texas

- religious significant names (saints, Biblical names)

  San Antonio, Santa Fe, St. Louis, Sacramento, Santa Barbara, Los Angeles; Bethany, AK; Canaan, VT; Jericho, IA; Shiloh, OK

- explorers and colonizers (French, English, Spanish, Dutch setters

  Columbus, OH; Coronado, CA; Balboa Park (San Diego), CA; Cadillac, MI; Hudson, NY; Bronx, NY; Raleigh, NC; Henrico County, VA

- famous persons (presidents, politicians, Native Americans)

  Lincoln, NE; Mount McKinley, WA; Austin, TX; Washington, D. C.; Jackson, MS; Tuscaloosa, AL; Pensacola, FL; Arizona, Mississippi, Utah

- culturally based names (immigrant's homeland, famous locations

  New Orleans; New Mexico; New Amsterdam, NY; Troy, NY; Rome, GA; Oxford, MS; Athens, GA; Birmingham, AL; Toledo, OH

- business oriented names (wealthy individuals, politicians, corporate sponsors)

  Sears Tower, Wrigley Stadium, Trump Towers, Gwinnett County, Dolby Theater, Verizon Center, Sun Trust Park

A classification scheme proposed by George Stewart (1982) focuses on ten basic themes which dominate North American toponyms. These include the following categories: descriptive (Rocky Mountains), associative (Mill Valley, CA), commemorative (San Diego, CA), commendatory (Paradise Valley, AZ), incidents (Battle Creek, MI), possession (Johnson City, TX), folk (Plains, GA), manufactured (Truth or Consequences, NM), mistakes (Lasker, NC), and location shift (Lancaster, PA).

5.4.2 Language and the Discourse of Advertising

Commercial advertisement occupies a noticeable expanse in the cultural landscape. An individual *text* (use and arrangement of specific language forms) is designed to promote or sell a product within a social context. A commercial text may be accompanied by music and visual depictions. The text may also be accompanied by paralanguage features of oral language (gestures, voice quality, facial expressions) and written language (choice of typeface, letter sizes, range of colors).

The advertisement itself brings up several of discourse concerns: Who (seller) is communicating with Whom (consumer) and Why (inform/convince/persuade about the product' importance/usefulness/ uniqueness)? The participants in the discourse may include various message senders/participants: the actor/s in the TV commercial along with the supporting role of the advertising agency and the studio production staff. The receivers may be a specific target group or anyone who sees the advertisement.

Highway billboards, store signs and product advertisements provide a visual representation of commercial language use in a community. Most billboard structures are located on public spaces and display advertisements to passing motorists and pedestrians. They can also be placed in other locations where there are many viewers (mass transit stations, shopping malls, office buildings and sports' stadiums). Some billboards may be static, while others may change continuously or rotate periodically with different advertisements. In addition, there are product promotions within a retail store, which often involve product placements at the end of aisles and near checkout counters.

Novelty ads can appear on small tangible items such as coffee mugs, t-shirts, pens and shopping bags. They can be distributed directly by the advertiser or as part of cross-product promotion campaigns. Advertisers use the popularity of cultural celebrities in the worlds of sports, music and entertainment to promote their products. Even aircrafts, balloons and skywriting are used as moveable means to display advertisements.

Store signs and highway billboards can be viewed as a *visual language trail*, stretching point A to point B on highway X in a specific geographic area. Depending on the population characteristics of a location, diverse forms of advertisement are used to convince the customer that a company's services or products are the best in quality and price, most useful and socially desired. A drive through various roads and highways across Gwinnett County Georgia, for example, might indicate how advertisers respond to the diverse population characteristics.

Ethnolinguistic Diversity in Gwinnett County Georgia[11]

The American Community Survey, aggregate data, 5-year summary file, 2006 to 2010, provides the following profile of ethnolinguistic diversity.

Language in Use	Ages 5 years and Above	Percentage of Population
English	484,134	67.80%
All languages other than English	229,932	32.20%
Spanish	124,331	17.41%
Korean	17,911	2.51%
Vietnamese .	12,692	1.78%
Chinese	9,184	1.29%
African languages	8,750	1.23%
Serbo-Croatian	5,097	0.71%
Guajarati	4,725	0.66%
French	4,713	0.66%
Other Indo-European languages	4,232	0.59%
Hindi	4,227	0.59%
Other Asian languages	4,181	0.59%
Urdu	4,088	0.57%
Russian	2,487	0.35%
French Creole	2,344	0.33%
German	2,127	0.30%
Arabic	2,038	0.29%
Tagalog	1,539	0.22%
Persian	1,289	0.18%
Mon-Khmer, Cambodian	1,265	0.18%
Japanese	1,198	0.17%
Other Slavic languages	1,038	0.15%
Polish	886	0.12%
Other Pacific Island languages	864	0.12%
Portuguese	781	0.11%
Laotian	645	0.09%
Thai	411	0.06%
Other West Germanic languages	395	0.06%
Hmong	379	0.05%

Some important questions regarding language use can be addressed within this multilingual context.

- What type of products are marketed to different ethnolinguistic communities?
- What type of services are advertised to different ethnolinguistic communities?
- What type of products are marketed bilingually or in the language of the ethnolinguistic community?
- What type of services are advertised bilingually or in the language of the ethnolinguistic community?

The visual content and design of an advertisement aimed to draw attention to a specific product might focus on customer needs such items as food, clothing, furniture, restaurants, home and garden, cosmetics and beauty care, auto maintenance, fitness and recreation, travel and hotels, communication and computers. The advertising style for product promotion often tends to be laudatory, positive and emphasizing the uniqueness. The vocabulary is usually vivid and concrete, involving play-on-words and commercial slogans in some cases. Ads rely primarily on language, and it is the visual content and design that creates an interest in the product and persuades people buy it.

The advertisement of services for the general population and targeted ethnolinguistic communities might encompass health services (doctors, dentists, hospital and emergency care), financial institutions (banks, credit unions, home and car loans, bail bonds), legal services (lawyers, notary public, public defenders) and community resources (schools, libraries, museums, parks). Customer needs usually dictate what services are available in a specific geographic area. Interest in niche marketing or ads targeted to a specific social group represents the strong relationship that exist between cultural and technological changes in contemporary US society.

5.4.3 Language and the Digital World

Social media are computer-mediated technologies which allows multiple senders and receivers to create and share information, ideas, career interests and other forms of expression via virtual com- munities and communication networks. Social media use relies on web-based technologies such as desktop computers, smartphones and tablet computers to create highly interactive formats which allow individuals, communities and organizations the possibility to share, co-create, discuss diverse topics and comment on content previously posted online. Social media allows for mass cultural exchanges and intercultural communication among people from different regions of the world.

The term social media is often used to indicate that many senders and receivers can communicate almost simultaneously across space and time. At the same time, the term social media is used to mean *social networks* (relationships and contacts among many individuals). If one is using the term to mean social networks (*who* interacts with *whom* in the linguistic community), then the researcher can "observe" and document the interactional patterns or the researcher can "interview" the participants to determine the type of social networks. Social networks, from a sociolinguistic perspective, can be differentiated on the basis of whether they are "dense" or "loose". In *dense networks* all members know each other. In *loose networks* not all members know each other. Networks can also be distinguished with the quality of *ties* (connections) that exist among the members. In *uniplex ties,* individuals are connected by one type of relationship (participate in the same swim club, take same courses at a college, work in the same business). In *multiplex ties,* members know each other in several different roles (student/friend/classmate; parent/co-worker/neighbor).

The term social media is usually associated with different networking sites such as the following:

- Facebook, an online social network which allows users to create personal profiles, share photos and videos and communicate with other users.

- Twitter, an internet service that allows users to post "tweets" (brief messages totaling 140 characters) for their followers to see in real-time.

- LinkedIn, a network designed for the business community allowing users to create professional profiles, post resumes, and communicate with other professionals and job-seekers.

- Pinterest, an online network that allows users to send photos of items found in the web by"pinning" them and sharing comments with others in the virtual community.

- Snapchat, an application on mobile devices that allows users to send and share photos of themselves performing daily activities.[12]

Social media takes many other forms including blogs, forums, product/services reviews and social gaming and video sharing. The social networking world has changed the way individuals and organizations use language to communicate with each other. Research findings indicate the significant impact that social media is having on society in the United States and elsewhere.

- Nearly 80 percent of American adults are online and nearly 60 percent of them use social media.

- Among the adolescent population, 84 percent have a Facebook account.

- Over 60 percent of 13 to 17-year-olds have at least one profile on social media, with some spending more than two hours a day on social network sites.

- Internet users spend more time on social media sites than any other type of web-based sites. The use from July 2011 (66 billion minutes) to July 2012 (121 billion minutes) represents a 99 percent increase.

- Young adults, some 33 percent, get their news from social media.

- More than half (52 percent) of internet users use two or more social media sites to communicate with their friends and family.

- In the United States, 81 percent of users look online to get news about the weather, 53 percent for national news, 52 percent for sports news, and 41 percent for entertainment or celebrity news.[13]

There are both positive and negative effects associated with social media. The positive effects include the ability to document memories, learn about and explore different topics, advertise oneself and form many friendships. On the negative side, social media often invades on personal privacy, fosters information overload, promotes isolation, affects users' self-esteem and creates the possibility for online harassment and cyberbullying.

Mapping actual language use in the context of the digital world is problematic. It is a complex communication universe. Unlike the geography of place names and the discourse of advertisement, social networking occurs in a virtual environment, involving many senders/receivers and different computer-mediated technologies. Data "mining" is a technique employed to analyze large-scale social media data fields to establish general patterns regarding the content/topics that emerge from people's actual online activities. Usage patterns in social media interest many advertisers, major businesses, government organizations and political parties. Research methods from the social sciences have been used to establish user's activities with different types of social media technologies. These include pencil-and-paper questionnaires, individual/group oral interviews and focus group sessions. These methods involve language-driven interactions with a limited number of users who may or may not reveal their actual social media patterns of behavior.

5.5 SUMMARY

Language is a mental capacity that allows members of a speech community to produce and understand countless number of utterances which include grammatical elements like words, phrases and sentences.

Language as a means of communication makes use of different communicative acts (orders, questions, apologies, suggestions) performed during conversational situations across varied social contexts. Language is a symbol of social identity and serves to express ideas, beliefs and attitudes shared by a cultural group. It is reflected in cultural stereotypes, notions about different languages, and behaviors during speech situations which presuppose the use of cultural schemata and cultural scripts.

Languages are commonly classified according to membership in a language family such as Indo-European, Sino-Tibetan, Indo-Pacific, Mayan, Niger-Congo. Members within a family are further subdivided into branches (Germanic, Slavic, Finno-Ugric, Indo-Iranian) and the branches into subgroups (English in the Germanic branch; Spanish in the Romance branch).

The distribution of languages around the world is influenced by numerous factors: settlement history, demographic strength, ways of living and contact with other ethnolinguistic groups. Some languages become more dominant and as a result displace others that may eventually become extinct, leading to language death. The world's ten most widespread languages include Chinese, Spanish, English, Hindi, Arabic, Portuguese, Bengali, Russian, Japanese, and Javanese.

The number of dialects or varieties of American English have changed over time due to settlement histories, political changes (Louisiana Purchase, Mexican American War, Spanish-American War, territory annexation). Language diversity and multilingualism continue to be prevalent in the United States. Recent 2015 Census data reveal extensive language diversity in states like California, Texas, New Mexico, New York, New Jersey, Nevada, Florida, Hawaii, Illinois and Massachusetts.

Place names provide us with cultural insights about the significance of geographic locations, important features of the landscape, the recognition of famous personalities, and local reference to distant places and times. Diverse forms of advertisement are used to inform and convince customers that the products and services offered are the worthiest in the marketplace.

The use of different social media technologies (Facebook, Twitter, LinkedIn, Snapchat, among others) allows for online interaction between many senders and receivers. Users can create and share information, ideas, photos, career interests, and other concerns via virtual communities and networks.

Geographic mapping of the use and users of web-based technologies (desktop computers, smart phones and tablet computers) is unattainable at this time. Research methods from the social science (questionnaires, oral interviews, focus group sessions) may reveal some insights about the pervasive ways individuals, communities and organizations communicate in the virtual world.

5.6 KEY TERMS DEFINED

Speech community: People who share a similar cultural background and language knowledge.

Langue: The internal mental capacity for language.

Parole: The external manifestation of ideas through speech.

Creole: A blended language differentiated from a pidgin language by its more complex grammar and its status as a first language.

Cultural schemata: A system of standards for perceiving, believing, evaluating and acting.

Speech situations: Locations and occasions requiring the use of different styles of language.

Cultural scripts: The "scripts" that guide social behavior and language use in everyday speaking situations.

Language branch: A group of languages which share common linguistic and have evolved from a common ancestor.

Language family: A collection of languages within a family with a common ancestral language.

Proto-language: An historic language from which known languages are believed to have descended by differentiation of the proto-language into the languages that form a language family.

Dialect: Variants of the single language.

Pidgin: A composite language with a simplified grammatical system and a limited vocabulary.

Lingua franca: A language used to make communication possible between people who do not share a native language.

Bilingual: Being able to use two languages with varying degrees of fluency.

Toponymy: The study of place names.

Text: The use and arrangement of specific language forms.

Typological classification: Classification based on the comparison of the formal similarities in pronunciation, grammar and vocabulary which exist among languages.

Official language: A language that is given a special legal status over other languages in a country.

Intergenerational language shift: A linguistic pattern of acculturation found in US immigrant groups in which a group shifts from being non-English monolingual to English monolingual.

5.7 WORKS CONSULTED AND FURTHER READING

Bloomer, Aileen, Patrick Griffths and Andrew John Merrison. 2005. *Introducing Language in Use: A Coursebook.* London: Routledge.

Chimombo, M.P.F. and Robert L. Roseberry. 1998. *The Power of Discourse: An Introduction to Discourse Analysis.* Mahwah, New Jersey: Lawrence Erlbaum Associates, Publishers.

Cook, Guy. 1992. *The Discourse of Advertising.* London: Routledge.

Crystal, David. 1987. *The Cambridge Encyclopedia of Language.* Cambridge: Cambridge University Press.

Downes, William. 1998. *Language and Society,* 2nd ed. Cambridge: Cambridge University Press.

Grosjean, François. 1982. *Life with Two Languages: An Introduction to Bilingualism.* Cambridge, Mass: Harvard University Press.

Hinton, Sam and Larissa Hjorth. 2013. *Understanding Social Media.* Thousand Oaks, CA: Sage Publications.

Kramsch, Claire. 1998. *Language and Culture.* Oxford: Oxford University Press.

Lightfoot, David. 1999. *The Development of Language: Acquisition, Change, and Evolution.* Oxford: Blackwell Publishers.

Meyerhoff, Miriam. 2006. *Introducing Sociolinguistics.* London: Routledge

Ostler, Nicholas. 2005. *Empires of the Word: A Language History of the World.* New York: Harper Collins Publishers.

Ramírez, Arnulfo G. 1995. *Creating Contexts for Second Language Learning*. White Plains, New York: Longman Publishers, USA.

Ramírez, Arnulfo G. 2008. *Linguistic Competence across Learner Varieties of Spanish*. Munich: LINCOM-EUROPA.

Romaine, Suzanne. 1994. *Language in Society: An Introduction to Sociolinguistics*. Oxford: Oxford University Press.

Spolsky, Bernard. 1998. *Sociolinguistics*. Oxford: Oxford University Press.

Stewart, George. 1982. *Names on the Land: A Historical Account of Place-Naming in the United States*. Available through Penguin Random House, paperback edition 2008.

Yule, George. 1996. *Pragmatics*. Oxford: Oxford University Press.

5.8 ENDNOTES

1. Engel, Matthew. *That's The Way It Crumbles: The American Conquest of the English Language*. London: Profile Books, 2017)

2. https://www.theguardian.com/us-news/2017/jul/13/american-english-language-study

3. https://www.nytimes.com/2012/10/11/fashion/americans-are-barmy-over-britishisms.html

4. Data adapted from https://www.kaggle.com/rtatman/world-language-family-map and http://jonathansoma.com/open-source-language-map/

5. Ethnologue. 2013. https://www.ethnologue.com/world

6. Noack, Rick. 2015. "The World's Languages, in 7 Maps and Charts." Washington Post. https://www.washingtonpost.com/news/worldviews/wp/2015/04/23/the-worlds-languages-in-7-maps-and-charts/

7. Nisen, Max. 2013. "Map Shows How American Speak 24 Different English Dialects." Business Insider. https://www.businessinsider.com/dialects-of-american-english-2013-12

8. "Languages." 2016. Accredited Language Services (blog). https://www.accreditedlanguage.com/category/languages/

9. United States Census Bureau. https://www.census.gov/programs-surveys/acs/news/data-releases/2015.html

10. Hallock, Jie Zong, Jeanne Batalova Jie Zong, Jeanne Batalova, and Jeffrey. 2018. "Frequently Requested Statistics on Immigrants and Immigration in the United States." Migrationpolicy.Org. February 2, 2018. https://www.migrationpolicy.org/article/frequently-requested-statistics-immigrants-and-immigration-united-states .

11. United States Census Bureau. https://www.census.gov/programs-surveys/acs/news/data-releases/2015.html

12. "What Is Social Media? - Definition from WhatIs.Com." 2018. https://whatis.

techtarget.com/definition/social-media .

13. "Social Media." 2018. Wikipedia. https://en.wikipedia.org/w/index.php?title=Social_media&oldid=855175226 .

6 Religion

David Dorrell

STUDENT LEARNING OUTCOMES

By the end of this section, the student will be able to:

1. Understand: the significance of religion as a historical spatial phenomenon
2. Explain: the significance of sacred spaces and places to understandings of culture locally, regionally, and globally
3. Describe: the hearths and diffusion patterns of the major religions of the world
4. Connect: religious belief and values to trade, colonialism, and empire

CHAPTER OUTLINE

6.1 INTRODUCTION

"I love you when you bow in your mosque, kneel in your temple, pray in your church. For you and I are sons of one religion, and it is the Spirit."

- Khalil Gibran

This chapter is an exploration of the geography of religion. Like language and ethnicity, religion is a cultural characteristic that can be closely bound to individual identity. Religion can provide a sense of community, social cohesion, moral standards, and identifiable architecture. It can also be a source of oppression, social discord, and political instability. Religion is more than metaphysics- magical explanations for natural phenomena; it is also a governing philosophy of behavior and an organizing cosmology of the universe.

The following pie chart gives us an idea of the relative size (in terms of adherents) of the world's major religions (**Figure 6.1**). Bear in mind that these numbers are estimates; there is no world governing body collecting detailed statistics of religion. Christians and Muslims make up over half of the world's population. The next category, the unaffiliated, are a large but diffuse body containing people who do not identify with any religion. Hindus, who cluster in the Indian subcontinent, are the next largest group. The category of Folk religion is similar to that of

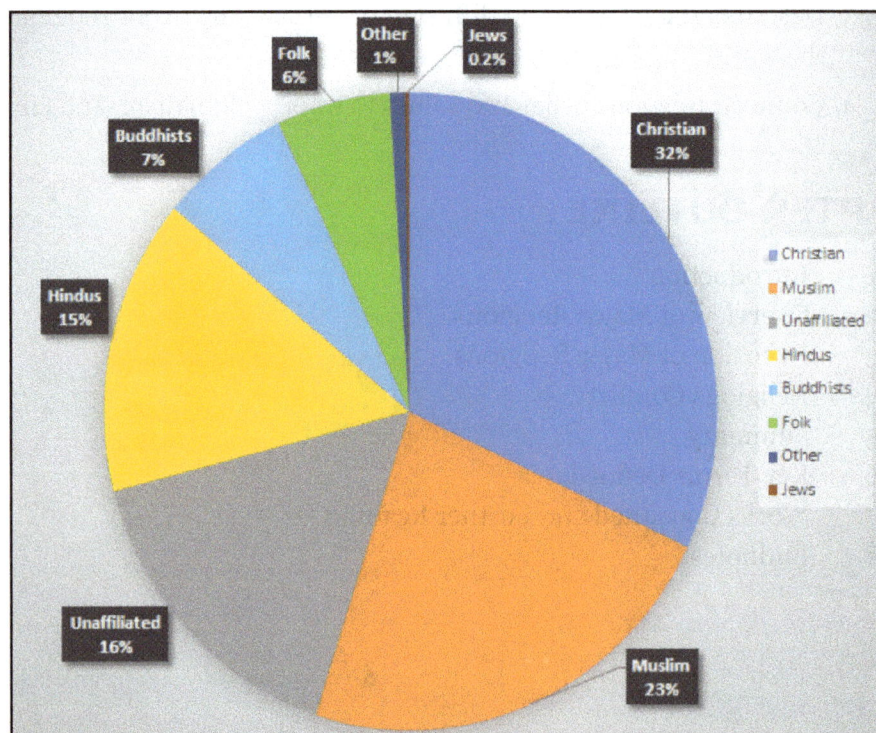

Figure 6.1 | Global Religious Percentages[1]
This pie chart shows the relative distributions of the World's Major Religions.
Author | David Dorrell
Source | Original Work
License | CC BY SA 4.0

Unaffiliated-it is a large group of religions that are bound into one category solely for logical consistency. Folk religions may consist of ancestor worship in China, animism in central Africa, or any other number of indigenous, local religions. The Other category contains newer religions that are just gaining their footholds and other religions that may be fading in the contemporary milieu. Judaism is included in the chart, although it has comparatively few adherents. It is included for two reasons. First, it provided the cultural spoor for both Christianity and Islam, and second, it is the predominant religion of the modern state of Israel.

In some places, religion can be considered a separate element of civil society, but in many others, religion cannot be meaningfully separated from daily life or governance.

Charts such as this can be somewhat misleading, as can maps of religion (**Figure 6.2**). All these methods of tabulating religion rely on estimates with varying degrees of accuracy. One problem is determining the predominance of a particular religion. Does predominance require over fifty percent? What if no religion in a country has a majority? In the case of this map, if no religion has a majority, but there are two large religions (for example Christianity and Islam in Nigeria) then the country is split between the two. If there are numerous fragmented groups, then the group with a plurality is used.

In some parts of the world, some forms of religious expression are discouraged or banned outright. For example, in North Korea, the state ideology is known as Juche, which is not a traditional religion with supernatural elements. The practice of Buddhism or Christianity in North Korea must necessarily be circumspect. In other places, religion has reached the status of being nominal (in name only) in which people identify with a religion, but the practices of that religion have little impact on their daily lives.

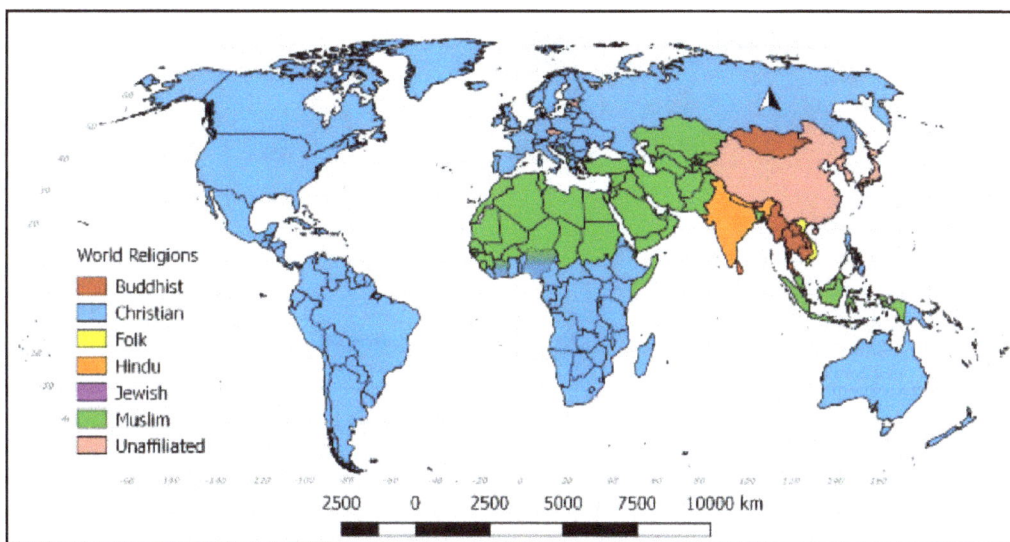

Figure 6.2 | World Religions by State 2012[2]
Author | David Dorrell
Source | Original Work
License | CC BY SA 4.0

State religions are religions that are recognized as the official religion of a country. In some places with a state religion, such as Denmark or the United Kingdom, the official status of one religion has little effect either on the practice of other religions or on the society at large. In places like Saudi Arabia or Pakistan, however, the official religion is closely connected to the power of the state, most obviously in the form of blasphemy laws which allow for state penalties (including death) for violations of religious statutes.

Maps such as this one can be very misleading in the sense that they present homogeneous, religious landscapes by country (**Figure 6.2**). This is, of course, untrue and it represents one of the problems with mapping anything- the level of aggregation. As any spatial phenomenon is aggregated into larger and larger groups, the details of those groups are often lost. An example of the importance of scale is seen in the following graphic. Although Mormons make up 88 percent of Utah County, Utah, they are only 61% of the state of Utah, and a mere 1.6% of the United States.

6.1.1 Scale and Predominant Religion

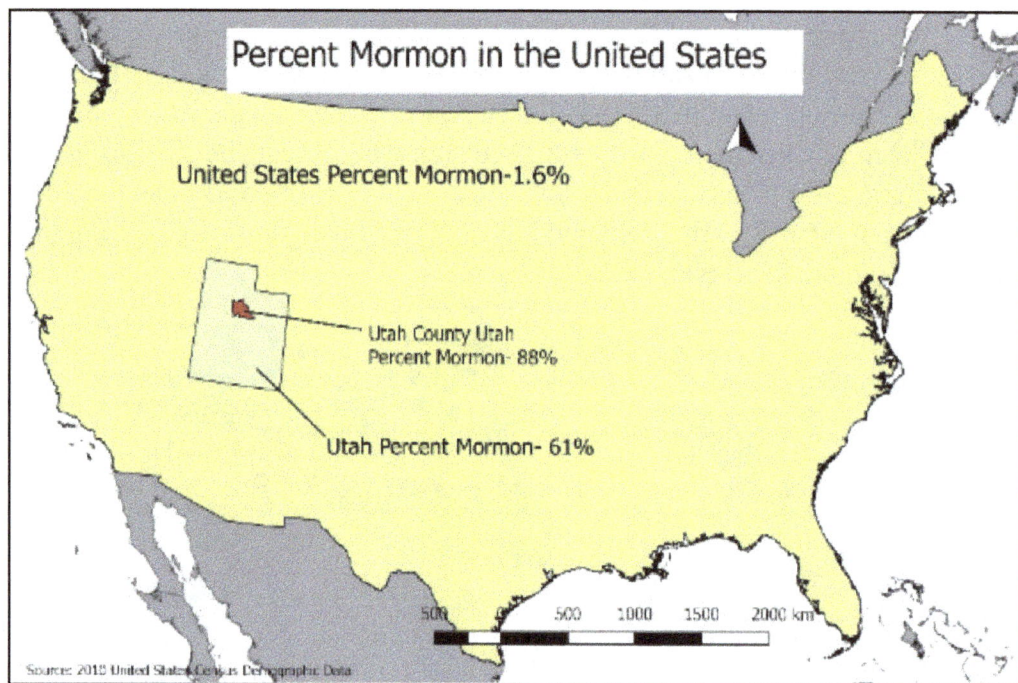

Figure 6.3 | Percent Mormon aggregated to the County, State, and National level [3]
Author | David Dorrell
Source | Original Work
License | CC BY SA 4.0

6.1.2 Religions in History

The current religious map of the world is best thought of as a snapshot. The religious landscape has been continuously changing throughout human history and will continue to change in the future. New religions are founded and old ones die out.

New religions are often made using pieces of older religions; Christianity and Islam deriving in part from Judaism and Buddhism deriving from Hinduism are not aberrations, but instead are examples of a common occurrence.

Within the relatively recent past, it has been possible to watch the creation and diffusion of several new religions just within the United States. Mormonism, the Jehovah's Witnesses, Seventh Day Adventism, and Scientology are all religions that were founded in the U.S. in the relatively recent past.

6.1.3 The Religious Contribution to Culture and Identity

Religions are not isolated social phenomena. They exist within a cultural complex that nurtures and sustains them, or conversely, demeans and undermines them. Religion can be closely bound to other elements of identity -language, or nationality. In many societies, the boundaries between religion and social life, family structure, and law and politics are nonexistent. Religion in those places is the center of all life and everything else revolves around religious concepts. A place that is purely governed by a religious structure is known as a **theocracy**. There are very few of these in the modern world, although many states have strong religious influence. Many modern societies have built barriers between religious influence and political life. These places are called **secular** and are much more common in the developed world.

Religion, along with ethnicity and language, are very often core components of an individual's identity. It can define the way a person sees the world, what clothing is appropriate, gender roles, employment, and even your position within the greater society. As such, it has tremendous cultural influence, and this influence is visible in the landscape.

6.1.4 Esthetics and religion

Religion has a motivating factor that few other social phenomena can match. When people are doing something for God, they generally have fewer limits than in other spheres of life. One of the ways this lack of limits is manifest in the landscape is through religious architecture.

Sacred spaces can be religious structures, but they can also be historic battlefields, cemeteries, mountains, or rivers. Anything that humans use to generate a sense of the divine can be considered a sacred space. Sacred spaces have expectations of behavior. In some places, it is still possible to claim **sanctuary** in a sacred place. The small altars that mark roadside fatalities in the United States can be considered sacred spaces, as could a closet that is used as a prayer room.

Elements of culture may be manifest in different types of churches, temples on the landscape, as well as clothing, the food grown, and small home devotionals. Another way that religion manifests in the visible sphere is through codes of acceptable dress as well as acceptable public behaviors.

A less obvious way that religions may influence the landscape is through religious influences on dietary choices. Food production can be influenced by religions. Many religions have doctrine regarding what is acceptable food, and what is not. Halal, Kosher and Ital are all representative of food restrictions. Religions that prohibit the consumption of pork will probably not have swine farms. Cattle wander through the countryside in India, since they are religiously protected from harm. Another effect that religion has on the landscape is the effect of **pilgrimage**. Many religions have an activity that requires gathering at a particular place. Probably the best-known pilgrimage is the Hajj of Islam, but this isn't actually the largest in the world. That would be the Hindu Kumbh Mela. The Camino Santiago is a well-known Christian pilgrimage that ends in Santiago de Compostela, Spain. Pilgrimage is not just visible through the pilgrims, but in the entire infrastructure that develops to support the pilgrims.

6.2 OVERVIEW OF MAJOR RELIGIONS

6.2.1 Types of Religions

We often break religions into one of two basic types: ethnic and universalizing. **Ethnic religions** are associated within one group of people. They make little to no effort at **proselytizing** (converting others), although that possibility may exist. The largest ethnic religion is Hinduism. Judaism is another well-known ethnic religion. Through migration, both of these religions have become dispersed around much of the world, but they are closely tied to their own ethnic groups.

Universalizing religions seek to convert others. For some religions, it is a requirement for practitioners to spend part of their lives in missionary work attempting to convert others.

Another way of dividing religions is into the categories of polytheistic (many gods) and monotheistic (one god). Although the difference between monotheistic religions and polytheistic religions seems unbridgeable, there are religions that have managed to combine elements of **polytheism** and **monotheism** into the same religion. For example, in Voudon (Voodoo), entities that had previously represented African gods are recast as Catholic saints, who themselves are semi-divine in the Catholic cosmology. Combining two religions to create a new religion is known as syncretism.

6.2.2 A Brief Description of the Major Religions

Christianity

Christianity is a monotheistic religion centered on the life and teachings of Jesus of Nazareth. It dates to some time in the first century AD since the Western world uses the Christian calendar. Christianity began as an offshoot of Judaism, and includes the Hebrew Bible (known to Christians as the Old Testament) as well as the New Testament as its canonized scriptures. It has three main branches: Catholic, Orthodox, and Protestant (**Figure 6.4**). Catholic and Orthodox Christianity split roughly one thousand years ago, while the Protestant/Catholic **Schism** began in the sixteenth century. The split between the Orthodox and Catholic hierarchies centered around whose authority in the church was final. The split between Protestantism and Catholicism mostly centered on practices conducted by the Catholic Church that the future Protestants did not believe were suitable for a religious organization.

The three **branches** of Christianity have their own spatiality, with a great deal of overlap between them. Orthodox Christianity is mostly seen in Russia, Eastern Europe, and Southern Europe with notable examples in Africa (Ethiopia) and in places where large numbers of people from these places have migrated (the United States, Canada). Catholic Christianity is seen in a wider range of places. It largely formed around the historic Roman Empire, then spread to the north and west of Europe. Catholicism did not stop there, however. The age of colonial expansion transplanted Catholicism to such widespread places as the Philippines, much of the Americas and Caribbean, and large parts of Africa. Protestantism is the most recently developed Christian branch, but it has also diffused widely. The initial Protestant countries were in northern Europe, but again due to colonialism, Protestant Christianity was exported to places like the United States, South Africa, Ghana, and New Zealand. The current expansion of Christianity, particularly in Asia, is largely due to the growth of Protestantism.

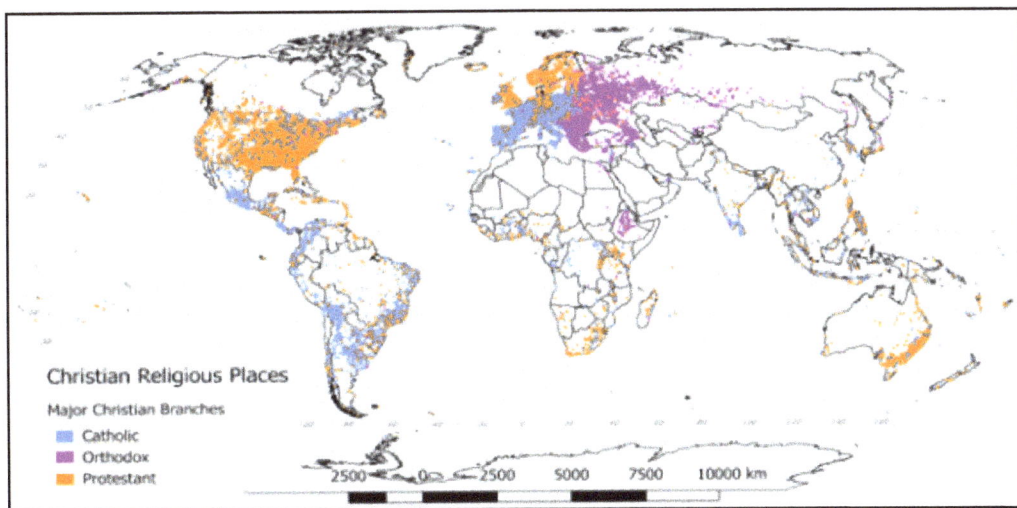

Figure 6.4 | Christian Places [4]
This dot map shows structures designated as Christian from OpenStreetMap. Note how it varies from Figure 6.2 in that it shows a religious landscape that is more fragmented and interspersed.
Author | David Dorrell
Source | Original Work
License | CC BY SA 4.0

Each Christian branch has developed a distinct appearance in the landscape. Orthodox churches are meant to invoke a sense of the divine. Buildings are elaborate, both inside and outside. Catholic churches also tend to be elaborate, in a similar vein to Orthodox churches, but with a different architectural tradition. This is understandable due to the fact that these two branches of Christianity arose in different places with different ideas of architectural grandeur and beauty. Protestant churches as a collective are less elaborate that their close relatives. This is a reflection of the early history of Protestant churches, which were often specific rejections of the elaborate ceremony and ostentatious display of the Catholic Church.

Islam

Islam is a monotheistic religion originating with the teachings of Muhammad (570-632), an Arab religious and political figure. The word Islam means "submission", or the total surrender of oneself to God. An adherent of Islam is known as a Muslim, meaning "one who submits (to God)." Both Islam and Christianity inherited the idea of the chain of **prophecy** from Judaism. This means that figures such as Moses (Judaism) and Jesus (Christianity) are considered prophets in Islam. Muslims believe that Muhammad is the very last in that chain of prophecy. Islam has two main branches, and many smaller ones. Of the two main branches -Sunni and Shi'a, Sunni is much larger, comprising roughly 80% of all Muslims (**Figure 6.5**). The split between the two largest branches of Islam centered around the question of succession, that is to say, who would be the rightful leader of the Muslim world. Currently, there is no single voice for the global Muslim community. Other forms of Islam include Sufi (mystical) Islam and Ahmadi Islam. India is the

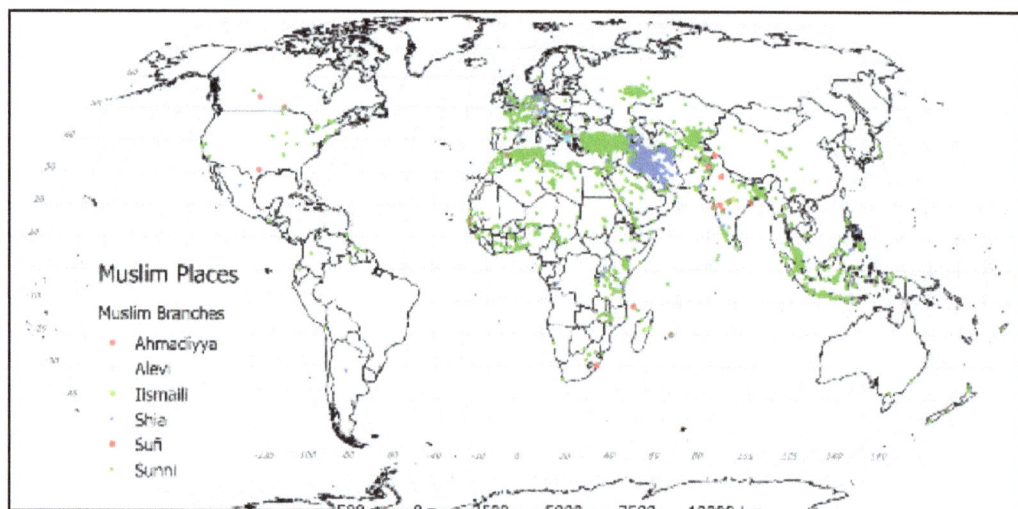

Figure 6.5 | Muslim Places [5]
This dot map shows structures designated as Muslim from OpenStreetMap. The distribution of the different branches (particularly Sunni and Shia) belie any hard lines in the religious landscape.
Author | David Dorrell
Source | Original Work
License | CC BY SA 4.0

number three Muslim country, but there are five times as many Hindus in India as there are Muslims.

The Muslim world is somewhat more contiguous than the Christian world. This is mostly due to the fact that the Muslim expansion did not occur in two phases in the same way that Christianity did. As can be seen in the following map, Sunni and Shi'a countries are somewhat spatially separated. Only the countries of Iran, Iraq, Azerbaijan, and Bahrain are majority Shi'a. There are sizable minority Muslim sects in the world. Many of these groups, such as the Ahmadiyya, are subject to discrimination by other Muslim populations and/or governments. The world's most theocratic governments are Muslim, particularly those of Iran and Saudi Arabia. This is notable in that these two countries are also regional rivals and the two most powerful states in the Muslim world.

Buddhism

Buddhism is an offshoot of Hinduism that dates to the fifth century BCE. It was founded by Siddhartha Gautama near the modern border between Nepal and India. The three largest branches of Buddhism are Theravada, Mahayana and Vajrayana (**Figure 6.6**). The main differences between the branches are their approaches to **canonized doctrine**.

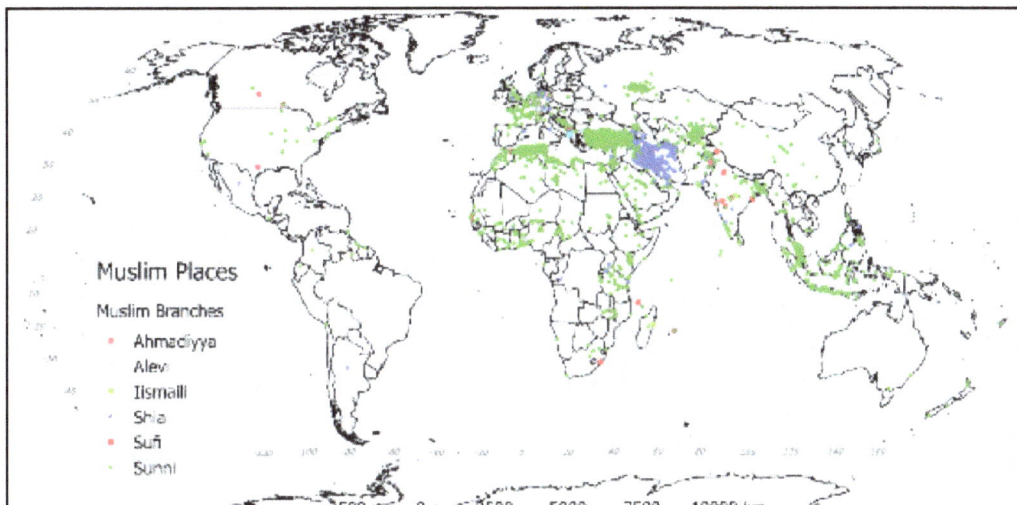

Figure 6.6 | Buddhist Places [6]
This dot map shows structures designated as Buddhist from OpenStreetMap. It does not break Buddhism into its major branches of Theravada, Mahayana, and Vajrayana.
Author | David Dorrell
Source | Original Work
License | CC BY SA 4.0

Hinduism

Hinduism is a religious tradition that originated in the Indian subcontinent (**Figure 6.7**). Its origins can be traced to the ancient Vedic civilization (1500 BCE), a product of the invasion of Indo-European peoples. A conglomerate of diverse beliefs and traditions that assembled organically over a period of centuries, Hinduism has no single founder. Due to its concurrent growth with Indian civilization, Hinduism has historically been tightly bound to the caste system, although the modern Indian state has worked to ameliorate the more damaging effects of this relationship.

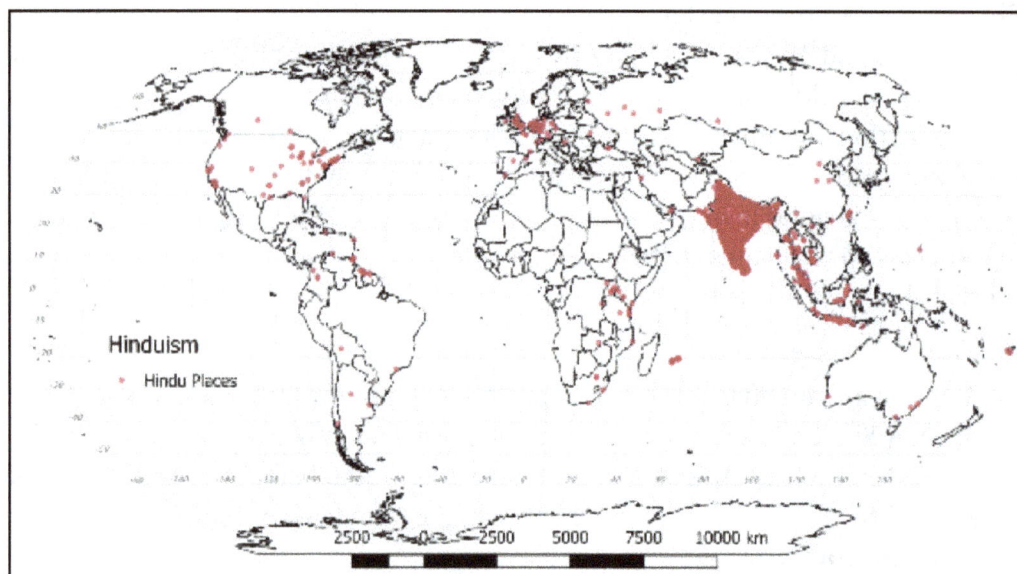

Figure 6.7 | Hindu Places [7]
This dot map shows structures designated as Hindu from OpenStreetMap. Although heavily associated with the modern Indian state, large numbers of Hindus now live elsewhere.
Author | David Dorrell
Source | Original Work
License | CC BY SA 4.0

Chinese Religions

Not strictly located in China, Chinese religions are closely tied to Daoism (a nature religion), Confucianism (a philosophy of living), and ancestor worship. Chinese religious structures are associated with people of Chinese descent within and external to China. Because of the diversity of religious practices and beliefs, this category is best thought of as a complex of beliefs, rather than a defined set of beliefs and practices.

Sikhism

Sikhism is a 15th Century amalgamation of Islam and Hinduism. It is in many ways emblematic of syncretic religions. **Syncretic religions** are created by the combination of two or more religions, with the addition of doctrinal elements to create cohesion between the disparate pieces. Founded by Nanak Dev Ji (1469 – 1539) Sikhism reconciles Hinduism and Islam by recasting Hindu gods as aspects of a single god, in a manner similar to the Catholic Trinity. Although heavily associated with the Punjab region of the Indian subcontinent, Sikhism has spread widely through relocation diffusion (**Figure 6.8**). It has about 26 million adherents.

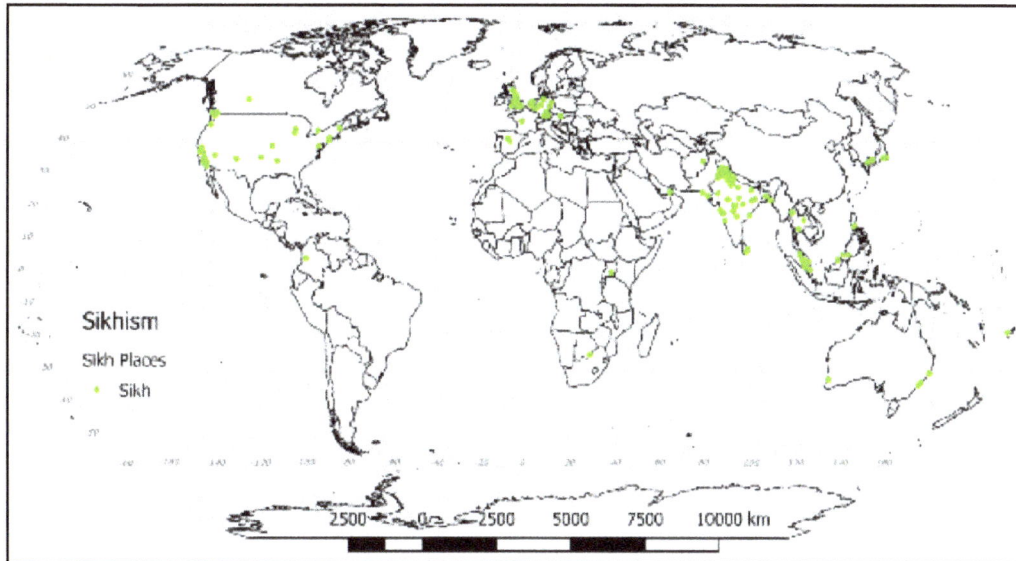

Figure 6.8 | Sikh Places [8]
This dot map shows structures designated as Sikh from OpenStreetMap. Although originating in the Punjab on the Indian Subcontinent, it has diffused widely.
Author | David Dorrell
Source | Original Work
License | CC BY SA 4.0

Judaism

Judaism is a monotheistic religion originating in the Bronze Age in the eastern Mediterranean (**Figure 6.9**). Although it has no single founder, it holds the Torah as its holy book. In the modern context of Judaism, there are three major forms—

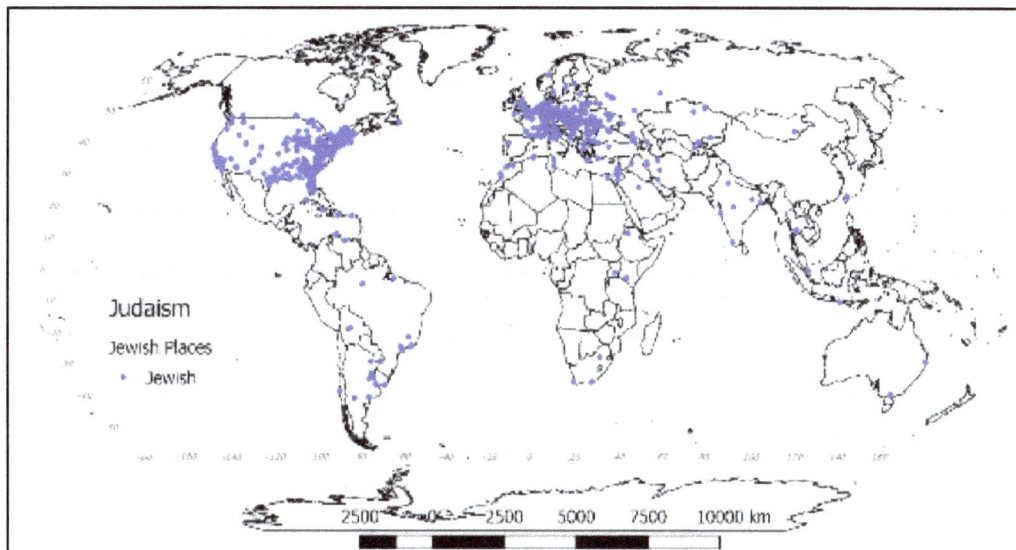

Figure 6.9 | Jewish Places [9]
This dot map shows structures designated as Jewish from OpenStreetMap. It does not break Judaism into Orthodox, Conservative and Reform.
Author | David Dorrell
Source | Original Work
License | CC BY SA 4.0

Orthodox, Conservative and Reform—each with their own set of interpretations of correct practice. Judaism, as the initial Abrahamic religion, influenced other religions (particularly Christianity and Islam).

Animism, Jainism, Bahai, Shinto and Others

This catch-all category combines together religions that are all quite different. They are here due to their similar ties to places or ethnicities, not because they share any doctrinal or historical connections. Before continuing on a discussion of the following religions, it is important to make a point clear. It is possible to practice more than one religion. Many people in the world practice two or more religions with no sense of contradiction. In many parts of the world, pre-Christian or pre-Islamic beliefs persist alongside the newer religions.

Animism is a broad category, found in a variety of environments (**Figure 6.10**). The underlying theme is the idea that almost anything in the environment-people, mountains, rivers, rain, etc. is alive and worthy of recognition as such. Animism is frequently practiced with other ideologies or philosophies.

Baha'i was founded by Mirza Husayn Ali Nuri (1817-1892) in 1863. Baha'i was an offshoot of another religion, Babism, that in itself was a derivative of Islam. Although traditional Muslims believe that Muhammad was the last of the prophets (the seal of the prophets) many religions have been founded on the idea that there could be other, later people who also spoke for god. Baha'is believe that new messengers would be sent to humanity to remind people of their universal relationship to god and one another. The late date and historic context of this religion informed a religion that explicitly rejected racism and nationalism. One

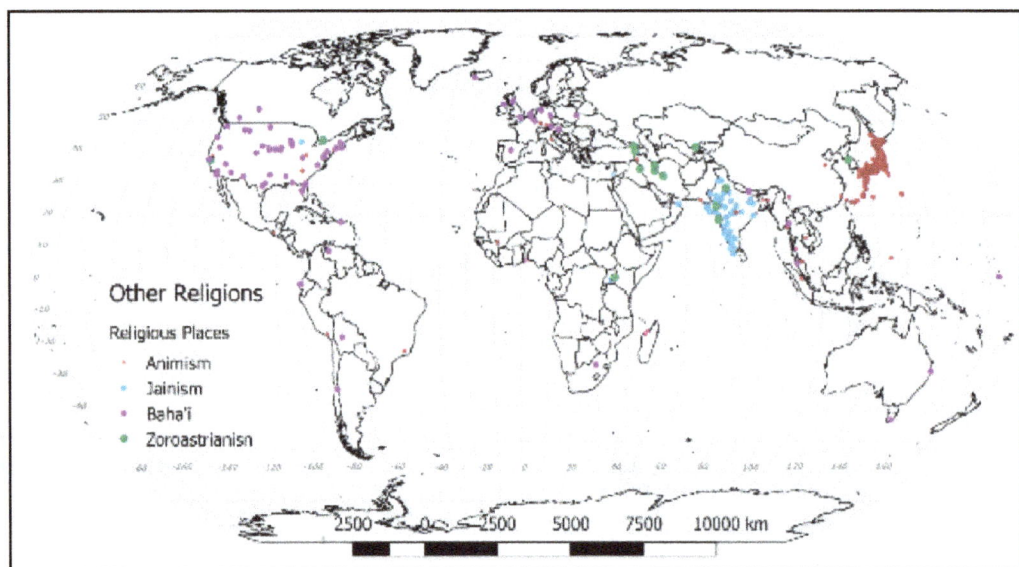

Figure 6.10 | Other Religious Places [10]
This dot map shows some structures designated as Animist, Jain, Baha'I, and Zoroastrian from OpenStreetMap.
Author | David Dorrell
Source | Original Work
License | CC BY SA 4.0

of the notable characteristics is that although Baha'is are not one of the larger religions on Earth, they have a temple on every permanently inhabited continent.

Jainism is another ancient religion that arose in India. It is best known for its concept of *ahimsa*, or nonviolence.

Shinto, the ethnic religion of Japan is often practiced in conjunction with Buddhism. It is polytheistic and dates back centuries. The most important consideration of Shinto is that the rituals are so ingrained in Japanese national identity that the religion can either be considered vibrant and relevant, or moribund and ritualistic, depending on the perspective of the viewer.

Syncretic Religions

Syncretic religions are formed by the combination of two or more existing religions to produce a new religion (**Figure 6.11**). Some of the larger syncretic religions have already been mentioned, such as Baha'i or Sikhism.

Cao Dai is a religion founded in twentieth century Vietnam. It has strains of Taoism and Buddhism and represents an attempt to reconcile many diverse religious traditions into a single religion.

Voodoo arose in French Caribbean colonies as a combination of Catholicism and the beliefs of another set of West African peoples, the Ewe and the Fon. Practitioners speak to God using intercessors called loa that function as saints do in both the Catholic and Sufi worlds.

Candomble is a syncretism formed from many West African religious traditions and Catholicism. It has existed in Brazil for centuries. It believes in a creator god (Oludumare) and a series of demi-gods (Orishas).

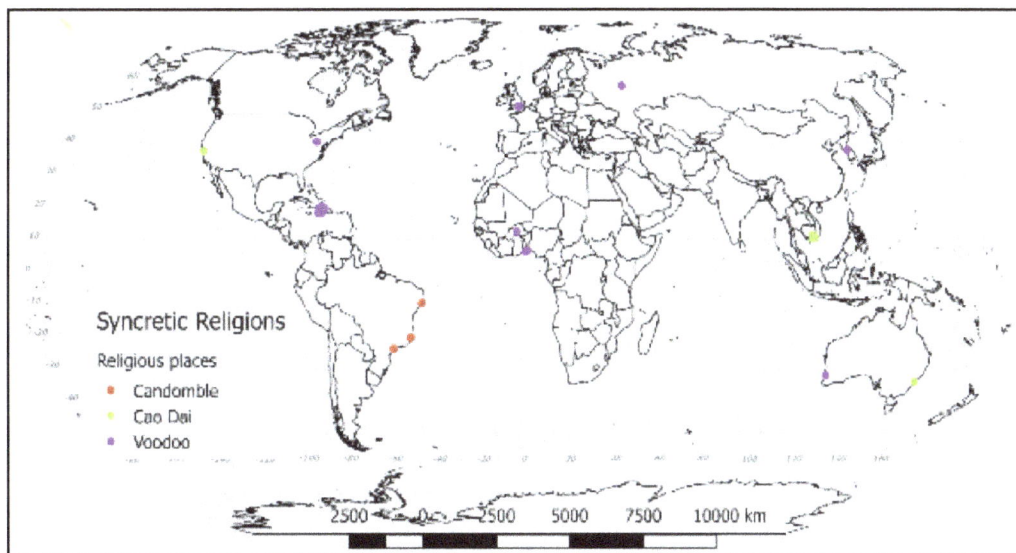

Figure 6.11 | Syncretic Religions [11]
This dot map shows structures designated as Candomble, Voodoo, or Cao Dai from OpenStreetMap. Notice that Cao Dai has diffused from its origin in Vietnam to Australia and the United States.
Author | David Dorrell
Source | Original Work
License | CC BY SA 4.0

New Religions

Much like any other human phenomena, new religions are formed continuously. They are usually adaptations or combinations of existing religions. It is impossible to list the most recent arrivals. This category includes such religions as Scientology, the Unification Church, Seicho no Ie and Wicca.

What about the Nonreligious?

Sometimes the nonreligious are considered a religion unto themselves. This is generally not true. The nonreligious category is amorphous. There are no documents of beliefs that all nonreligious people must abide by. There is no over-arching nonreligious creed. It is another catch-all category that contains a large, diverse population with divergent beliefs and practices. Within these categories, however, there are notable manifestations. First, there are those places which are officially **atheistic** or non-religious. This label is problematic. It provides only the perspective of the government of these places. In many places that officially have no religions, practitioners simply do not advertise their religious affiliations. In other places, religious attendance has declined to a point that many people have no connection to a particular religious tradition. The label **agnostic** refers to the idea that the existence of a higher power is unknowable. It is important to point out that religions do not necessarily require the existence of a god-like force. Daoism relies on nature as its driving force.

6.3 DIFFUSION OF MAJOR RELIGIONS

Religion uses nearly all forms of diffusion to reproduce itself across space. Hierarchical diffusion generally involves the conversion of a king, emperor, or other leader who then influences others to convert. Relocation diffusion, often through missionary work, brings "great leaps forward" by crossing space to secure footholds in far-away places. Contagious diffusion is most often seen in a religious context as the result of direct proselytizing. All these forms of diffusion produce patterns of diffusion that are complex. It is impossible to know why certain religions have appeal in particular times or places, but they do, and that appeal can wear off over time.

Another important thing to remember is that the religious landscape is just a snapshot. In the same way that it has always changed, it will continue to do so. These maps can reinforce this idea in that they demonstrate the historical nature of current religious distributions. Remind yourself that these religious expansions occurred at different times. This will help explain why some places will become Buddhist or Christian at one point in history, but become Muslim at another.

6.3.1 The Diffusion of Buddhism

Buddhism originated near the current Nepalese-Indian border. Like many other religions, it spread in other directions, particularly to the south and east (**Figure 6.12**).

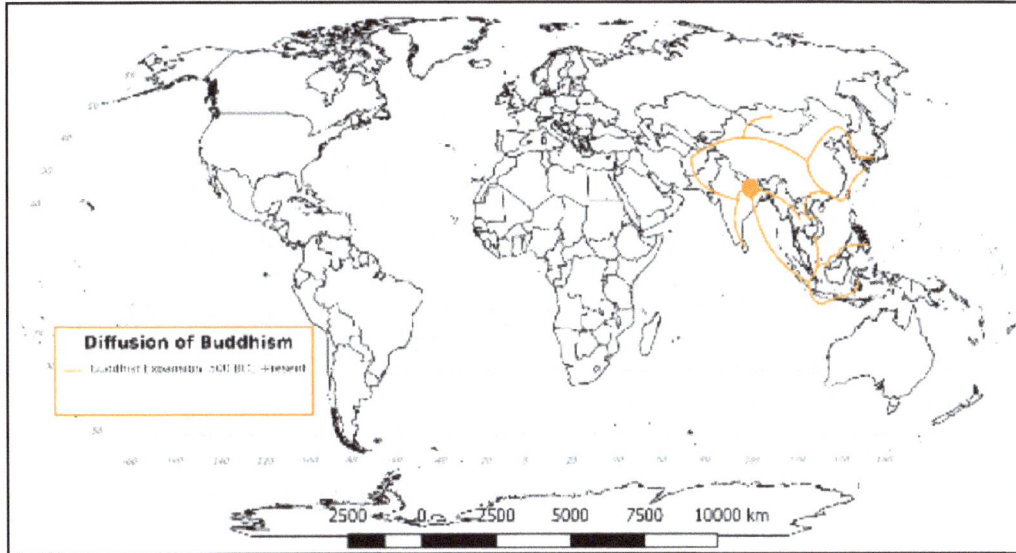

Figure 6.12 | Diffusion of Buddhism [12]
This map shows the distinct waves of the diffusion of Buddhism.
Author | David Dorrell
Source | Original Work
License | CC BY SA 4.0

Due to its position as the oldest large, universalizing religion, Buddhism is a good example of the lifecycle of a religion. From its origins, the religion spread across what is now India and Nepal. It spread in all directions but looking at a current religious map reveals that the process did not end 1500 years ago. Much of its territory on the Indian subcontinent would become mostly Hindu or Muslim. To the east and south, however, the religion continued and expanded. It is not unusual for a religion to prove popular far from its place of origin. In fact, that is the key to a successful universalizing religion.

6.3.2 The Diffusion of Christianity

Christianity was founded in the eastern Mediterranean, although much like Buddhism, its greatest successes were found in other parts of the world (**Figure 6.13**). Christianity initially grew in areas dominated by the Roman Empire, but it would adapt and thrive in many places. With the collapse of the Empire, Christianity became the only source of social cohesion in Europe for centuries. Later, Christianity was promoted through the process of colonialism, and as such, it was modified by the process that distributed it. The spread of Christianity helped drive the process that created the modern world.

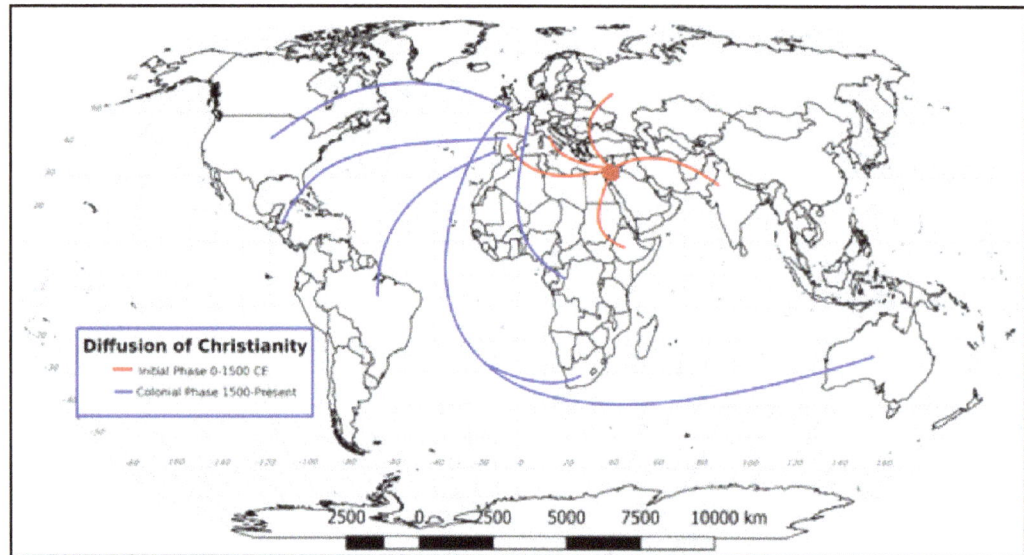

Figure 6.13 | Diffusion of Christianity [13]
This map shows the distinct waves of the diffusion of Christianity, the initial wave from its foundation, and the subsequent wave associated with European Colonialism.
Author | David Dorrell
Source | Original Work
License | CC BY SA 4.0

6.3.3 The Diffusion of Islam

The most recent of the world's largest religions, Islam is also the one that is expanding the fastest (**Figure 6.14**). This is not necessarily through conquest or conversion, but mostly through current demographics. Islam provides a blueprint

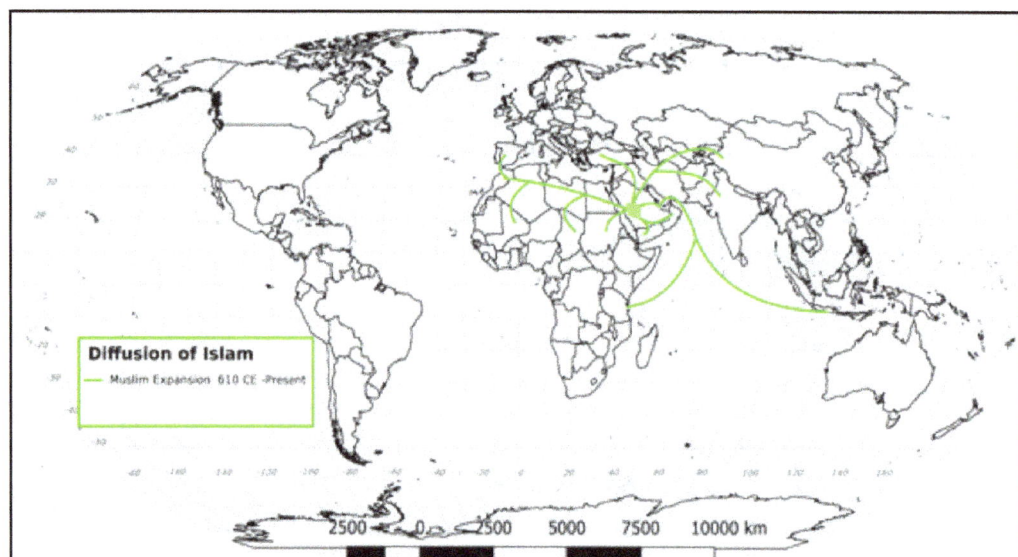

Figure 6.14 | Diffusion of Islam [14]
This map shows the distinct waves of the diffusion of Islam.
Author | David Dorrell
Source | Original Work
License | CC BY SA 4.0

for most aspects of life and as such, has often been associated with rapid expansion driven by military conquest. Although military conquest occurred in the past, military campaigns have been rare since the fall of the Ottoman Empire. The relative distributions of Buddhists, Christians, and Muslims have in fact changed little in half a millennium. Although there has been some migration of Muslims into western Europe, the percentages of Muslims in each country is small. France has the largest percentage of Muslims at 7.5%. To keep this in perspective, that is much lower than the percentage of Muslims in Spain in 1492.

6.4 RELIGIOUS CONFLICT

Human beings have struggled against one another for a variety of reasons. Religious disagreements can be particularly intense. **Sectarian violence** involves differences based on interpretations of religious doctrine or practice. The struggles between the Catholic and Orthodox churches, or the wars associated with the Protestant Reformation and Counter Reformation, are examples of this form of conflict. The current violence seen between Sunni and Shia Muslims is also in this category. Closely associated with this kind of conflict is **religious fundamentalism**. Religious fundamentalism rests on a literal interpretation and strict and intense adherence to the basic principles of a religion. The conflict arises when religious fundamentalists see their coreligionists as being insufficiently pious. Extremism is the idea that the end of a religious goal can be justified by almost any means. Some groups that are convinced that they have divine blessing have few limits to their behavior, including resorting to violence.

Another form of religious violence is between completely different religions. Wars between Muslims and Christians or Hindus and Buddhists have been framed as wars for the benefit or detriment of particular religions. What is described as religious strife, however, is often not. Although some religions are fighting over doctrinal differences, most conflict stems from more secular causes- a desire for political power, a struggle for resources, ethnic rivalries, and economic competition.

The Israel/Palestine conflict is a struggle over territory, resources, and political recognition. The Rohingya crisis in Myanmar has less to do with religion and more to do with differences in ethnicity, national origin, and post-colonial identity. Massacres in Sahelian Africa are better framed as farmers versus herders. The long running violence between Protestants and Catholics in Northern Ireland is better framed as a violent dispute between one group who holds allegiance with the Republic of Ireland and the other who holds allegiance with the United Kingdom.

This is not to say that religious violence does not exist. It does. The most obvious example of this in recent years has been the emergence of Islamic State. This organization carries all the worst examples of religious extremism- sectarianism toward other Muslims (the Shi'a), attempted genocide of religious minorities (Yazidis and Christians), and brutal repression through the apparatus of the state.

6.5 SUMMARY

Religion can be key to a person's identity. It manifests both as an internal sentiment as well as in structures in a landscape. The religious world is always changing, but at a pace that is generally very slow. New religions are created, while other religions may fade away, or change. The general historical trend has been toward a small number of universalizing religions gaining ground over local, ethnic religions. Religious differences can lead to conflict, although many conflicts presented as religious in nature have their roots elsewhere. Just like language, religion is another way of sorting people into groups, either for good or bad. Another such way of sorting people is ethnicity, the subject of the next chapter.

6.6 KEY TERMS DEFINED

Agnostic: The belief that existence of the supernatural is unknown or   unknowable.

Atheistic: The belief that there is nothing supernatural.

Branches: A large division of a religion.

Canonized doctrine: The officially recognized documents or ideas of a religion.

Ethnic religions: A religion associated with a particular ethnic group.

Monotheism: The belief in one god.

Pilgrimage: A journey to a sacred place.

Polytheism: The belief in many gods.

Prophecy: Communication with a supernatural power.

Proselytizing: Seeking converts to a religion.

Religious fundamentalism: The belief in the absolute authority of a religious text.

Sacred spaces: Places associated with a sense of the divine.

Sanctuary: A haven or place of safety, often defined by law.

Schism: The fracturing of an organization.

Sectarian violence: Violence between different sects of the same religion.

Secular: A condition of separation between a state and any religion.

State religion: The official religion of a state. This is not the same as theocracy.

Syncretic religion: A religion formed by the combination of other religions.

Theocracy: A state ruled by religious principles.

Universalizing religion: A religion that seeks converts.

6.7 WORKS CONSULTED AND FURTHER READING

Alcalay, Ammiel. 1992. *After Jews And Arabs: Remaking Levantine Culture*. 1 edition. Minneapolis: University Of Minnesota Press.

Black, Jeremy. 2000. *Maps and History: Constructing Images of the Past.* Yale University Press.

"Buddhist Pilgrimage Tours | Pilgrimage Tour | Buddha Darshan Tour | Buddha Tours | Buddhist Destination | Holy Buddha Places |Hotel In Buddhist Destinations | Buddhist Temples | Buddhist Monasteries." n.d. Accessed March 17, 2013. http://buddhistpilgrimagetours.com/.

Cao, Nanlai. 2005. "The Church as a Surrogate Family for Working Class Immigrant Chinese Youth: An Ethnography of Segmented Assimilation." *Sociology of Religion* 66 (2):183–200.

"Catholic Pilgrimages, Catholic Group Travel & Tours By Unitours." n.d. Accessed March 16, 2013. http://www.unitours.com/catholic/catholic_pilgrimages.aspx.

Dorrell, David. 2018. "Using International Content in an Introductory Human Geography Course." In *Curriculum Internationalization and the Future of Education.*

"Global-Religion-Full.Pdf." n.d. Accessed August 22, 2017. http://assets.pewresearch.org/wp-content/uploads/sites/11/2014/01/global-religion -full.pdf.

Gregory, Derek, ed. 2009. *The Dictionary of Human Geography.* 5th ed. Malden, MA: Blackwell.

"Hajj Umrah Packages 2011 | Hajj & Umrah Travel, Tour Operators | Umra Trips 2011 : Al-Hidaayah." n.d. Accessed March 17, 2013. http://www.al-hidaayah.travel/ .

Knowles, Anne Kelly, and Amy Hillier. 2008. *Placing History: How Maps, Spatial Data, and GIS Are Changing Historical Scholarship.* ESRI, Inc.

"Modern-Day Pilgrims Beat a Path to the Camino | Travel | The Guardian." n.d. Accessed March 16, 2013. http://www.guardian.co.uk/travel/2011/may/02/camino -pilgrims-route.

"OpenStreetMap." 2018. OpenStreetMap. Accessed January 5, 2018. https://www.openstreetmap.org/ .

Thompson, Lee, and Tony Clay. 2008. "Critical Literacy and the Geography Classroom: Including Gender and Feminist Perspectives." *New Zealand Geographer* 64 (3):228–33. https://doi.org/10.1111/j.1745-7939.2008.00148 .x.

"World Religions Religion Statistics Geography Church Statistics." n.d. Accessed January 5, 2018. http://www.adherents.com/ .

Bank, World. 2017. "Metadata Glossary." DataBank. Accessed August 20. http://databank.worldbank.org/data/glossarymetadata/source/all/concepts/ .

6.8 ENDNOTES

1. Data source: Pew Research Center's Forum on Religion & Public Life 2012. http://www.globalreligiousfutures.org/explorer#/

2. Data source: Pew Research Center's Forum on Religion & Public Life 2012. http://www.pewforum.org/2015/04/02/religious-projections-2010-2050/

3. Data source: United States Census Bureau http://www2.census.gov/geo/tiger/
    TIGER_DP/2016ACS/ACS_2016_5YR_COUNTY.gdb.zip

4. Data source: OpenStreetMap. http://www.geofabrik.de/data/

5. Data source: OpenStreetMap. http://www.geofabrik.de/data/

6. Data source: OpenStreetMap. http://www.geofabrik.de/data/

7. Data source: OpenStreetMap. http://www.geofabrik.de/data/

8. Data source: OpenStreetMap. http://www.geofabrik.de/data/

9. Data source: OpenStreetMap. http://www.geofabrik.de/data/

10. Data source: OpenStreetMap. http://www.geofabrik.de/data/

11. Data source: OpenStreetMap. http://www.geofabrik.de/data/

12. Adapted from Proliferation of Buddhism. Spiegel Online http://www.spiegel.de/
    international/spiegel/grossbild-460247-779134.html

13. Adapted from https://www.ed.ac.uk/divinity/research/resources/animated-maps

14. Adapted from http://www.oxfordislamicstudies.com/article/opr/t253/e17

7 Ethnicity and Race

David Dorrell

STUDENT LEARNING OUTCOMES

By the end of this section, the student will be able to:

1. Understand: the differing bases of ethnic identity
2. Explain: the relationship between ethnicity and personal identity
3. Describe: the degrees of relevance of ethnicity in a society
4. Connect: ethnicity, race, and class as they relate to political power

CHAPTER OUTLINE

7.1 WHAT ARE ETHNICITY AND RACE?

A common question asked in introductory geography classes is "What is ethnicity and how is it different than race? The short answer to that question is that ethnicity involves learned behavior and race is defined by inherited characteristics. This answer is incomplete. In reality, both race and ethnicity are complex elements embedded in the societies that house them. The relationship between race, ethnicity and economic class further complicates the answer.

Other students have asked, "How is this geography? Ethnicity and race have strong spatial dimensions. Both races and ethnicities have associated places and spatial interactions. A person's ability to navigate and use space is contingent upon many factors- wealth, gender, and race/ethnicity. Anything that sets limits on a person's movement is fair game for geographic study. Numerous geographic studies have centered on the sense of place. Race and ethnicity are part of a place. Signs are written in languages, houses have styles, people wear clothing (or not!) and all of these things can indicate ethnicity.

7.1.1 The Bases of Ethnicity

Ethnicity is identification through language, religion, collective history, national origin, or other cultural characteristics. A cultural characteristic or a set of characteristics is the constituent element of an ethnicity. Another way of thinking of an ethnicity is as a **nation** or a people. In many parts of the world, ethnic differences are the basis or political or cultural uprisings. For example, in almost every way the Basque people residing on the western border between France and Spain are exactly like their non-Basque neighbors. They have similar jobs, eat similar foods, and have the same religion. The one thing that separates them from their neighbors is that they speak the Basque language. To an outsider, this may seem like a negligible detail, but it is not. It is the basis of Basque national identity, which has produced a political separatist movement. At times, this movement has resorted to violence in their struggle for independence. People have died over the relative importance of this language. The Basques see themselves as a nation, and they want a country.

The ethnicities of dominant groups are rarely ever problematized. **Majority** ethnicities are considered the default, or the normal, and the smaller groups are in some way or another marginal. Talking about ethnicity almost always means talking about minorities.

There are three prominent theories of Ethnic Geography: amalgamation, acculturation, and assimilation. These theories describe the relation between majority and minority cultures within a society. Amalgamation is the idea that multiethnic societies will eventually become a combination of the cultural characteristics of their ethnic groups. The best-known manifestation of this idea is the notion of the United States as a "melting pot" of cultures, with distinctive additions from multiple sources.

Acculturation is the adoption of the cultural characteristics of one group by another. In some instances, majority cultures adopt minority cultural characteristics (for example the celebration of Saint Patrick's Day), but often acculturation is a process that shifts the culture of a minority toward that of the majority.

Assimilation is the reduction of minority cultural characteristics, sometimes to the point that the ethnicity ceases to exist. The Welsh in the United States have few, if any, distinct cultural traits.

When we looked at the previous chapters- Language, Religion, and now Ethnicity, we have explored subjects that are often the core of a person's identity. Identity is who we are and we, as people, are often protective of those who share our collective identity. For example, ethnicity, and religion can be closely tied, and what can appear as a religious conflict may be in fact a politicized ethnic disagreement or a struggle over resources between ethnicities that has become defined as a religious war. Muslim Fula herders and Christian farmers in Nigeria aren't battling over religious doctrine; they're two different peoples fighting for the same land and water resources.

One of the enduring ideas of modern political collectives is that we consider everyone within the boundaries of our country as "our group." The reality has not lived up to that concept, however. Many modern countries are wracked by ethnic struggles that have proven remarkable resistant to ideas of ethnic or racial equality.

7.1.2 Race

The central question around race is simple: "Does race even exist?" Depending on how the question is framed, the answer can be either yes or no. If race is being used in a human context in the same way that species is used in an animal context, then race does not exist. Humans are just too similar as a population. If the question is rephrased as, "Are there some superficial differences between previously spatially isolated human groups?" then the answer is yes. There are genetic, heritable differences between groups of people. However, these differences in phenotype (appearance) say very little about genotype (genetics). Why is that? The reality is that human beings have been very mobile in their history. People move and they mix with other groups of people. There are no hard genetic lines between different racial categories in the environment. As a consequence of this, racial categories can be considered socially constructed.

7.1.3 How are ethnicity and race different?

People tend to have difficulties with the distinctions. Let's start with the easiest racial category in the United States- African American. Most people understand that the origin of the African American or Black population of the United States is African. That is the race part. Now, the ethnic part appears to be exactly the same thing, and it almost is, but only for a particular historical reason. If Africans had been forcibly migrated by group, for example large numbers of BaKongo or Igbo people were taken from Africa and brought to Virginia and settled as a group, then we would be talking about these groups as specific ethnicities in the same way we talk about the Germans or Czechs in America. The Germans and Czechs came in large groups and often settled together, and preserved their culture long enough to be recognized as separate ethnicities.

That settlement pattern did not happen with enslaved Africans. They were brought to the United States, sold off effectively at random, and their individual ethnic cultures did not survive the **acculturation** process. They did, however, hold onto some general group characteristics, and they also, as a group, developed their own cultural characteristics here in the United States. Interestingly, as direct African immigration to the United States has increased, the complexity of the term African American has increased, since it now includes an even larger cultural range.

7.1.4 Specifically Ethnic

The United States is a multiethnic and multiracial society. The country has recognized this from the very beginning, and the U.S. Census has been a record of ethnic representation for the U.S. since 1790. Here are the current racial categories **(Figure 7.1)**.

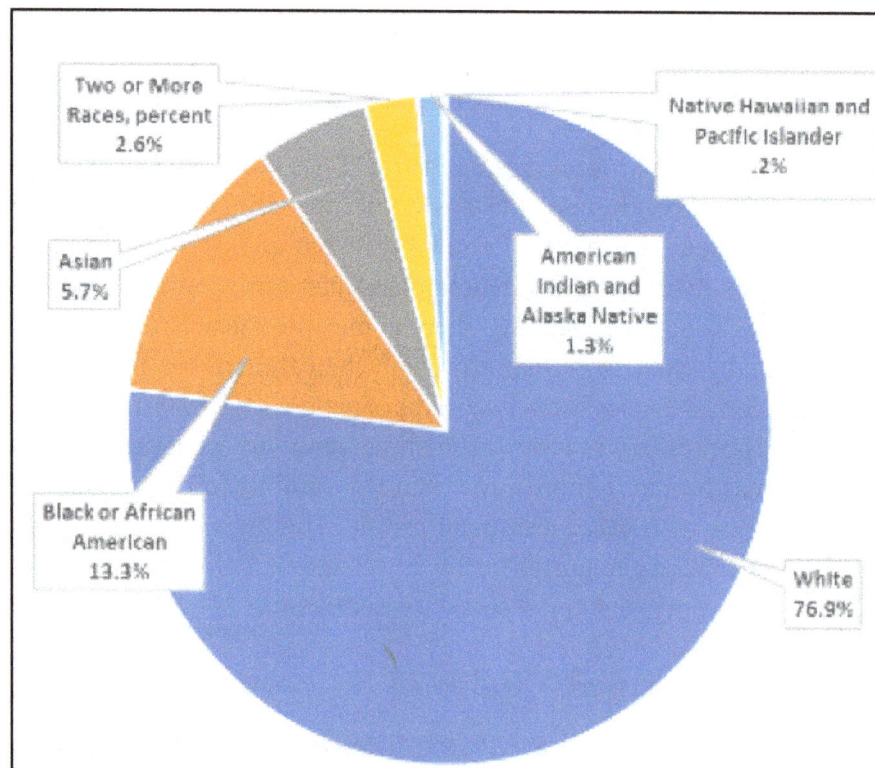

Figure 7.1 | U.S. Racial Makeup according to the United States Census of 2016 [1]
Author | David Dorrell
Source | Original Work
License | CC BY SA 4.0

There are many ethnicities in the United States, and data are collected to a granular level, but in many ways, the ethnic categories are subsets of the racial categories. The idea is that race is a large physical grouping, and ethnicity is a smaller, cultural grouping. Thinking about the data this way helps understand why African American is both an ethnicity and a race (Remembering that there are African-Americans who come directly from Africa). Another, more complete example is the numerous ethnicities within American Indian. Within the race category of American Indian and Alaska Native are dozens of individual nations (**Figure 7.2**).

American Indian Nations			
Apache	Comanche	Menominee	Seminole
Arapaho	Cree	Navajo	Shawnee
Assiniboine Sioux	Creek	Osage	Shoshone
Blackfeet	Crow	Ottawa	Sioux
Cherokee	Delaware	Paiute	Tohono O'Odham
Cheyenne	Hopi	Pima	Ute
Chickasaw	Houma	Potawatomi	Yakama
Chippewa	Iroquois	Pueblo	Yaqui
Choctaw	Kiowa	Puget Sound Salish	Yuman
Colville	Lumbee		

Figure 7.2 | Federally recognized American Indian Nations [2]
Author | David Dorrell
Source | Original Work
License | CC BY SA 4.0

7.1.5 Hispanic Ethnicity in the United States

Since 1976, the United States government has required the collection and analysis of data for only one ethnicity: "Americans of Spanish origin or descent." The term used to designate this ethnicity is Hispanic. It is a reference to the Roman name for what is now modern Spain. Hispanics, however are generally not Spanish; they are people who originate in one of the former colonies of Spain. Another term that is used is Latino, which is another reference to the Roman Empire. Both of these labels are very vague. Generally, people identify with the country of their ancestors (Mexico, Thailand), and not with a label generated by the Census Bureau for the purposes or recordkeeping.

Hispanics can be of any race. It is important to note that all racial and ethnic information is self-reported. This means that the person who decides if you are African American, Hispanic, or any other category is you. One final detail is that native people of hispanophone countries, even if they themselves do not speak Spanish, will often be considered Hispanic.

7.2 RELEVANCE OF RACE AND ETHNICITY IN THE UNITED STATES

The importance of race and ethnicity is variable both across space and across time. Historically, divisions in the United States along ethnic or racial lines have been the norm. but now these divisions based on race or ethnicity are not as prevalent as they have previously been. From the earliest days of the country and codified in the US Constitution, slavery created a profoundly divided society, particularly between the free, white population and the enslaved Black population. Free people of color provided a small degree of linkage between the groups.

These were not the only divisions in the US, however. The dominant group were people of English descent. In geographical terms, we refer to them as a charter group. The charter group does not refer to the first people to come to a place; they are people with the **first effective settlement**. This is an academic way of saying that they are the first group with political dominance. English settlers produced laws that furthered their own interests. They promoted their own language (English), religion (Protestant Christianity) and governance. Groups coming in later found themselves in a place where many of the cultural questions had already been answered. The pressure to assimilate in the United States applies to everyone. There can be political pressure; for example, during World War One German Americans largely stopped speaking German. The pressure can be social; for example, young children at school can feel isolated when they cannot speak the majority language. Particularly, the pressure can be economic. Without conforming to general social (majority) norms, it can be difficult to navigate the employment market. A lack of English, unawareness of the norms of formal dress or behavior, or just the inability to recognize social cues can make life difficult for those who have not acculturated.

The charter group also changed. For example, the definitions of "whiteness" and "blackness" have not been historically constant. Consider the history of the United States. Initially, the U.S. population was made largely of Protestant British white people and African black people. Adding people from other places required that definitions be amended.

Would Catholic Italians be considered "White?" In the past, many Americans would have said no. For that matter, neither would the Irish (because of their Catholicism) or Jews (because they aren't Christian), but over time, these groups were generally included into the white category. Whiteness broadened to include more people. It became less of an ethnic category and more of a racial category.

Definitions of blackness evolved as well. In the American South, there eventually arose a legal framework that defined blackness as having any African ancestry. It would be possible (and relatively common) to be phenotypically white and legally black. Historically, mixed-race Creole people in Louisiana did not consider themselves to be black or white; they were another category altogether.

People attempting to emigrate from Asia, particularly China, were subject to their own set of exclusionary laws, which severely limited their migration to the United States. As late as World War II, it was considered acceptable for the government to intern (imprison) American citizens of Japanese descent over questions of their racial origins and loyalty.

One of the current interesting ethnic questions in the United States is the status of Hispanic people in the existing racial categories. Since Hispanic is not itself a racial category, people within this ethnicity can choose what label they feel is most appropriate. It appears now that Hispanics are identifying themselves as white in the U.S. census. This has an impact on projections for the future U.S. population. If Hispanics identify as white, then the U.S. will remain majority white for quite some time. If they do not, the U.S. will have no racial majority in a few decades.

Although race and ethnicity in the U.S. were largely associated with state-mandated identification, restrictive laws, and onerous obligations, today both race and ethnicity are self-identified for the census. Whereas at one time being Irish could be enough to deny someone employment, now it is a slogan to place on your welcome mat and celebrate once a year in March.

7.2.1 Racial Identifiers

The language used to identify racial groups has changed as well. For example, in broad terms of ethnicity, people of Asian descent who were born in the United States are now referred to as Asian Americans, although the census racial category is still Asian. The term Asian implies a relationship with Asia and no relationship with America. Asian American explicitly ties this group to America.

People who trace their ancestry to Africa have a different problem. This problem is a function of American history. The first census label for this group was simply Black. Over time other labels were used, such as Negro (which means black), and eventually the term African American was adopted. This term is meant to provide a relationship between a population of people and a place of origin. In other words, it explicitly ties a group to their ancestral origin.

Although Native American is used in common speech in the U.S., the Census category is still American Indian, which is not the same as Indian-American (peoples associated with South Asia). The continued use of American Indian is somewhat outside the trend toward more descriptive categories.

7.2.2 Racism in the United States

Although racism and ethnic **discrimination** are similar, they are not the same thing. Although ethnic markers (generally) diminish over time, physical differences do not.

Exclusionary racial policies existed in the United States from the very beginning and have continued beyond the **Jim Crow** era of the twentieth century. From the US Constitution that counted slaves as 3/5 of a person to restrictive housing covenants in the 1960s, the country has had a history of racism that did not end in the Civil Rights era. This exclusion has not solely been limited to African

Americans. Many groups have been subject to racist laws and acts. The indigenous people of the United States were not fully considered citizens until 1924. In the past, voting rights, access to housing and even union membership had racialized politics directed at many marginalized groups.

This is not to say that ethnically-based discrimination does not exist. Such discrimination has been prevalent in United States history, but it tends to subside as the host population absorbs the immigrant population.

7.2.3 Housing

Historically, ethnic groups tended to live near one another in spatially contiguous areas. Many cities have a Chinatown or Little Italy. These are known as ethnic **enclaves**. There are many reasons why groups cluster; some reasons are voluntary and some are not. In the United States, it was not uncommon for cities to restrict where African American citizens could live. These restrictions were either through the force of law, or through unwritten behavioral norms that resisted renting or selling houses to African American families outside of certain areas. This residential, spatial **segregation** was accompanied with educational, social, and economic segregation. African American communities were often known as **ghettos**, places where a certain population is forced to live. The word ghetto is older than the United States itself. Ghetto was an Italian name for the area that Jews were forced to live in. Although the word is Italian, the idea of forcing **minority** populations to live in designated areas has unfortunately had wide historical appeal. Legal housing segregation ended in the United States in 1968, but behaviors change more slowly than laws.

Many ethnic communities have arisen from less coercive means. There are numerous reasons that an ethnic community would choose to live close together. Mutual support networks, the ability to develop schools and businesses catering to their own needs, a sense of safety, and the ability to retain their own cultural connections are examples of positive reasons. Institutionalized poverty, marginalized political representation, and active discrimination are negative reasons.

7.2.4 Environmental Justice

One of the spatial manifestations of racism and ethnic discrimination is the difference in levels of political representation. Another one is the location of unpleasant environmental activities. Landfills and airports tend to be built in places inhabited by less-powerful groups, while dominant groups rarely if ever have to organize to prevent such things being built in their neighborhoods. Some groups find their economic situations limited by underfunded schools or inadequate infrastructure. The idea that different groups should have access to decent places to live called **environmental justice**.

7.2.5 Ethnic diversity in the United States

Like all predominantly immigrant countries, the United States is ethnically diverse, but the range of ethnicities has varied over time as new groups arrive and previous groups acculturate and eventually assimilate. A male of Italian descent in the United States will sometimes just say, "I'm Italian." This may be a person who speaks no Italian, isn't Catholic, and never been in Italy in his entire life. What then, does this statement mean? It just signals an historic connection with an ethnicity, even if the connection has faded over time. This isn't to single out Italian-Americans. Generally, as groups assimilate, their distinctive ethnic markers fade. Comparing Polish-Americans with Mexican-Americans may involve people who speak the same language (English), have the same Catholic religion, and live very similar lifestyles. The label has faded to a marker, with food being the one of the last cultural elements.

7.2.6 Foodways

One of the ways groups demonstrate ethnicity is through food. One of the most obvious hallmarks of the arrival of an ethnicity into the United States, or any other country, is the diffusion of a food from the group of origin. Pizza in the United States, curry in the United Kingdom, and doner kebab in Germany all exemplify the degree to which a food brought by immigrants can reach the status of adopted national cuisine. Food is also the cultural element that is most accessible to outsiders. Foodways are used to construct a spatial sense of one location as a reflection of the entire world.

Foodways refer to the types of food that people eat, the ways they are prepared, and the cultural factors that surround and contextualize the food. Food is the most resilient cultural artifact. In countries undergoing language unification, foods can define ethnic groups. In mostly monolingual countries like the United States, foods may indicate geographical origins or social class. Food is easily bought, tried and accepted, or rejected. As such, it is the most accessible cultural element.

In many ways, the consumption of a food and its production have been divorced from its roots by the modern restaurant industry and international food conglomerates. Americans have eaten foods they consider Chinese or Mexican for generations, while few know the histories of said foods. Questions of whether or not a food is authentic are difficult to answer when the cooks in a restaurant are of a completely different ethnicity from the stated cuisine.

We can compare foodways between places and groups. Quantities of food, the ratio of prepared foods, and consumption of tobacco and alcohol all help us get inside the lives of people in different places, at different states of technological development, and different socioeconomic classes.

7.2.7 The Ethnic Landscape

Urban ethnic landscapes are often immediately recognizable. Signs in other languages advertising exotic products, ethnic architecture, and even local tourism reveal the ethnic fabric of a place. Most people in the US do not live in large cities with obvious ethnic architecture. The majority of Americans, including many ethnicities, live in the suburbs and smaller towns. Instead of obvious population clusters, ethnic populations here can be widely dispersed. Instead of living within walking distance of their local store or religious structure, people will simply drive to such a place. Ethnicity has sprawled along with the rest of America. Waves of migration to U.S. cities and suburbs have created landscapes of tremendous ethnic difference embedded in architectural homogeneity.

7.2.8 Ethnic Festivals and the Idealized Homeland

One of the ways that ethnicities represent themselves is through festivals. People wear traditional clothing, play music from the old country, eat food previously reserved for holidays, dance the old dances, and promote their culture to others. Festivals are a way of reproducing a sense of home in emigrant communities. They are also a way of keeping children participating in activities that would otherwise forget.

Places represented in ethnic festivals in the United States are often not representative of those places now. Traditional Czech clothing at a Kolache Festival in Oklahoma represents a place/time that no longer exists, except perhaps to market "Czech-ness" to tourists.

7.3 ETHNICITIES IN THE UNITED STATES

The distribution of ethnicities in the United States follows patterns that have been in place for some time. Historic migrations (some of which were forced) produced the patterns we still see in the United States.

7.3.1 African Americans in the United States

African Americans in the United States are still heavily southern. Their distribution (**Figure 7.3**) dates to the beginning of the United States and the forced importation of millions of Africans. Starting in the early twentieth century, many African Americans migrated out of this region, but most did not. In the last decades of the twentieth century and the beginning of the twenty-first, there has even been a reverse migration of African Americans back to Southern cities and suburbs.

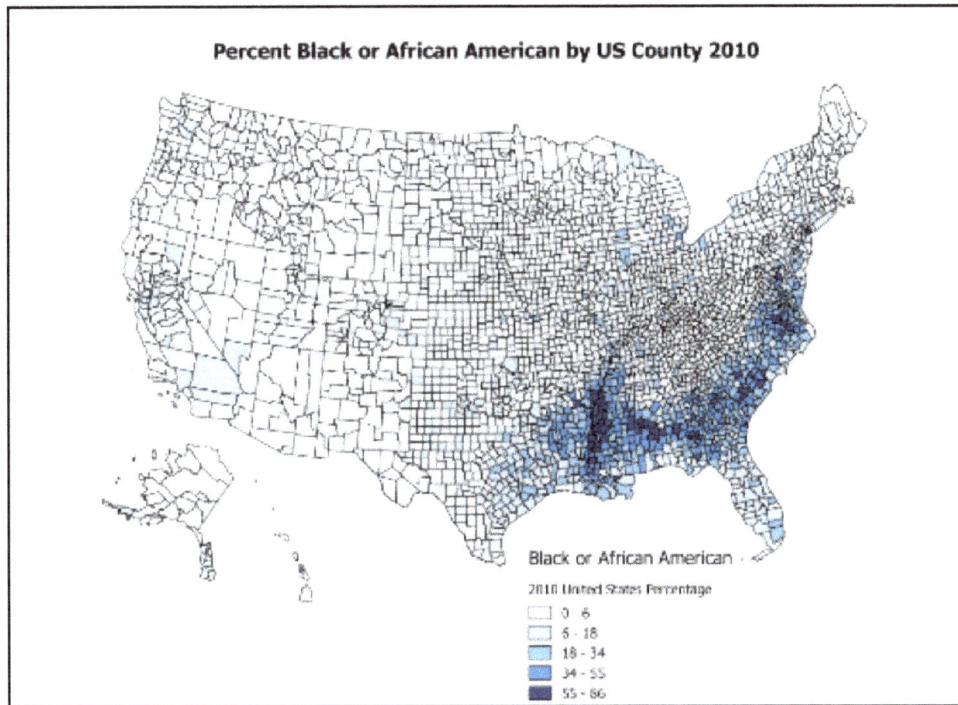

Figure 7.3 | The Distribution of the African American Population [3]
Author | David Dorrell
Source | Original Work
License | CC BY SA 4.0

7.3.2 Hispanics in the United States

Many of the states with large Hispanic populations were states taken from Mexico by the United States in the Mexican American War. In some ways, these places didn't come to the United States, the United States came to them. Certainly, a pattern is apparent. Generally, the parts of the U.S. closest to Mexico or the Caribbean are the most Hispanic (**Figure 7.4**). There are other areas with high Hispanic populations. These places have been attractive to immigrants for their employment prospects.

7.3.3 Asian Americans in the United States

Asian Americans also have a distinctive distribution based in history. The western United States, and in particular, Hawaii, are physically the parts of the United States that are closest to Asia. A proximity effect similar to that of Hispanics is in play here. **Figure 7.5** shows their distribution.

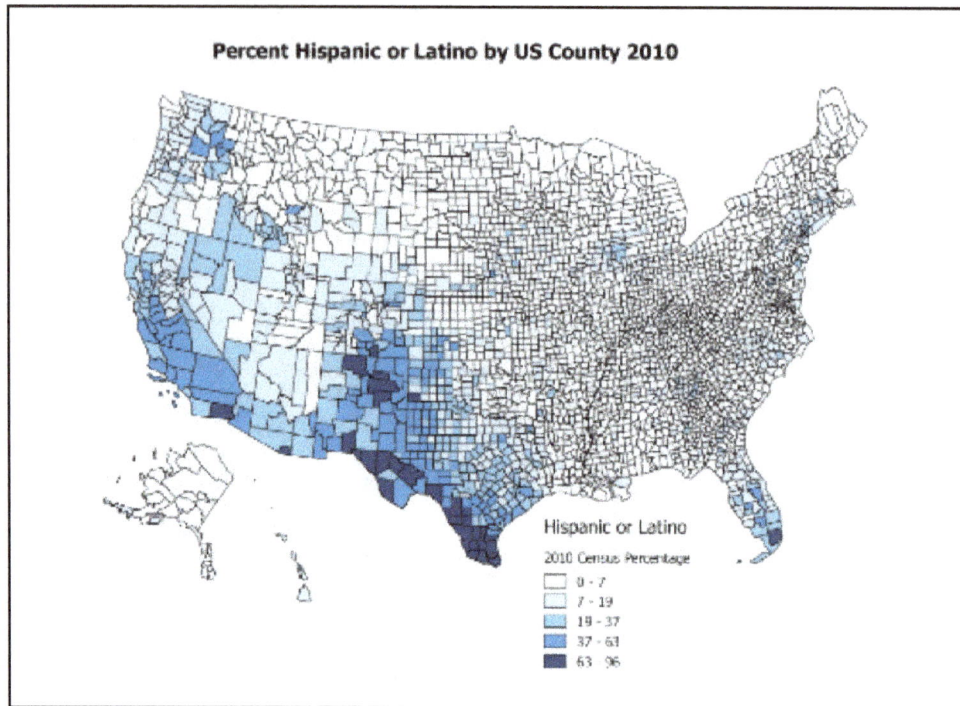

Figure 7.4 | The Distribution of the Hispanic Population [4]
Author | David Dorrell
Source | Original Work
License | CC BY SA 4.0

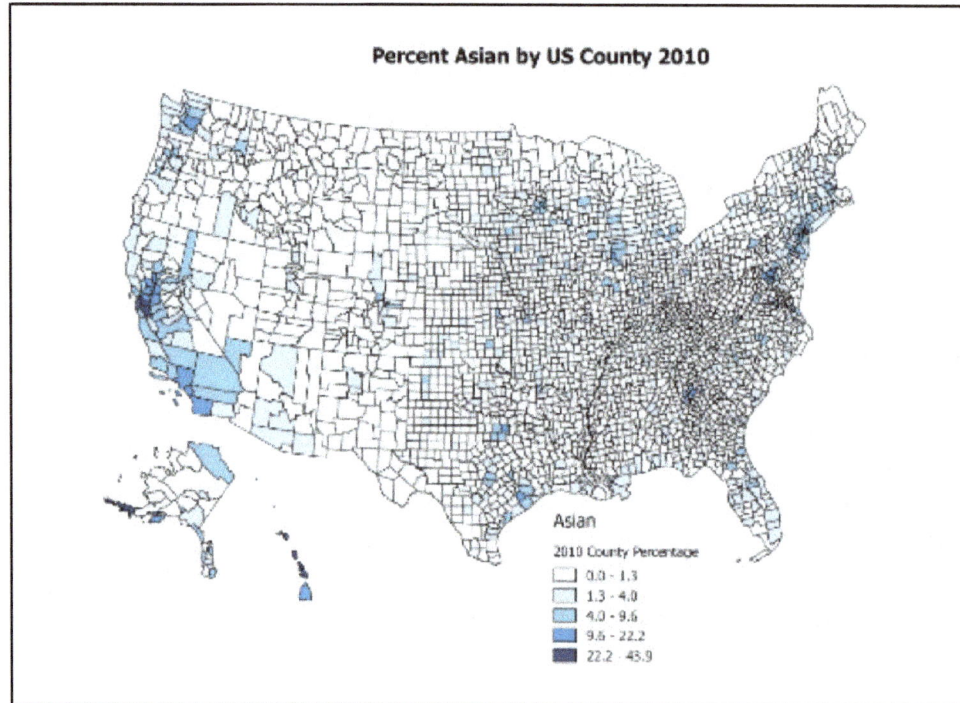

Figure 7.5 | The Distribution of the Asian American Population [5]
Author | David Dorrell
Source | Original Work
License | CC BY SA 4.0

7.3.4 American Indian and Alaska Natives

At one time, all of the current territory was occupied by Native Americans. Due to the influence of disease, **genocidal** wars, and poverty they have been reduced to roughly 2 percent of the overall population of the United States. Some live on reservations, but most do not (**Figure 7.6**).

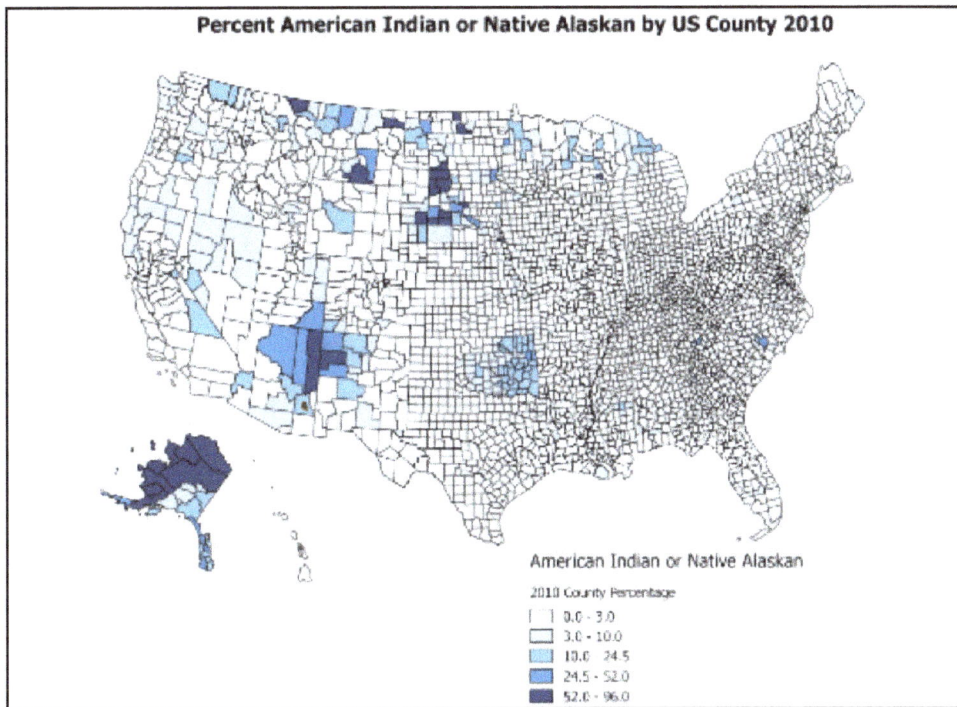

Figure 7.6 | The Distribution of the American Indian Population [6]
Author | David Dorrell
Source | Original Work
License | CC BY SA 4.0

7.3.5 Native Hawaiian and Other Pacific Islander

Hawaii at one time had been an independent kingdom. Other territories in the Pacific were taken during wars with other dominant regional powers. Many of these groups have migrated to the mainland of the United States (**Figure 7.7**). In the same way that American Indians are a minority in every state, Native Hawaiians are a minority in Hawaii.

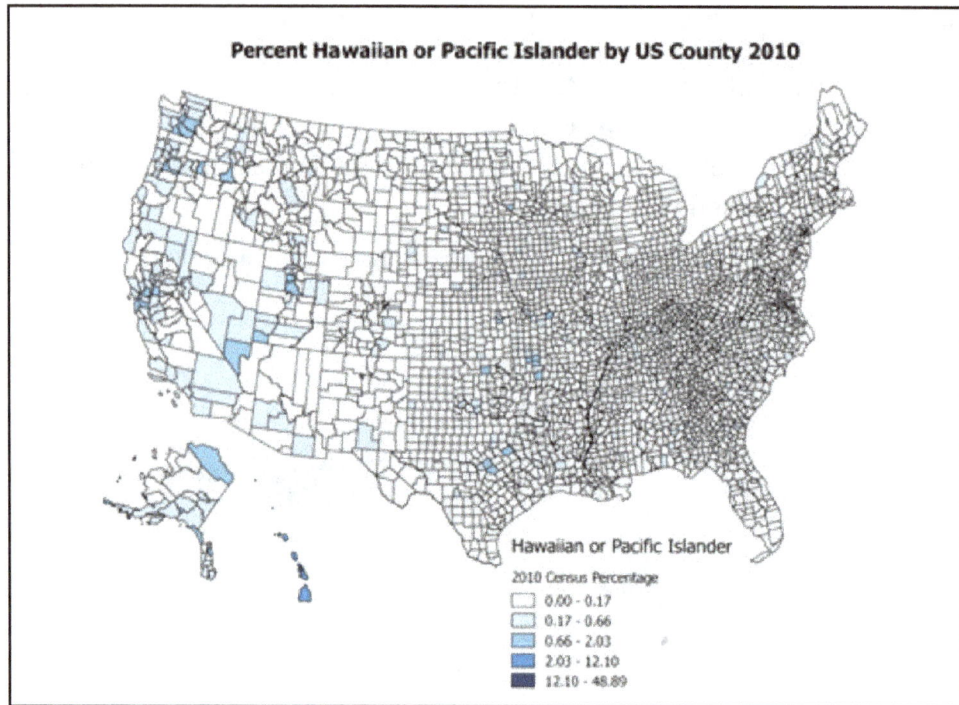

Figure 7.7 | The Distribution of the Native Hawaiian or Pacific Islander Population [7]
Author | David Dorrell
Source | Original Work
License | CC BY SA 4.0

7.4 RELEVANCE OF RACE AND ETHNICITY IN OTHER PLACES

Although the social implications of race or ethnicity in the United States have eroded over time, this does not mean that that is no longer relevant. It also does not mean that ethnicity is not relevant anywhere else. In many places, it is still very important. In much the same way as race defined the early United States, it defined South Africa, Brazil and other settler colonies. There were important differences between these places. Whereas white people who made up the racial majority ruled the United States, South Africa was ruled by a white racial minority. In order to preserve power for themselves, South African whites developed a system known as apartheid, which divided the population into a number of legally-defined categories. Similar to the U.S. development of Indian reservations, the South African state also developed ethnically-based "Homelands" which were used as a means of denying citizenship to black South Africans.

Sustaining such a system required the use of a police state that eventually became unsustainable. In 1994, full and open elections were held, and the black majority gained political power. The state policy of separating people ended, but this did not immediately transform South Africa into a new kind of state. It has continued to negotiate the relationship between the outside world and internal political and economic struggles between differing factions in the country.

Brazilian society was far more racially mixed from the beginning. This simply changed the social equation from a binary black/white relationship to a society stratified by skin tone and migration status. As was the case in many colonies, people born in the colonizing state (in this case, Portugal) continued to enjoy elevated social standing well after the colonial era ended. In the same manner of the United States and South Africa, social standing was related to being part of the charter group.

In other places, purely ethnic differences have had violent consequences. In the 1990s in Rwanda and in Yugoslavia, ethnic tensions flared into open warfare and genocidal massacres. A new term was coined—**ethnic cleansing**—which denoted an attempt to complete expunge traces of another population from a place. In both of these places, it would have been difficult for an outsider, and sometimes even a local, to tell the differences between the two groups. Remember that ethnic differences can be based on historical groups that may now be very similar.

It should be noted that the massacre of opposing ethnicities and the appropriation of their territories was not a product of the twentieth century. The colonial phase of world history was largely defined by the massacre and marginalization of indigenous people around the world by people of European descent.

7.4.1 Ethnicities and Nationalities

Some countries have only one ethnicity and are called nation-states (remember that an ethnicity can also be called a nation) Most places are not like this and contain many ethnicities. Some ethnicities are minorities solely by a political boundary. Many groups have found themselves on the wrong side of an imaginary line. Sometimes this is due to outside forces imposing a boundary, for example the Hausa in Nigeria and Niger, but sometimes it is a product of state creation itself. When the state of Germany was created, there were pockets of ethnic Germans scattered all over Europe. It would have been impossible to incorporate them all, since they were spatially discontiguous.

In the former Union of Soviet Socialist Republics (USSR), attempts were made to make political boundaries match ethnic boundaries. Kazakh people had the Kazakh SSR, Uzbek people had the Uzbek SSR, and so forth. It was an idea based on ease of administration, but it wasn't based on the actual distribution of the ethnic groups. The distributions were far too messy to draw clear, neat lines between them. This didn't matter as long as the USSR was still functioning, but when it collapsed, it created another landscape of minorities on the wrong side of a boundary. Irredentism is when your ethnic group has people on the wrong side of a boundary, and it's necessarily destabilizing. The following chapter on political geography will go into greater detail regarding this, but suffice it to say that split nations do not like being split.

Some ethnicities are numerous, but find themselves minorities in several countries. Kurds, Balochs (**Figure 7.8**), and Sami are all nations who are distributed across several countries. Such groups often harbor strong desires to create their own independent political entities to the detriment of currently existing states.

Figure 7.8 | Pakistan Major Ethnic Groups, 1980
This map details the politically fragmented spatiality of the Baloch people, separated into Iran, Pakistan, and Afghanistan.
Author | U.S. Central Intelligence Agency
Source | Wikimedia Commons
License | Public Domain

As places unify politically and develop industrially, ethnicity often declines. Moving populations into cities and stirring them around in schools, militaries, and jobs fosters intermarriage and acculturation to the larger, national norms. Rural places tend to be more diverse, and somewhere like Papua New Guinea is probably the most diverse, due to the fact that smaller villages still predominate. Somewhere like South Korea, which was already relatively ethnically homogenous, has become almost fully so due to economic development.

7.4.2 Is Diversity Good?

Diversity in developed countries is often promoted as a self-evident benefit, but there are some downsides to increasing diversity. Studies have shown that ethnic diversity decreases political participation. This is likely due not only to factors such as difficulties in communication, but also simple mistrust of other groups of people, known as **xenophobia**. This mistrust can apply to all parties in the relationship. There are places in the world with very low levels of diversity. South Korea, Japan and Finland are all highly productive economies with very little cultural or ethnic variability and high levels of social cohesion.

There are benefits to diversity. Aside from the benefits to genetic diversity (a reduction in recessive-gene disorders) ethnic diversity opens citizens to a wider range of experiences. Without pizza, sushi, tacos, stir-fry, or hamburgers, the United States would be a cultural wasteland forced to subsist on our British inheritance of boiled lunch and steak and kidney pie. Diversity has made our lives more pleasant, and it has made our ability to relate to others broader.

7.4.3 Immigration and Ethnicity

The United States is not the only place to receive immigrants or to have ethnic diversity. In fact, many places have far more ethnic diversity, even places that have little history of immigration. India, China, and Russia are all countries that have had diverse populations speaking different languages and living different lifestyles for a very long time.

In many ways, the impact of immigration on the ethnic fabric in Europe is the same as it has been in the United States. Due to the relative strength of their economies, European countries have been receiving large numbers of immigrants for some time. These immigrants are usually culturally distinct from the host population. In many instances, the immigrants come from places that had previously been colonies of European powers. The increase in immigrants with backgrounds dissimilar to the host country has triggered a rise in nationalistic or xenophobic activities, and in some places, a rise in political parties dedicated to reducing immigration or even repatriating current immigrants. The separate category of guest worker has created an even more complicated ethnic relationship. Guest workers are temporary workers who are contracted for a set period of time with the understanding that they will leave when the period of work has ended. By and large, that is not what happens due to the economic realities of short-term employment. People are reluctant to return to poverty.

7.4.4 Models of Ethnicity

Different places have different conceptions of ethnicity. In the United States, we separate race from ethnicity, and we have exhaustive lists of ethnicities collected by the census. France collects neither racial nor ethnic data, under the belief that every French citizen is ethnically French. This doesn't include linguistic minorities such as the Bretons, the Basques or the Alsatians, all of whom are indigenous to France and whose ideas about their own ethnicities are different from that of the state.

In other places, ethnic identity is the most significant impediment to state cohesion. In Nigeria, no less a person than Nobel Prize winner Wole Soyinka wrote," There is no such thing as a Nigerian." He wasn't saying that Nigerians are a figment of the imagination. He was stating that in his country, few people would identify first as a Nigerian, but instead as Igbo, Yoruba, Hausa or many others. This is another concept that will be addressed in the next chapter.

7.5 SUMMARY

Ethnicity is key to our identity. It can be formed around a variety of nuclei-historical ties, national origin, language, religion, or any admixture thereof. Ethnicity creates a feeling of belonging to a group. Ethnicity, of course can also be a source of exclusion to those who do not belong to the group that holds power in a place. Discrimination and prejudice often have a root in ethnicity, although other factors, like economics generally play a part as well. Race (and racism) are closely related to ethnicity, in that both ethnicity and race have been used to separate people, and some ethnicities can be associated with particular races. The next chapter also deals somewhat with human identity and delves further into nations and nationality.

7.6 KEY TERMS DEFINED

Acculturation: Cultural change, generally the reconciliation of two or more culture groups.

Discrimination: Mistreatment due to perceived difference.

Diversity: Having a range of different people.

Enclave: Self-enforced separation for a racial or ethnic group.

Environmental Justice: The concept that environmental benefits and burdens should be equally shared across different socio-economic groups.

Ethnic cleansing: An attempt to complete expunge or remove traces of another population from a place. May or may not relate to genocide.

Ethnicity: group of people sharing a common cultural or national heritage and often sharing a common language or religion.

First effective settlement: Doctrine in which the first group able to assert dominance provides the template for the future society.

Foodway: The cultural, social, and economic practices relating to the production and consumption of food.

Genocidal: having the purpose of exterminating an entire people.

Ghetto: Area of externally forced and legally-defined ethnic or racial separation.

Immigration: Incoming migration to a place.

Jim Crow: A set of laws enforcing racial segregation and disenfranchisement in the southern United States in the poet Civil War era.

Majority: A group making up more than half of a population.

Minority: A group making up less than half of a population.

Nation: An ethnicity or a people.

Race: The categorization of humans into groups based physical characteristics or ancestry.

Segregation: The spatial and/or social separation of people by race or ethnicity.

Xenophobia: fear of the different.

7.7 WORKS CONSULTED AND FURTHER READING

Alcalay, Ammiel. 1992. After Jews And Arabs: Remaking Levantine Culture. 1 edition. Minneapolis: University Of Minnesota Press.

Aluisio, Faith, and Peter Menzel. 2005. Hungry Planet. New York: Material World Books.

Baerwald, Thomas J. 2010. "Prospects for Geography as an Interdisciplinary Discipline." Annals of the Association of American Geographers 100 (3):493–501. https://doi.org/10.1080/00045608.2010.485443 .

Banerjee, Sarnath. 2004. Corridor: A Graphic Novel. New Delhi ; New York: Penguin Books.

Black, Jeremy. 2000. Maps and History: Constructing Images of the Past. Yale University Press.

Brown, Dee, and Hampton Sides. 2007. Bury My Heart at Wounded Knee: An Indian History of the American West. 1st edition. New York: Picador.

Cao, Nanlai. 2005. "The Church as a Surrogate Family for Working Class Immigrant Chinese Youth: An Ethnography of Segmented Assimilation." Sociology of Religion 66 (2):183–200.

Dorrell, David. 2018. "Using International Content in an Introductory Human Geography Course." In Curriculum Internationalization and the Future of Education.

Gillespie, Marie. 1995. Television, Ethnicity and Cultural Change. Psychology Press.

Gupta, Akhil, and James Ferguson, eds. 1997. Culture, Power, Place: Explorations in Critical Anthropology. N edition. Durham, N.C: Duke University Press Books.

Halter, Marilyn. 2002. Shopping for Identity: The Marketing of Ethnicity. 58081st edition. New York, NY: Schocken.

Hear, Nicholas Van. 1998. New Diasporas: The Mass Exodus, Dispersal, and Regrouping of Migrant Communities. 1 edition. Seattle, Wash: University of Washington Press.

Hutchinson, John, and Anthony D. Smith, eds. 1996. Ethnicity. 1 edition. Oxford ; New York: Oxford University Press.

Kelley, Robin D. G. 1996. Race Rebel : Culture, Politics, and the Black Working Class. New York: Free Press.

Olsson, Tore C. 2007. "Your Dekalb Farmers Market: Food and Ethnicity in Atlanta." Southern Cultures 13 (4):45–58.

Que Vivan Los Tamales! n.d. University of New Mexico Press.

Sacks, Oliver. 1998. The Island of the Colorblind. First edition. Vintage.

Seal, Jeremy. 1996. A Fez of the Heart: Travels Around Turkey in Search of a Hat. Houghton Mifflin Harcourt.

Sorkin, Michael, ed. 1992. Variations on a Theme Park | Michael Sorkin. 1st ed. Macmillan. http://us.macmillan.com/variationsonathemepark/michaelsorkin .

Lemann, Nicholas. 1992. The Promised Land: The Great Black Migration and How It Changed America. Vintage

Van Mechelen, Niki, Debra De Pryck, Niki Van Mechelen, and Debra De Pryck. 2009. "YouTube as a Learning Environment." https://www.learntechlib.org/p/33021/ .

Yang, Gene Luen. 2008. American Born Chinese. Reprint edition. New York: Square Fish.

7.8 ENDNOTES

1. Data source: United States Census Bureau 2016 http://www2.census.gov/geo/tiger/TIGER_DP/2016ACS/ACS_2016_5YR_COUNTY.gdb.zip

2. Data source: United States Census Bureau 2016 http://www2.census.gov/geo/tiger/TIGER_DP/2016ACS/ACS_2016_5YR_COUNTY.gdb.zip

3. Data source: United States Census Bureau http://www2.census.gov/geo/tiger/TIGER_DP/2016ACS/ACS_2016_5YR_COUNTY.gdb.zip

4. Data source: United States Census Bureau http://www2.census.gov/geo/tiger/TIGER_DP/2016ACS/ACS_2016_5YR_COUNTY.gdb.zip

5. Data source: United States Census Bureau http://www2.census.gov/geo/tiger/TIGER_DP/2016ACS/ACS_2016_5YR_COUNTY.gdb.zip

6. Data source: United States Census Bureau http://www2.census.gov/geo/tiger/TIGER_DP/2016ACS/ACS_2016_5YR_COUNTY.gdb.zip

7. Data source: United States Census Bureau http://www2.census.gov/geo/tiger/TIGER_DP/2016ACS/ACS_2016_5YR_COUNTY.gdb.zip

8 Political Geography

Joseph Henderson

STUDENT LEARNING OUTCOMES

By the end of this section, the student will be able to:

1. Understand: how political space is organized.
2. Explain: how states cooperate in military and economic alliances.
3. Describe: the various types of boundaries and how boundary disputes develop.
4. Connect: the electoral process in the United States to ethnic, urban/rural, and regional affiliations.

CHAPTER OUTLINE

8.1 INTRODUCTION

When most people think of geography, they think about memorizing the states, state capitals, and perhaps learning the location of various countries around the world. These facts deal with the subdiscipline of political geography because they show how politics is reflected on the surface of the Earth, but political geography is much more than a "trivial pursuit" of such information. Although having knowledge about the locations of countries and states is fundamental and an important foundation for the study of political geography, the subject matter deals with many other topics such as military and economic alliances, boundaries between countries, terrorism and other civil-military conflicts, and the geography of the electoral process.

8.2 HOW POLITICAL SPACE IS ORGANIZED

The fundamental unit of political space is the **state**, and this type of state is different than the states that make up the United States. A state is basically synonymous to a country and represents a formal region in which the government has sovereignty or control of its own affairs within its territorial boundaries. The number of states in the world is currently 196, but this number changes through military conquests or the devolution, or breakup, of states. For example, the United Kingdom has devolved over the past 70 years as the Republic of Ireland has broken away from the UK, and a new referendum may occur in the next few years to decide whether or not Scotland will become independent. Another prime example of the creation of new states occurred after the breakup of the Soviet Union, when fifteen states were created in Eastern Europe. Even a terrorist group, the Islamic State, has tried to establish its own state in portions of Syria and Iraq, even though their legitimacy is not recognized by the international community.

States in which the territorial boundaries encompass a group of people with a shared ethnicity are known as **nation-states**. These states are generally homogeneous in terms of the cultural and historical identity of the people, and these groups of people are referred to as a **nation**. A few current examples of nation-states are Japan, Finland, and Egypt. Nation-states are actually in the minority compared to **multinational states**, which are states that have more than one nation within their borders. With international migration being a significant phenomenon worldwide, more states become multinational. In contrast, some nations exist but do not have their own state, and those nations that desire to become nation-states are known as **stateless nations.** In the United States, a prime example of stateless nations are the many Native American tribes scattered throughout the countries. Other examples include the Palestinians living in Israel, Syria, Lebanon and Jordan, and the Kurds found in Iraq, Turkey, Syria, and Iran (**Figure 8.1**). Both the Kurds and the Palestinians are actively seeking statehood, but serious obstacles must be overcome because the countries where they live are reluctant to grant them independent territories.

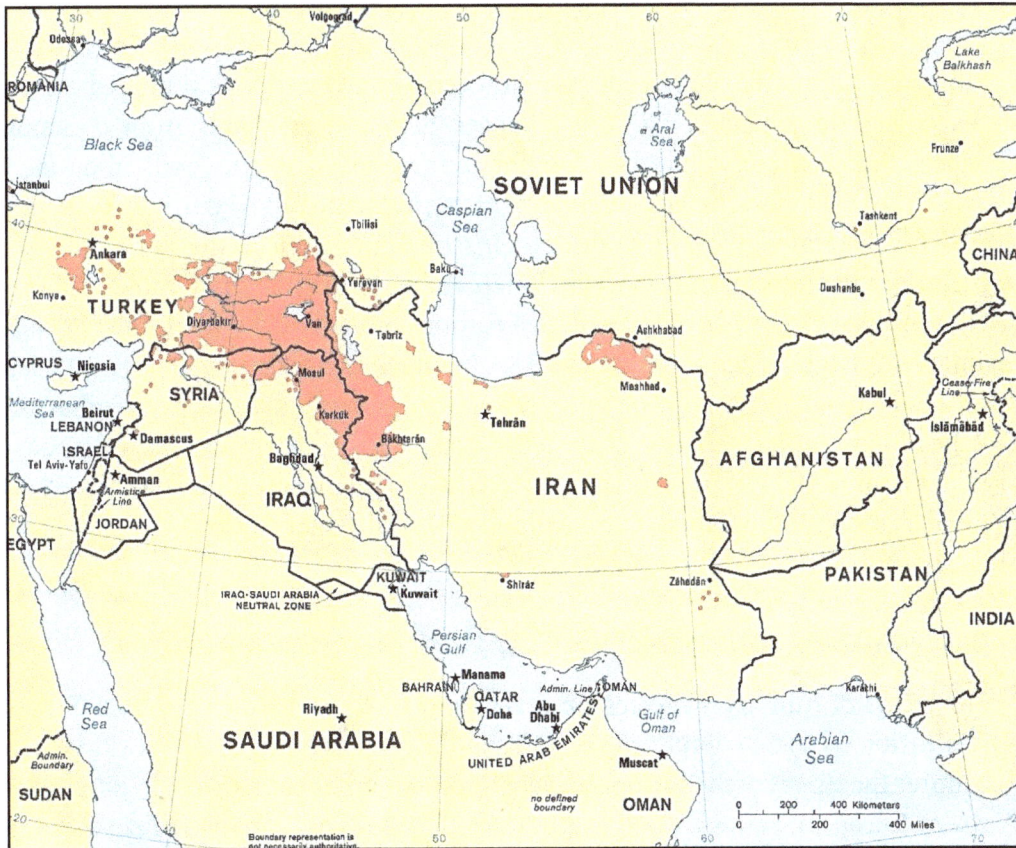

Figure 8.1 | Kurd-majority areas in Turkey, Syria, Iraq, and Iran
Author | U.S. Central Intelligence Agency
Source | Wikimedia Commons
License | Public Domain

The solidarity and unity of a state is influenced by both centripetal and centrifugal forces. **Centripetal forces** tend to bind a state together, and **centrifugal forces** act to break up a state. Examples of centripetal forces include nationalism, economic prosperity, and strong, ethical security forces. Centrifugal forces include wars, ineffective or corrupt governments, and market failure. Other factors that can influence the solidarity of a state include types of boundaries, ethnic differences (which may result in unity or discord), and the compactness of a state.

The compactness of a state is related to the shape of a state, and a **compact state** is one that is ideally circular in shape, where the distance from the center to any border is roughly equal. In contrast, a **fragmented state** is one that is discontinuous in nature and may consist of a number of islands. A few examples of fragmented states include Indonesia and the Philippines. Indonesia consists of over 17,000 islands, and in order to increase the solidarity of the state, the government actively encouraged migration to less populated islands in order to assimilate indigenous populations. In the Philippines, control of its southern islands such as Mindanao is problematic because of terrorist groups that are active in those areas.

8.3 COOPERATION BETWEEN STATES

In order to provide shared military and economic security as a unified entity, states engage in alliances. Military alliances help protect states from common enemies, and economic alliances allow for the free exchange of goods in a larger market. These alliances are also referred to as **supranational organizations**, and they all involve states giving up some of their sovereign power for the common good.

The largest supranational organization in the world is the United Nations (UN). Formed originally as the League of Nations after World War II, the UN now includes 193 states. The work of the UN includes peacekeeping, humanitarian relief, and the establishment of internationally approved standards of behavior. The headquarters of the UN is in New York City, and important subsidiary organizations of the UN include the World Health Organization (WHO), UNESCO (United Nations Educational, Cultural and Cultural Organization) and the Food and Agriculture Organization (FAO).

8.3.1 Military Alliances - NATO and Warsaw Pact

In terms of military alliances on a regional scale, the North Atlantic Treaty Organization (NATO) comprises 28 states and was developed after World War II to counter the threat of the former Soviet Union. Member states include numerous Western European states as well as the United States and Canada (**Figure 8.2.**).

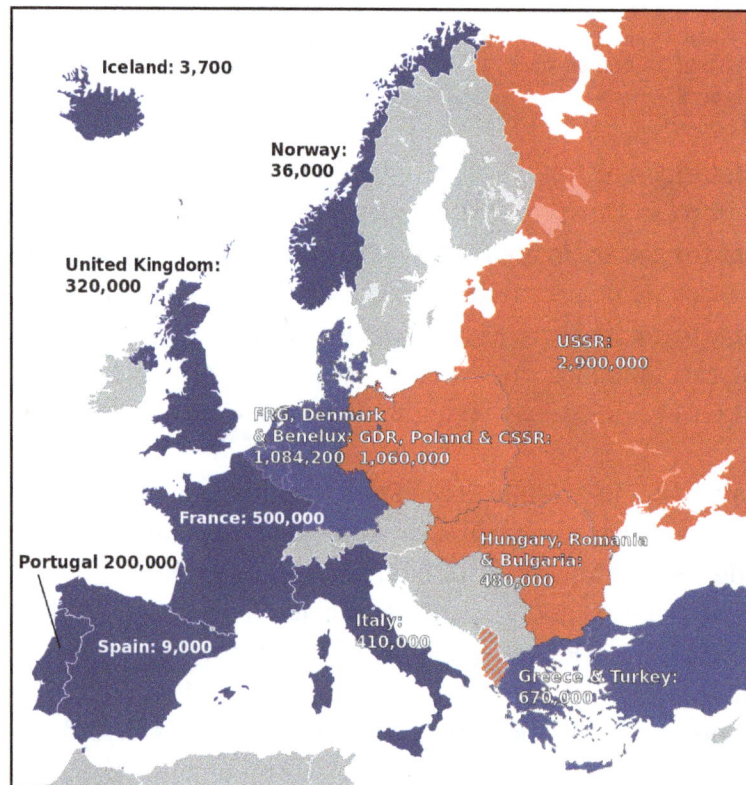

Figure 8.2 | Current NATO map and former Warsaw Pact
Author | User "Alphathon"
Source | Wikimedia Commons
License | CC BY SA 3.0

When the Soviet Union existed, the Warsaw Pact was a military alliance between the Soviet Union and seven satellite states of Eastern and Central Europe (**Figure 8.2**). The Warsaw Pact disbanded in 1991, and several of the former Soviet states as well as satellite states have subsequently joined NATO. As a result, Russia has felt isolated and vulnerable, and as a result, has been aggressively acting to seize or control territories in states close to its borders.

For example, in 2008, Russia engaged in a military conflict in Georgia, one of the former Soviet states, in order to support a separatist movement allied with Russia. In 2014, Russia invaded the peninsula of Crimea, within the territorial boundaries of Ukraine, one of the former states in the Soviet Union. Furthermore, Russia has intervened militarily against the rebel forces fighting in eastern Ukraine. In response to these provocations, NATO commenced Operation Atlantic Resolve, an ongoing series of training exercises between the United States and other NATO countries in former Warsaw Pact countries such as Poland, Romania, and Latvia.

8.3.2 Military Alliances – Terrorism

Although terrorists are characterized as non-military, non-state actors, they have a tremendous impact on states around the world and involve allied groups in many countries. **Terrorism** is the intimidation of a population by violence in order to further political aims. The first terrorist group with global influence is Al Qaeda, formed in 1988 by Osama bin Laden. Although its influence has waned in the past decade, Al Qaeda was responsible for the 9/11 attacks and had several affiliates including Boko Haram in Nigeria and Abu Sayyaf in the Philippines. With the rise of the Islamic State (ISIS/L) in Iraq and Syria in 2013, terrorist groups such as Boko Haram and Abu Sayyaf have consequently declared allegiance to ISIS/L. ISIS/L is an extremist Muslim group that intends to seize as much territory as possible in the Middle East and force their subjects to adhere to their strict version of Islamic fundamentalism. To facilitate their goal, ISIS/L conducts an extensive campaign on social media to recruit fighters to come to Syria and also conduct individual attacks in their home country. Their media campaign has been successful in recruiting numerous militants to come to Syria and to inspire or instigate attacks in the United States, France, Belgium, England, Sweden, Turkey, Afghanistan, Yemen, and Saudi Arabia (**Figure 8.3**). Furthermore, ISIS/L is conducting regular military operations in Libya and Egypt.

To combat the Islamic State and other terrorist groups, interesting military alliances have developed, and the situation is exceptionally complex. The United States cooperates with numerous NATO allies to train and equip local military forces in Iraq and Syria. Iran, in alliance with Iraq, has provided assistance in driving ISIS/L out of Iraq. Kurdish forces in northern Iraq and Syria, in concert with NATO forces, have conducted much of the military action, and they hope for an independent state because of these efforts. Complicating this situation is the objection of the central governments of Iraq, Syria, and Turkey to an independent Kurdish state. Further clouding the situation is the alliance between Russia and the Syrian central government. Although these two countries do fight against ISIS/L, they are also opposed to other Syrian rebel forces, which the United States supports. The problem of ISIS/L propaganda is being addressed by numerous governments and social media outlets such as Facebook, Twitter, and YouTube. More importantly, though, military success on the ground against ISIS/L forces in Iraq and Syria is decreasing the influx of foreign fighters as well as the proliferation of internet propaganda.

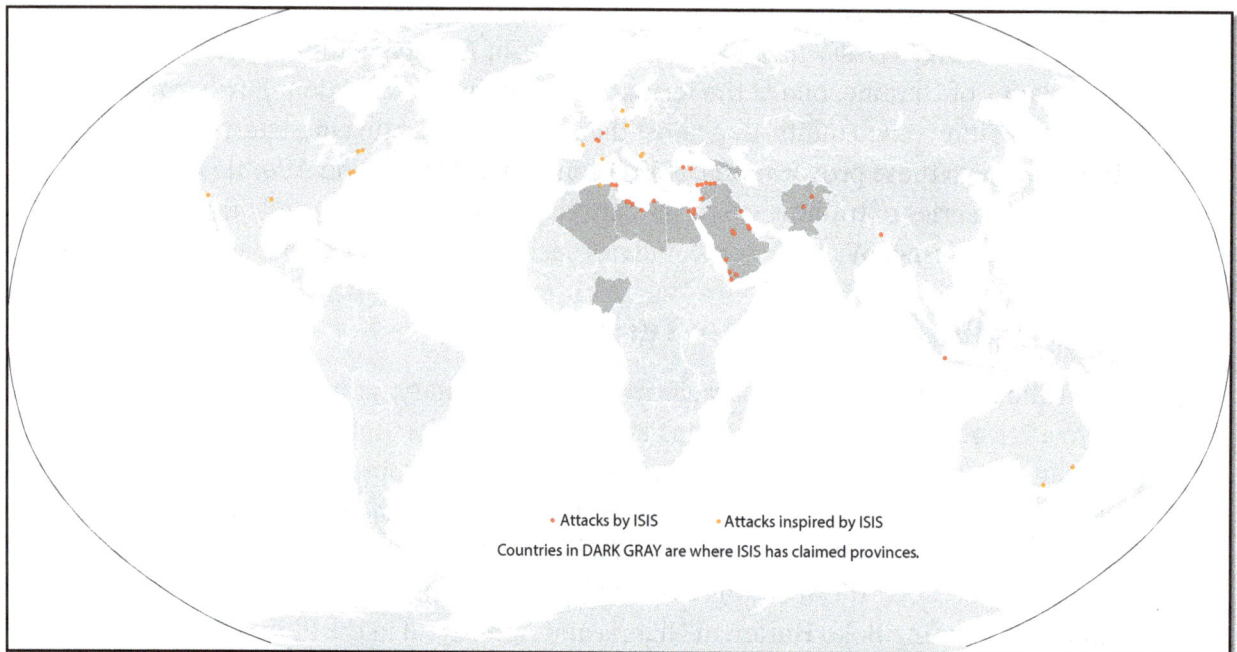

Figure 8.3 | Map of ISIS/L global influence [1]
Author | Corey Parson
Source | Original Work
License | CC BY SA 4.0

8.3.3 Economic Alliances – European Union and NAFTA

One of the most prominent economic alliances in the world is the European Union (EU), which consists of 28 member states (**Figure 8.4**). What began as the European Community (EC) in 1958, the European Union has grown significantly from the original six members and now includes seven Eastern European states that were formerly in the Soviet Union. The EU has developed a common currency, the euro, for all member countries and a European Central Bank. Furthermore, at most boundaries, a passport is not required to enter another country.

One of the weaknesses of the EU is the need to subsidize poorer countries, creating financial difficulties for the more wealthy members. For example, Greece has experienced large debts that have required rescue funding from the EU. Another issue confronting the EU is whether or not to allow Turkey to join, as Greece has long-standing disputes with Turkey over territory in Cyprus, and the Turkish central government has been accused of anti-democratic practices. Perhaps most concerning for the EU is the imminent departure of the United Kingdom (UK) from the alliance in an action termed "Brexit." In 2016, the UK voted by referendum to leave the EU, and is on schedule to formally break away in 2019. The UK's decision to leave the EU is not solely related to economics, as not only is the EU an economic alliance, but agreements on social and political policies are involved as well. The majority of British citizens are generally against the subsidizing of poorer states and the increasing number of immigrants who use scare public resources, and they generally desire greater autonomy. With the exit of Britain from the EU, EU members are concerned that other states may follow suit.

Figure 8.4 | European Union
Author | U.S. Central Intelligence Agency
Source | Wikimedia Commons
License | Public Domain

An important economic alliance for the United States is the North American Free Trade Agreement (NAFTA). Established in 1992, this alliance integrates the United States, Mexico and Canada and facilitates the flow of goods and services across borders. The Trump administration has repeatedly criticized this agreement, as manufacturers have relocated production to Mexico which has resulted in the loss of manufacturing jobs in America. Whether or not the United States withdraws from NAFTA or simply renegotiates the agreement remains to be seen.

Another significant alliance that is being considered is the Trans-Pacific Partnership, an agreement between 11 countries that border the Pacific Ocean, and originally included the United States. The proposal was signed in February 2017, but the United States had already withdrawn from the agreement in January of the same year, making ratification virtually impossible. The goal of the agreement is to promote economic prosperity by lowering tariff barriers, but also promote environmental and labor protections as well as protect intellectual property. Critics say that it would result in the loss of U.S. jobs and weaken the sovereignty of the U.S.

8.4 BOUNDARIES AND BOUNDARY DISPUTES

"Good fences make good neighbors."

-Robert Frost

As mentioned in Section 13.4, boundaries can influence the solidarity of a state, as boundaries disputes can result in conflict. A **boundary** is essentially an invisible, vertical plane that separates one state from another, so it includes both the airspace above the line on the surface and the ground below. Boundaries can be both physical and anthropogenic, and while it is difficult to categorize all boundaries, some prominent boundary types exist.

Physical boundaries are natural features on the landscape such as rivers, lakes, and mountains. The Rio Grande is an important physical boundary on the southern border of the United States. Like most rivers, the Rio Grande shifts gradually (and sometimes abruptly) through time. As a result of the fact that the course of a river is not fixed, a river boundary can be problematic. In fact, because of the gradual shift in the Rio Grande in the vicinity of El Paso, the United States and Mexico established the Chamizal Treaty which reestablished the boundary and included a more permanent relocation of the river channel by engineering (**Figure 8.5**). Some examples of mountain ranges as boundaries include the Zagros Mountains between Iraq and Iran, the Pyrenees between Spain and France, and the Andes Mountains between Chile and Argentina.

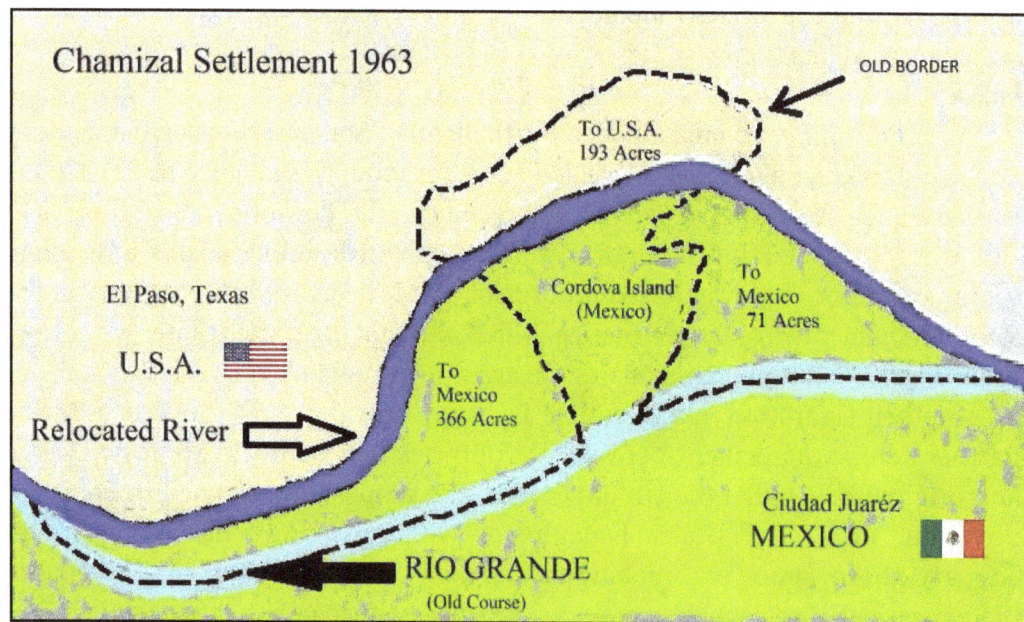

Figure 8.5 | Chamizal Treaty map
Author | Mike Hayes
Source | Wikimedia Commons
License | Public Domain

In contrast to physical boundaries, **geometric boundaries** and **ethnic boundaries** are not related to natural features. Instead, in the case of geometric boundaries, they are straight lines. These straight lines could coincide with latitude or longitude, as is the case with the northwestern boundary of the United States with Canada along 49° north latitude. Likewise, Indonesia and Papua New Guinea is separated by another geometric boundary along the 141st meridian.

For ethnic boundaries, they are drawn based on a cultural trait, such as where people share a language or religion. The border between India, which is predominantly Hindu, and Pakistan, which is predominantly Muslim, is one example. Some borders split ethnic groups that are more closely related to the people on the other side of the border. For example, in eastern Ukraine, the majority of the population speaks Russian and is sympathetic to Russians on the other side of the border. As a result, the current conflict between Russia and Ukraine has been problematic for the Ukrainian central government because of the Russian affiliation with eastern Ukraine. Russian influence in eastern Ukraine is an example of **irredentism**, or an effort to expand political influence of a state on a group of people in a neighboring state.

Another prime example of where boundaries do not coincide closely with ethnic groups is in Africa. Almost 50 percent of the boundaries in Africa are geometric, and at least 177 ethnic groups are split in two or more states. If all ethnic groups in Africa were to be enclosed in their own boundaries, Africa would have over 2,000 countries (1). Because ethnic groups straddle many boundaries in Africa, this situation has led to considerable cross-border trade, but also has created numerous conflicts. For instance, several wars have occurred because the Somali ethnic group is split between five different countries.

8.5 THE ELECTORAL PROCESS

8.5.1 Gerrymandering

In the United States, boundaries play an important role in the electoral process, but in this case, district, and precinct boundaries are significant in contrast to the country boundaries that have been previously discussed. Political parties in power will sometimes rearrange the boundaries of voting districts in order to ensure victory in elections, and this practice is called **gerrymandering**. Gerrymandering strategies can involve drawing the boundaries so that the majority of voters in a district favor the party in power. Another method is to segregate the opposition voters into several different districts (**Figure 8.6**). While gerrymandering is not generally illegal in the United States, it can be challenged in court when it appears to clearly discriminate against minority populations. For example, when legislative districts where redrawn in 2015 in Gwinnett and Henry counties in Georgia, the NAACP filed a federal lawsuit because of the perception that the adjustment violated the rights of minority black voters. A 2013 Supreme Court decision, however, declared that a requirement in the 1965 Voting Rights Act for

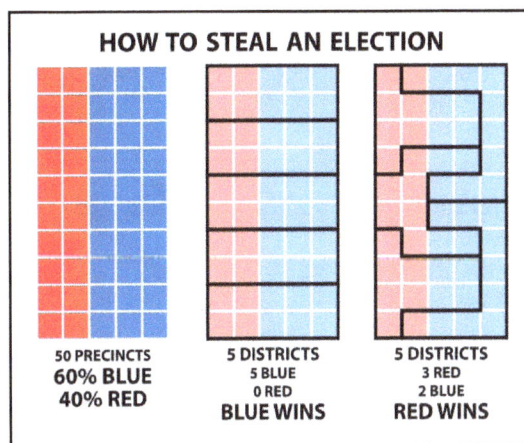

HOW TO STEAL AN ELECTION

50 PRECINCTS
60% BLUE
40% RED

5 DISTRICTS
5 BLUE
0 RED
BLUE WINS

5 DISTRICTS
3 RED
2 BLUE
RED WINS

Figure 8.6 | Gerrymandering Methods
Author | Steven Nass
Source | Urban Milwaukee
License | CC BY SA 4.0

federal oversight of redistricting is not constitutional, and this decision may have an impact on such lawsuits.

8.5.2 Presidential Elections

One of the more intriguing aspects of political geography in the United States is presidential elections and how the Electoral College process has a decidedly spatial component. An examination of the presidential results from the 2016 election reveals interesting regional affiliations (**Figure 8.7**). Northeastern and Mid-Atlantic states, as well as far Western states, all tended to vote Democrat, while the rest of the electoral map is Republican. Although it appears that the Republican victory was decisive from a spatial and Electoral College perspective, from the standpoint of the popular vote, Democrats actually had more votes than Republicans. The dichotomy is explained by the very large populations in both California and New York, predominantly Democrat states. California's large population is reflected in a high number (55) of Electoral College votes, more than any other state.

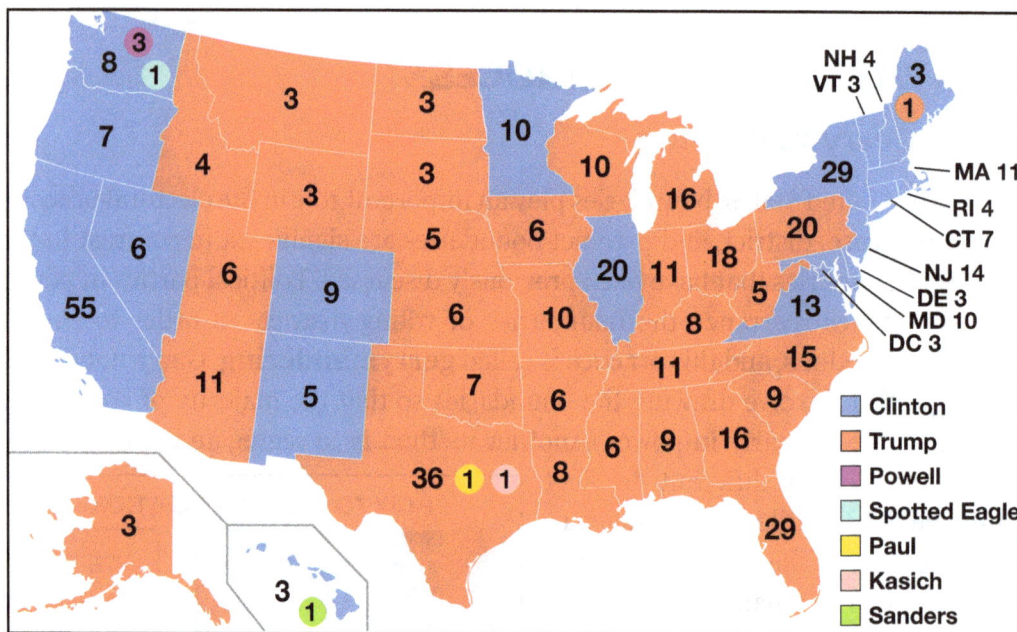

Figure 8.7 | Presidential Election of 2016
Author | User "Gage"
Source | Wikimedia Commons
License | CC BY SA 4.0

Another marked trend in the voting map is evident when the election results are examined by county (**Figure 8.8**). Ethnic groups tend to vote in particular voting blocks, and the impact of ethnic groups can be seen more clearly by looking at individual counties. For example, a crescent of blue (Democrat) can be seen running through south-central Alabama and into eastern Mississippi. African Americans tend to vote Democrat, and these blue counties contain predominantly African American voters.

Similarly, the blue counties in western Mississippi have a majority African American population. Another example of the influence of ethnic groups is evident in extreme southern Texas. Hispanics generally vote Democrat, and most counties in southern Texas are dominated by Hispanics.

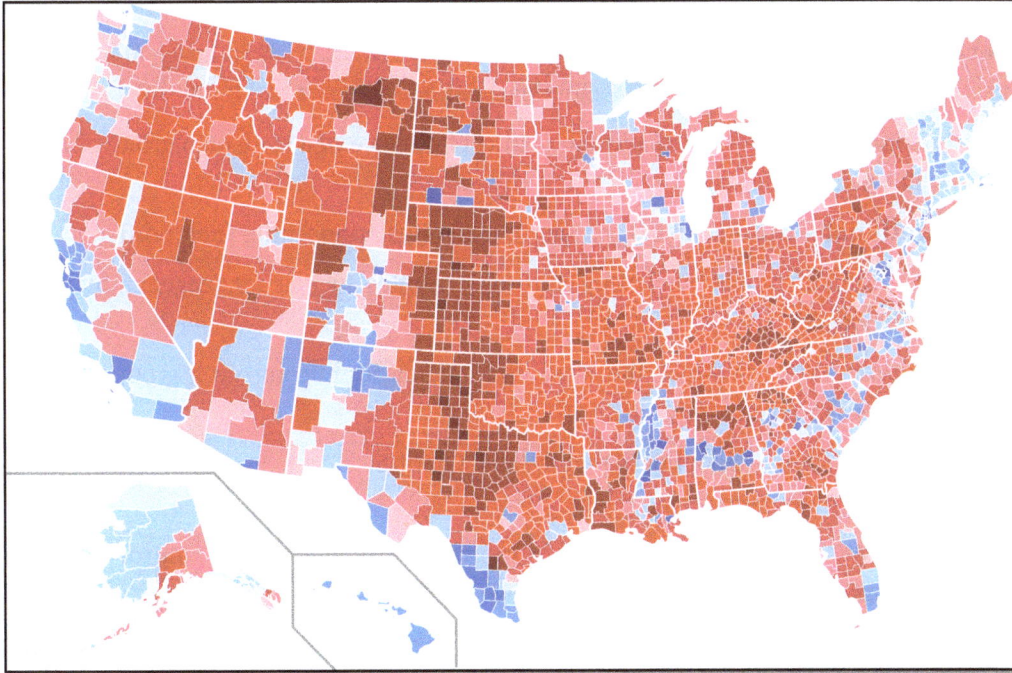

Figure 8.8 | 2016 Presidential Election by County
Author | Users "Ali Zifan" and "Inqvisitor"
Source | Wikimedia Commons
License | Public Domain

Related to ethnic affiliations by county is the trend for rural voters to lean toward Republican and urbanites to vote Democrat. Rural areas tend to have white majorities, and whites tend to vote Republican. Rural voters also have lower numbers of college graduates compared to urban areas. In the 2016 presidential election in both Georgia and Texas, urban counties tended to vote Democrat, and rural counties went to the Republicans. With increasing urbanization and diversity in the United States, the electoral map will continue to undergo significant changes in the future.

8.6 KEY TERMS DEFINED

Boundary – an invisible, vertical plane that separates one state from another, which includes both the airspace above the line on the surface and the ground below.

Centripetal force – a force that tends to bind a state together.

Centrifugal force – a force that tends to break a state apart.

Compact state – a state where the distance from the center to any border does not vary significantly; roughly circular.

Ethnic boundary – a boundary that encompasses a particular ethnic group.

Fragmented state – a state whose territory is not contiguous, but consists of isolated parts such as islands.

Geometric boundary – a boundary that follows a straight line and may coincide with a line of latitude or longitude.

Gerrymandering – the process of redrawing legislative districts in order to benefit the party in power and ensure victory in elections.

Irredentism – an effort to expand the political influence of a state on a group of people in a neighboring state.

Multi-national state – state that has more than one nation within their borders.

Nation – group of people bonded by cultural attributes such as language, ethnicity and religion.

Nation-state – state in which the territorial boundaries encompass a group of people with a shared ethnicity.

Physical boundary – a boundary that follows a natural feature on the landscape such as a river, mountain range, or lake.

State – a formal region in which the government has sovereignty or control of its own affairs within its territorial boundaries.

Stateless nation – a nation that aspires to become a nation-state but does not yet have their own territory.

Supranational organization – an alliance involving three or more states who have shared objectives that may be economic, political/military, or cultural.

Terrorism – intimidation of a population by violence in order to further political aims.

8.7 WORKS CONSULTED AND FURTHER READING

Acemoglu, Daron, and James A. Robinson. 2013. Why Nations Fail: The Origins of Power, Prosperity, and Poverty. Reprint edition. New York, NY: Currency.

Flint, Colin, and Peter Taylor. 2011. Political Geography: World-Economy, Nation-State and Locality. 6 edition. London New York: Routledge.

Frost, Robert. "Mending Wall by Robert Frost." Poetry Foundation. Accessed April 30, 2018. https://www.poetryfoundation.org/poems/44266/mending-wall.

Gallaher, Carolyn, Carl T. Dahlman, Mary Gilmartin, Alison Mountz, and Peter Shirlow. 2009. Key Concepts in Political Geography. 1 edition. London ; Los Angeles: SAGE Publications Ltd.

Glassner, Martin Ira, and Chuck Fahrer. 2003. Political Geography. 3rd edition. Hoboken, NJ: John Wiley and Sons.

Jones, Martin, Rhys Jones, Michael Woods, Mark Whitehead, Deborah Dixon, and Matthew Hannah. 2015. An Introduction to Political Geography: Space, Place and Politics. 2 edition. London ; New York: Routledge.

Mann, Michael. 2012. The Sources of Social Power: Volume 1, A History of Power from the Beginning to AD 1760. 2 edition. New York: Cambridge University Press.

Marshall, Tim. 2015. Prisoners of Geography: Ten Maps That Explain Everything About the World. First Edition edition. New York, New York: Scribner.

Mungai, Christine. "Africa's Borders Split over 177 Ethnic Groups, and Their 'real' Lines Aren't Where You Think." MG Africa. January 13, 2015. Accessed April 30, 2018. http://mgafrica.com/article/2015-01-09-africas-real-borders-are-not-where-you-think.

Painter, Joe, and Alex Jeffrey. 2009. Political Geography. 2nd edition. Los Angeles: SAGE Publications Ltd.

8.8 ENDNOTES

1. Adapted from https://www.nytimes.com/interactive/2015/06/17/world/middleeast/map-isis-attacks-around-the-world.html

9 Development and Wealth

Todd Lindley

STUDENT LEARNING OUTCOMES

By the end of this section, the student will be able to:

1. Understand: the primary measures and terms associated with wealth, poverty, economic growth, and social improvement for gauging levels of development by region and country

2. Explain: the two paths to development and the attendant strengths and weaknesses of each.

3. Describe: Regional differences in economic and social development across the world

4. Connect: the processes of globalization to development efforts and outcomes that vary dramatically by place

CHAPTER OUTLINE

9.1 DEVELOPMENT AND GEOGRAPHY: AN INTRODUCTION

If you could choose anywhere on the planet, where would you like to live? Would you choose a place with mountains or with beaches? A place with high taxes or few regulations? Do you love your community/state/country? Could you make more money somewhere else or might you be happier with warmer/colder weather? Do you think that you are able to realize your full personal potential in the place where you live now? Why or why not? Your answers might vary greatly from other people around the world based upon what language you speak, your religious preferences, your cultural framework, and your own persona values. Nonetheless, there are certain indicators that geographers can use to categorize places according to how developed they are in terms of technology, infrastructure, wealth, and opportunity. As you might guess, the differences between places can be quite stark, but it's important to understand the dynamics and geography of the patterns and processes associated with income, well-being, and opportunity. This chapter explains how those distinctions are made and how they vary across time and place using the concept of development (the processes related to improving people's lives through access to resources, technology, education, wealth, opportunity, and choice).

Which places on earth are the most developed? There is no simple answer to this question. San Francisco is often voted as the most beautiful city in the U.S. but the cost of living makes this place unaffordable for all but the wealthiest of residents. Ancient cities like Jerusalem, Athens, and Baghdad contain amazing architecture and the roots of western civilization, but streets also tend to be narrow and housing crowded. China has experienced the fastest economic growth of any other country in the past 30 years, but it is accompanied by catastrophic levels of air pollution, poor working conditions, and severe limits on personal liberty. Within the United States, people in Colorado tend to be the healthiest while those in Utah have the largest houses and people in Texas are the most loyal to their home state. There are multiple ways to measure development, and geographers spend a great deal of time and effort studying, measuring, and quantifying the differences and commonalities.

A few general truths about development and wealth in our 21st century world can simplify the complexity:

1. The world continues to be divided into the **Global North** and **Global South** by the Brandt Line (See **Figure 9.1**). Levels of wealth, well-being, access to technology, and health tend to be higher in northern countries than in southern ones. The line is problematic because it over-generalizes, but it remains a meaningful starting point to understanding development from a global scale.

2. The majority of people living in the 21st century have a higher standard of living, earn more money, are healthier, and live longer than was the case for people 50 years ago.

3. In spite of #2, the wealth disparity between those at the bottom and those at the top remains greater than ever before. According to Credit Suisse's global wealth report, the globe's richest 1% (of people) control more than half of the world's wealth, up from 42.5% in 2008.

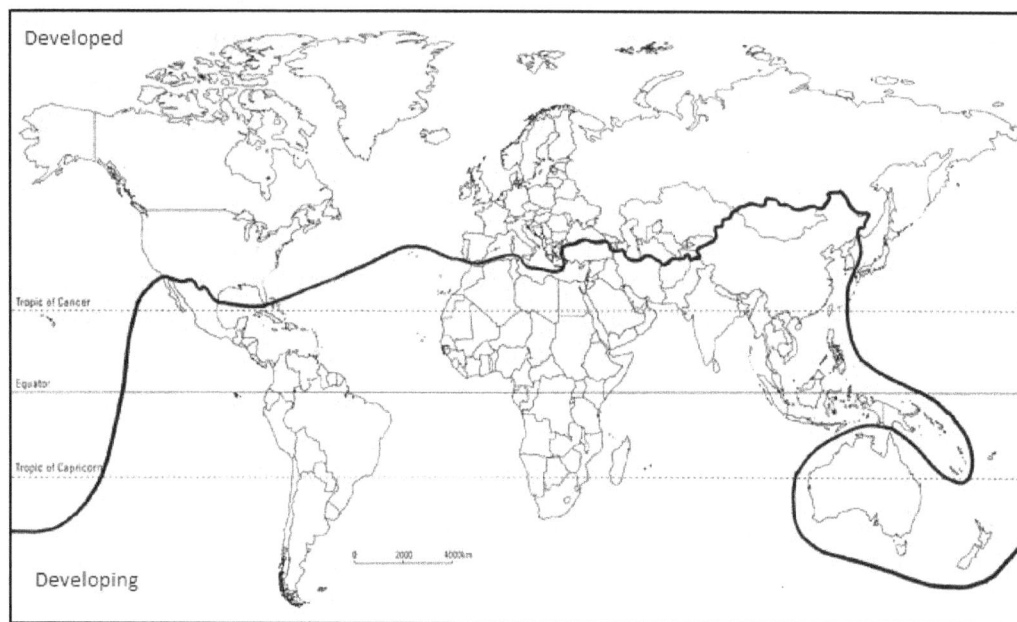

Figure 9.1 | The Global North as defined by the Brandt Line in 1980
Author | User "Jovan.gec"
Source | Wikimedia Commons
License | CC BY SA 4.0

9.2 IMPORTANT TERMS & CONCEPTS

People in **more developed countries** (MDC) and **less developed countries** (LDC) experience the world very differently, but let's think about the meaning of such terms. MDC's tend to have higher incomes, levels of technology, infrastructure development, and life expectancies while LDC's tend to have lower levels of those and other indicators. The simplest measure to consider is income. The World Bank (more on this institution later in the chapter) places countries into 4 categories. In 2017, low-income economies were defined as those with a **gross national income** (GNI) per capita, of $1,025 or less in 2016; lower middle-income economies are those with a GNI per capita between $1,026and $4,035; upper middle-income economies are those with a GNI per capita between $4,036 and $12,475; high-income economies are those with a GNI per capita of $12,476 or more. The map in **Figure 9.2** displays the general geographic patterns associated with the 4 categories. Take a moment to look at the map to see if any of the countries or regions surprise you. For example, are all African countries in the same category? What about countries in Latin America or Asia? Which country is in a higher category, China or Russia?

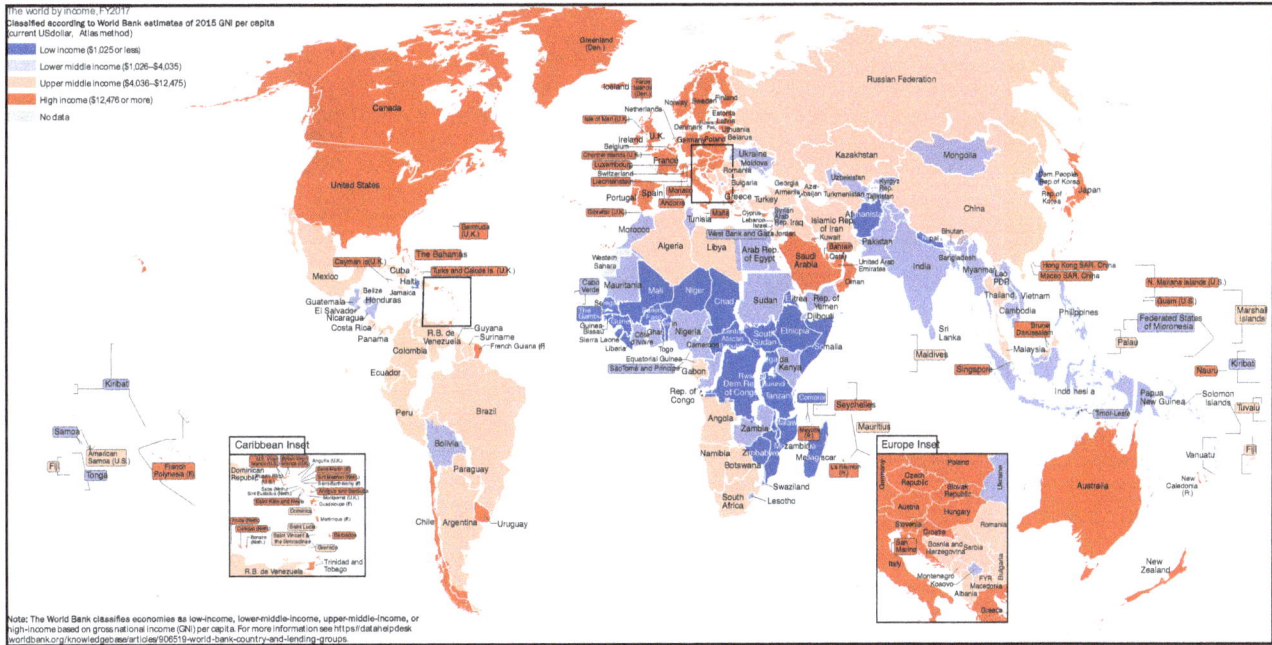

Figure 9.2 | Income Categories for Countries as Defined by the World Bank in 2017
Author | The World Bank
Source | The World Bank
License | CC BY 4.0

Now that you have pondered these categories and spatial patterns, let's take a step back. What exactly is meant by **Gross National Income** per capita? It is the sum of a nation's **gross domestic product** (GDP) and the net income it receives from overseas. **GDP** includes all of the goods and services produced with a country in a given year and is calculated according to the following formula: Consumption + Investment + Government Spending + Net exports. So if a government increases spending on the military or on road construction in a given year that will increase GDP (and GNI). If a country imports a lot of things but doesn't export many things, it will decrease that country's GDP (and GNI). Finally, if consumers buy a lot of alcohol for New Year's Eve, that will also increase GDP (and GNI). GNI also includes income derived outside of the country by members of the country. For example, Coca Cola, headquartered in Atlanta, Georgia, earns profits in 200 different countries around the world. Those profits outside of the U.S. are not included in GDP, but they are included in GNI. It's a little confusing, but this measure provides a valuable way to compare economic outputs between multiple countries and is the metric that is most often used in recent years. The total GNI for a country is divided by the total population of that country to arrive at a per capita figure, which includes children, the elderly, those imprisoned, etc. So, the figure of $12,476 does NOT mean that the average adult earns that amount annually as income, but rather it represents total income and economic output divided by the total population. Take a look at the cartogram in **Figure 9.3,** which shows countries represented, not by their physical size, but by the size of their respective GNI. **Table 11.2** shows the top 15 countries by GNI, with the

Figure 9.3 | Gross National Income by Country
Author | © Copyright Worldmapper.org / Benjamin D. Hennig
Source | WorldMapper
License | CC BY NC ND 4.0

United States ranked at the top and China second. The next table shows countries ranked according to GNI per capita (per person). Notice how the list of the top 15 countries changes dramatically. Can you think of reasons to explain why this might be? First, small oil-producing countries tend to have high incomes that are spread across a small population (e.g. Qatar, United Arab Emirates, Brunei). Second, countries with high incomes, high exports, and few children (e.g. Singapore and Switzerland) have high GNI. Third, certain small countries attract very wealthy member of society to live there in order to avoid paying higher tax rates elsewhere (e.g. Switzerland, Isle of Man, and Bermuda).

The difference between GNI and GNI per capita tends to be more pronounced in large, highly populated countries. The U.S. and China, for example, are the top two countries for total GNI, but neither appears in the top 15 for GNI per capita, in which the U.S. ranks 18th ($58,700 per person) and China drops all the way to 103 ($15,500 per person). A final important term useful to understand economic differences is **Purchasing Power Parity (PPP)**. The PPP accounts for differences in the cost of living and goods between 2 places to allow for a more meaningful comparison. Think about it this way. If you earned $60,000 per year in New York City or in Lincoln, Nebraska, where would your salary be more valuable? Most items are cheaper in Lincoln, so your salary would be worth more there than in NYC. Similarly, the cost of living in Mexico is much lower than in Sweden, so the PPP figures in **Table 11.1** account for such differences. If you are considering a move, you can check on salary equivalencies here: http://money.cnn.com/ calculator/pf/cost-of-living/index.html.

Ranking	Economy	(millions of US dollars)	Ranking		Economy	Purchasing power parity (international dollars)			
1	United States	18,357,322	1		Monaco				
2	China	11,374,227	2		Qatar	125,000			
3	Japan	4,816,892	3		Liechtenstein				
4	Germany	3,624,638	4		Channel Islan				
5	United Kingdom	2,778,488	5		Macao SAR,	98,650			
6	France	2,590,030	6		Isle of Man				
7	India	2,212,306	7		Singapore	85,190		World	16,161
8	Italy	1,923,095	8		Kuwait	83,310		East Asia & Pacific	17,023
9	Brazil	1,835,993	9		Brunei Darus	83,170		Europe & Central Asia	30,825
10	Canada	1,584,301	10		Gibraltar			Latin America & Caribbean	15,027
11	Russian Federation	1,425,702	11		United Arab	72,980		Middle East & North Africa	19,619
12	Korea, Rep.	1,414,400	12		Luxembourg	70,430		North America	57,163
13	Australia	1,313,016	13		Bermuda			South Asia	6,063
14	Spain	1,281,828	14		Switzerland	63,660		Sub-Saharan Africa	3,613
15	Mexico	1,153,529	15		Norway	62,550			

Figure 9.4 | Gross National Income & Gross National Income per capita (PPP) in 2016
Author | The World Bank
Source | The World Bank
License | CC BY 4.0

9.3 GLOBAL, NATIONAL, REGIONAL, AND LOCAL PATTERNS

9.3.1 How Fast Does the Global Economy Grow and How Much Money Is There in the World?

With 78 of the world's more than 200 countries and territories categorized by the World Bank as *high* and 64 are *low* or *lower middle income*, it is tempting to believe that wealth distribution might be relatively balanced in the world. After all, more people than ever are connected to the global economy. Instead, inequalities have been exacerbated by global capitalism in recent years. Jeff Bezos, the founder of Amazon, has a net worth of $105 billion, a figure that is larger than the annual GDP of 150 countries! Meanwhile, 800 million people earn less than $2.00 per day. Such click-bait worthy headlines, while fascinating, can also be misleading. Thirty-five percent of the world lived in extreme poverty in 1990. By 2013 that figure dropped to 11%, representing a shift of nearly 1.1 billion people out of extreme poverty. Nonetheless, economic differences between most wealthy countries compared to most poor countries have widened over that same period rather than narrowed.

The number of those entering the formal economy rose dramatically during that time period, so categorizing somebody as not in 'extreme poverty' simply because they earn more than $2.00 a day is also quite problematic. The majority of those that rose out of extreme poverty were in just two countries (India or in China), where hundreds of millions of people left unpaid work on subsistence farms and moved to cities, where they earned just a little bit of money. Does that make a country more developed or a person better off? In economic terms, the answer is yes, but in more qualitative terms, the answer is not quite so clear.

The global economy has grown persistently since 1960 as evidenced by **Figure 9.5**, with the most dramatic growth occurring since 2000. As a matter of fact, global GDP nearly doubled between 2002 ($34.6 trillion) and 2016 ($76 trillion)! So it took all of the economies of the planet tens of thousands of years to go from zero to $34 trillion, but then only 14 years to double that figure. Hmmm. Does this represent **sustainable development** – one that can continue into the future? Moreover, can the planet handle the effects of continued growth, consumption, CO2 emissions, and water pollution across the world at such a persistent growth rate? Such questions are difficult to assess and will be dealt with more substantially in the last chapter of this text. For now, it's important simply to understand that economic growth is just one of many indicators used to understand development and well-being, and the implications for economic growth come with complications. Finally, economic growth across the planet is uneven and difficult to predict. In spite of what appears to be a fairly even upward movement in global GDP, one

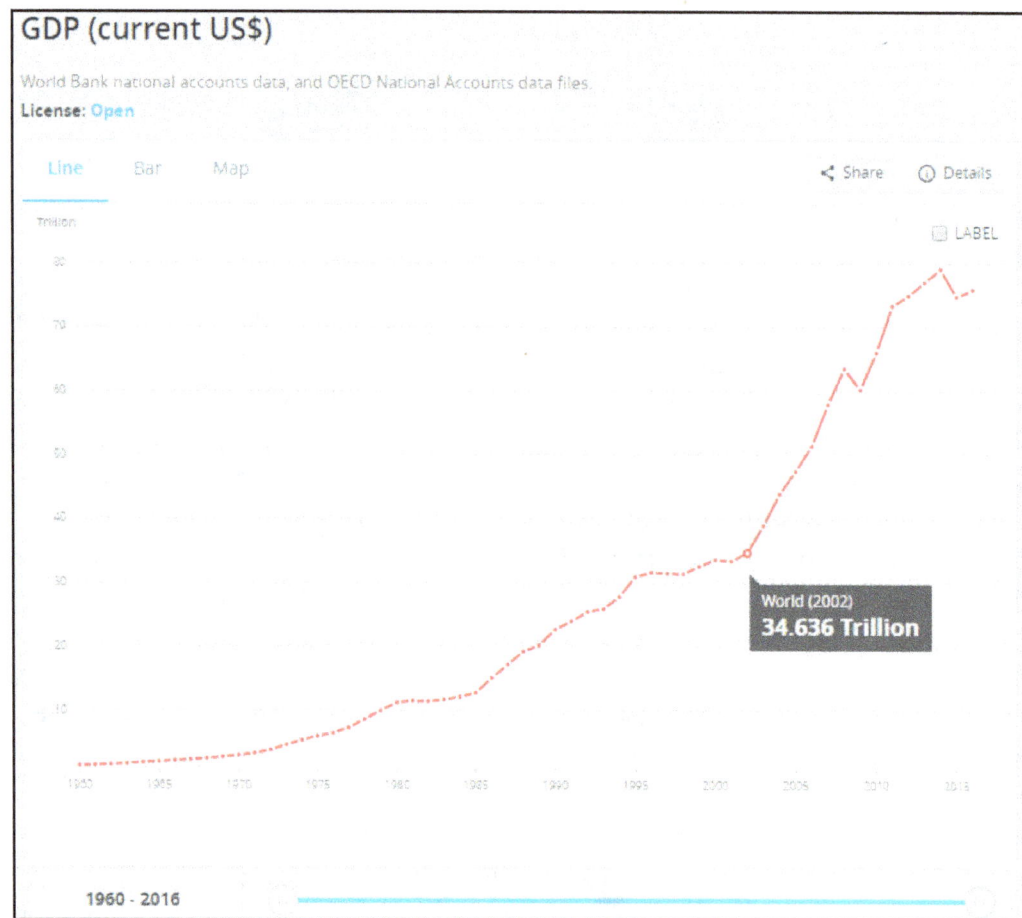

Figure 9.5 | Gross Domestic Product 1960-2016 in current US$
Author | The World Bank
Source | The World Bank
License | CC BY 4.0

can see more clearly the complex nature of GDP growth in **Figure 9.5** that shows the wild changes in growth that vary dramatically by place and time. In the 20 years from 1996-2016, Russia (an upper tier country) experienced many years of negative growth, while Rwanda (a lowest tier country) experienced rates of growth up to 14%, much higher than that of any other country. U.S. growth was negative in 2009 and has hovered around 2-3% in subsequent years. China's annual growth, that had averaged 12% for several years dropped to around 7%, where it is expected to remain into the 2020's.

The most basic spatial patterns of wealth and income can be easily observed in the maps and figures presented thus far in this chapter. Wealthier countries tend to be those in the **Global North** (North America, Europe, Japan, Australia, New Zealand), while poorer countries tend to be in the **Global South** (everywhere else). However, such generalizations are problematic in truly understanding wealth and well-being around the planet. Let's take a brief look at Latin America, for example. Mexico, by most accounts, is considered a **developing country** (another name for 'less developed country' that is on a pathway to improving). It falls south of the Brandt line, and is considered poor by most American standards. However, its per capita income places it in the top third of all countries and its economy is the 15th largest in the world. Carlos Slim, once the wealthiest person in the world, is Mexican and its economic performance far outpaces all of its neighboring countries to the south. The difference in economic indicators between Mexico and Haiti, for example, is greater than the difference between Mexico and the U.S. As such, it's important to be wary of simplistic categorization schemes in terms of wealth and development.

Moreover, in recent decades dozens of **newly industrialized countries (NIC**'s) have emerged that have reached or approached MDC status. Such places have moved away from agriculture-based economies to more industrial, service, and information-based systems. One example is a group of places known as the **Asian Tigers** or **Asian Dragons** (Singapore, South Korea, Hong Kong, and Taiwan), where massive investment in infrastructure and education facilitated an equally massive transformation of the economy in a very short period of time. South Korea, for example, lay in ruins at the climax of its civil war (1953), but has miraculously risen to a level of wealth similar to that of Italy. Another group of countries termed the **BRICS** (Brazil, Russia, India, China, and South Africa) fall outside of the Global North, but have experienced dramatic economic growth, raising its collective share of the global economy from 11% to 30% in just 25 years. Those countries continue to wield more political power in direct relation to the rise in economic might, and this could shift the economic, social, and political landscape of the world in the coming decades. Another group of NIC's are the oil-rich Gulf States of Qatar, United Arab Emirates, Saudi Arabia, Kuwait, Iran, and possibly a few others. Such entities have accumulated massive amounts of wealth as a result of **absolute advantage**, the abundance of rare and high-valued commodities. Other countries enjoy absolute advantage due to climatic conditions for growing coffee, tobacco, tropical fruit, etc. The high price of oil and its concentrated supply, however, have facilitate massive economic growth in places that were traditionally poor and less developed than in recent years. Perhaps you are familiar with some of the recent development projects in this region of the world. The tallest building in the world (Burj Khalifa) and the world's most ambitious set of artificial island construction projects are both located in the United Arab Emirates (UAE) (**Figure 9.6**) as evidence of the Gulf States development efforts in the 21st century.

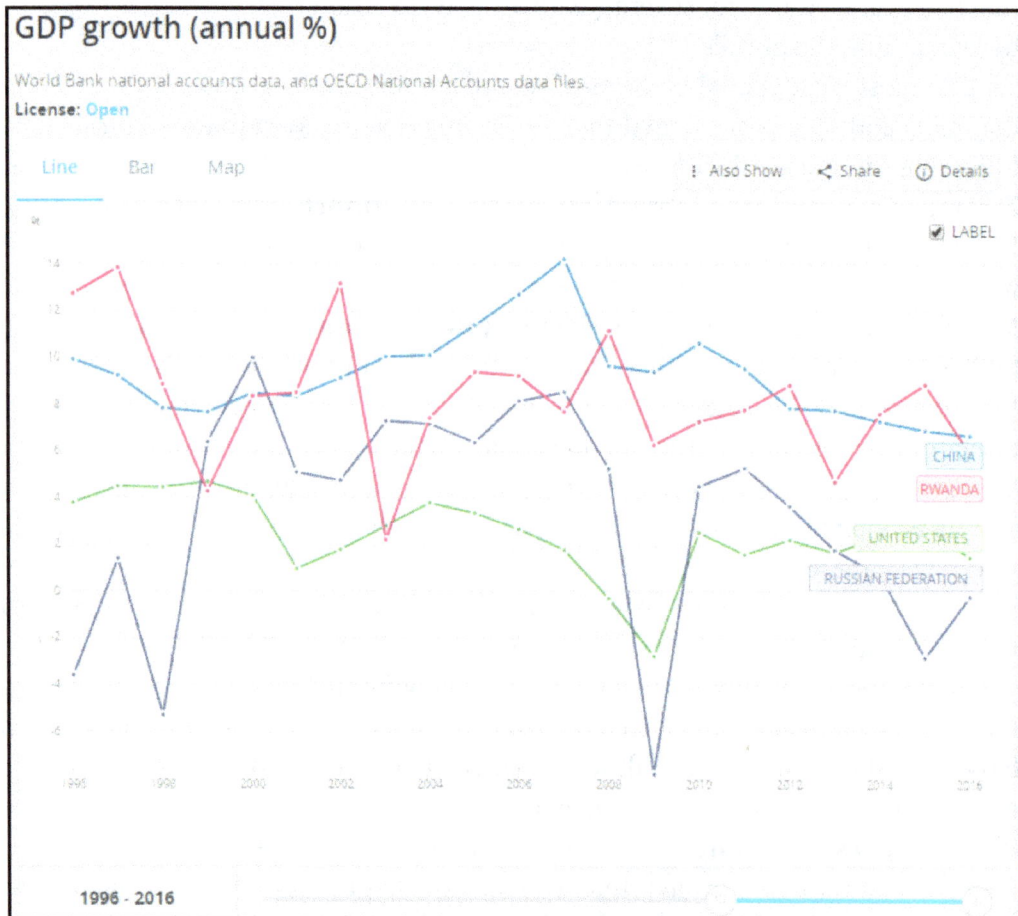

Figure 9.6 | Annual Growth Rates in GDP, selected countries, 1996-2016
Author | The World Bank
Source | The World Bank
License | CC BY 4.0

In spite of the obvious wealth benefits that accrue in oil-rich or other resource-laden countries, they can also suffer what's termed the **Resource Curse (aka Dutch Disease)**, as the benefits of a highly valuable commodity do not spread to other members of society and violence/conflict emerge as groups fight over the resource. While income may be very high, millions of workers continue to face horrendous work conditions directly related to development efforts. In Qatar, for example, thousands of workers have died during the massive construction of new stadiums and other infrastructure required to host the World Cup in 2022. Other examples of Dutch Disease can be found in Nigeria (oil), South Africa (diamonds), and the Democratic Republic of Congo (coltan).

Figure 9.7a | Burj Khalifa, Dubai, United Arab Emirates
Author | User "Donaldytong"
Source | Wikipedia
License | CC BY SA 3.0

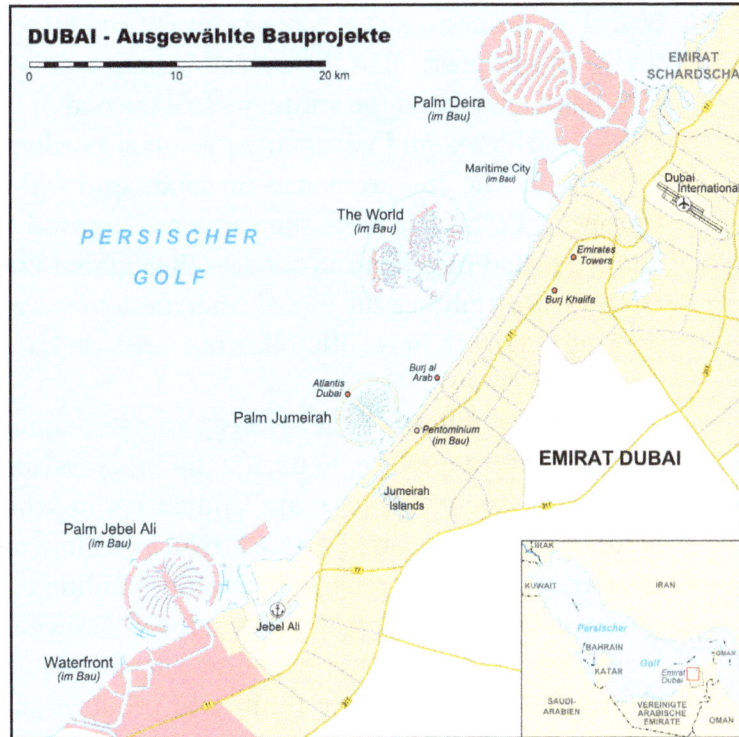

Figure 9.7b | Artificial Island Construction, Dubai, United Arab Emirates
Author | User "Lencer"
Source | Wikimedia Commons
License | CC BY SA 3.0

9.3.2 How Does a Country Improve Its Wealth and Well-Being?

The International Trade Model

The characteristics of this model are quite simple on the surface. In this strategy, a country embraces **free trade** (removing barriers to all imports and exports in a country) and willingly participates in all facets of the global economy. The basic benefits of the strategy are as follows:

1. Potential High ROI (Return on Investment)
2. Increased specialization leading to technological advantages
3. Simplified development strategy
4. Less government involvement

Libraries of books have been written about the transformative power of **capitalism** and other libraries of books have criticized the system. The international trade model asserts that trade between nations is the best way to bring about mutual prosperity for all. As a country removes barriers to trade, there will invariably be winners and losers, but classical **macroeconomic** (the branch of economics that focuses on entire systems rather than individuals or firms) theory posits that the overall benefit will be greater than the losses. The U.S. has championed the strategy since the end of WWII (1945) as it encouraged allies, neighbors, and adversaries to open borders, allow imports, and reduce controls on the free exchange of goods and services between countries. The European Union has done the same as it moved towards a common currency and a free flow of goods and services throughout Europe. The **Asian Tigers** and **BRICS** also embrace the global trade system to various degrees, having gained enormous growth in wealth following a recipe that calls for an intensive export-oriented economy.

Countries following the strategy remove domestic producer subsidies and allow global competition to decide the 'winners' and 'losers'. As such, countries must find specific services and industries in which to gain **specialization.** South Korea, for example, elected to focus on low-end electronics initially before moving into other sectors such as ship-building and automobiles. Initially, the products were inferior to those produced elsewhere, but with each generation it improved its workforce, technical knowledge, and facilities until it gained a **comparative advantage** (ability to produce particular items/services more efficiently than competitors given all the alternatives) in those industries. Specialization requires a lot of practice with an intensity of focus, investment, and time to gain price and quality competitiveness on the global market, but if done correctly the rate of return on investment can be very high. For example, in the post-Korean War era (circa 1953), South Korea transitioned away from an economy based mostly on farming to become the 7th leading exporter in the world, specializing in cars, auto parts, ships, and integrated circuits. Moreover, the technological knowledge can then be used to foster other industries. Under this strategy, governments don't try to protect certain companies over others, and since **tariffs** are removed, it simplifies the development strategy.

As countries gain comparative advantages in certain areas, they also tend to relinquish efforts in other areas. For example, as Japan focused developing its industrial sector after WWII, it focused less on agriculture, depending increasingly upon imports from other countries. Such a trade-off is termed **opportunity cost** in that choosing to do one thing prohibits you from doing something else. During the 1980's era, the United States aggressively pursued an international trade model that allowed for more manufactured goods to be imported into the country. People working in areas like steel production, and coal mining began to see their job opportunities diminish as more foreign goods entered the U.S. economy. Proponents of the system argue that such workers need to adapt and become re-trained in other high paying professions in order to escape the pain that comes with economic transformation. Can you think of other industries or jobs that suffer as a result of a country's choice to follow the international trade model?

A few other negative aspects to the strategy are:

- Susceptibility to unpredictable global markets
- Loss of local control
- Uneven benefits to the population.

Take a look at **Figure 9.8**. The dramatic price shift (first upward and then downward) represents the dramatic risk associated with global trade. For a farmer that has shifted away from food production and into palm oil (as did millions of farmers in Indonesia and Malaysia in the 2000's), this price drop is more than just economic theory. It can be the difference between living and dying. Focusing on a few key industries or products within a country comes with serious risks. What if prices drop unexpectedly or global preferences for certain products change without warning? Somebody gets left holding the bag and such shifts are very common in many commodity chains, with those at the bottom suffering the most serious consequences.

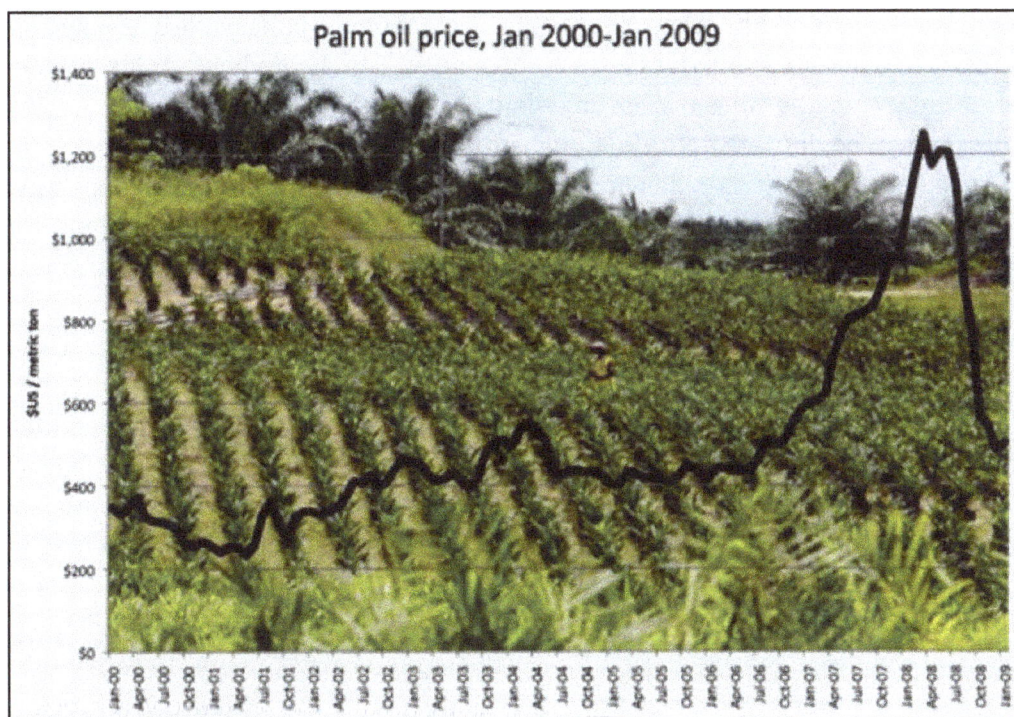

Figure 9.8 | Palm Oil Prices, 2000-2009
Author | Todd Lindley
Source | Original Work
License | CC BY SA 4.0

The second risk, loss of local control, occurs when countries have entered into free-trade agreements such as the North American Free Trade Agreement (NAFTA) between Mexico, the U.S., and Canada. As U.S. corporations increasingly began to re-locate production to Mexico, American workers' calls for politicians to intervene went unanswered because the agreement prevented state or local governments from taking intentional actions to protect jobs or to keep companies from moving.

Finally, international trade has generated significant growth in production and wealth, but it also brings new competition that disrupts local economies. While customers usually benefit from lower prices of imported products, many local producers lose their livelihoods entirely. As the U.S. lost manufacturing jobs to Mexico, more than 2 million small-scale Mexican farmers also lost their jobs, as Mexican corn could not possibly compete with the low-cost, mass produced crops from the U.S. and Canada. Corn exports to Mexico in 2016 were 5 times higher than in the year before NAFTA. It is not an accident that rates of immigration from Mexico to the U.S. increased dramatically during the same period. As farmers lost their jobs, the moved in search of new ones.

The Protectionist Model

As you probably have guessed from its name, the protectionist model (also known as import substitution) requires that countries sustain themselves without significant trade with other countries. Protectionist policies are applied to safeguard domestic companies from foreign ones. The theory behind this strategy is grounded in the idea that over-reliance on foreign labor, products, and/or services can be detrimental to national sovereignty and/or security. As such, the strategy became very popular among countries that gained independence after many years of imperialism and colonization. India, Jamaica, the Philippines, the former Yugoslavia, the former Soviet Union, and most countries in the Caribbean and in Africa embarked up this strategy from the 1950's-1980's. Yugoslavia, for example, produced its own car, called a Yugo, and India did the same. These automobiles were designed to be affordable for domestic consumers, but in both cases the strategy failed to produce a reliable, affordable product and left consumers without access to better alternatives imported from Germany, Japan, or the U.S. Protectionism is closely tied to nationalism in many cases as politicians and/or consumers ask the basic question, "Why don't we take care of ourselves, by ourselves, rather than depending on workers and producers from other parts of the world?" How does this question relate to your own opinions about whatever country you live in or were born in? Does the argument make sense to you?

By the 21st century, the vast majority of countries in the world turned away from protectionism as a strategy in favor or adopting an international trade oriented economy. Nonetheless, protectionism does offer certain advantages worthy of consideration such as:

1. More controlled decision-making
2. Benefits spread to more members of society
3. More government involvement
4. Food security

Decision-making by governments can be more controlled and policy decisions can be easier to make and policies can be designed to benefit and spread to more members of society under protectionism. Until 1993, for example, corn farmers in Mexico were protected from cheap corn produced in the U.S. and Canada, because of restrictions on imports. Most of the millions of small-scale corn farmers in Mexico could grow enough corn for their own families and produce a small surplus to be sold locally, providing a small but meaningful wage for millions of families. Protecting farmers from outside competition offered a measure of predictability, political stability, and food security from one year to the next. Mexico is also blessed with generous reserves of oil and traditionally the government has controlled domestic oil and gas prices, which again can bring a certain level of predictability to consumers as opposed to the wild shifts in prices that can often accompany imported oil.

Disadvantages to this isolationist strategy are probably apparent to you as you read and think about this topic. A few of them are listed here:

- Susceptibility to corruption, inefficiency, and slow response to market conditions
- Lacking in creativity and innovation
- Absence of natural resources in some places necessitate trade to meet demand

When a government protects a company from outside competition, there is a tendency for that company to become susceptible to corruption. Since 1938 a single oil company called Pemex had exclusive rights to drill, process, and sell oil within Mexico, which provided a huge source of revenue, protected jobs, and helped build national pride. In recent decades, however, the company increasingly found itself entangled in one scandal after another involving bribes, kickbacks, and various schemes that have kept gasoline prices at the pump artificially high. In 2013, however, Mexico began reforms known as **market liberalization** (a process of removing barriers to foreign-owned companies from operating and competing with domestic ones), allowing BP, Shell, and other companies to import petroleum and open gas stations in certain parts of Mexico. Such a shift is having a serious impact upon the national landscape both symbolically and materially as residents have a choice between competitors for the first time in 80 years!

Figure 9.9 | Nationally Owned/Operated Gas Stations in Mexico
Author | User "Diaper"
Source | Flickr
License | CC BY 2.0

Aside from corruption, another drawback to isolationism is the lack of creativity and innovation that otherwise comes from fierce competition. Venezuela, a country that has resisted international trade in favor of protectionism, currently is suffering from massive shortages in food and other necessary consumer items as domestic producers have failed to innovate and respond to consumer demand. Meanwhile, global supply chains of rice, beans, corn, tomatoes, mangoes, coffee, sriracha sauce, and fidget spinners have constantly evolved, innovated, and changed to efficiently provide for the demands of consumers internationally whether they happen to be in Sydney, Seoul, Santo Domingo, or Sao Paulo! It is very difficult for governments to plan for and anticipate the needs of an entire society for any length of time. Protectionism requires a government that can match consumers and producers efficiently, a task often handled much better by free-market forces than by public officials.

The final element that makes protectionism so problematic rests in simple geography. All countries have different conditions related to site, situation, climate, or natural resources, so trade isn't just advantageous but rather it is essential to survival. An extreme case can be found in North Korea, a country with notoriously tightly controlled borders. Without trade the country routinely fails to produce enough rice, leaving millions desperate and undernourished. Likewise, China faced similar droughts, food shortages, and periods of starvation when it favored protectionism for most of the 20th century, but it began to reverse its strategy as it cautiously opened its economy to the rest of the world in the early 1980's.

The 2 Models in Global Context

After reading the brief summaries of the 2 models above, you might ask yourself why any country would choose a protectionist model, given the obvious advantages and drawbacks of each. Most economists say the same thing. We need to exercise caution with overly simplistic conclusions, however. Here is where geography and history are worth careful consideration. Let's begin with Caribbean islands (most of them anyway). The majority of Caribbean countries were colonies until just very recently (1960's onward). During the colonial period, European powers established a monoculture (agricultural system heavily focused on a single item) plantation economic system in which particular islands or territories focused exclusively on one item such as sugar, bananas, or pineapple. Upon independence, many of these former colonies found that they had very little diversity in their own domestic economy, meaning that even after earning their freedom they were still almost entirely dependent upon the former colonizer economically because the plantation economy created a relationship of dependency. The concept of **dependency theory** explains the economic problems experienced by former colonies as a function of the disadvantageous terms and patterns of trade established over hundreds of years. Look at it this way. If all of the grocery stores around you closed permanently, would you be able to head into the wilderness tomorrow to find your own food? Would you know which berries are safe and which are poisonous? Probably not. This was the situation in which many former colonies found themselves, so it made sense for them to attempt to protect themselves from outside competition while domestic companies could emerge and develop. Moreover, **world systems theory** suggests that the global system of trade only works as long as there are a persistent set of winners (more developed countries), who mostly benefit from low-cost production in poor countries, and losers (less developed countries) that provide a readily available supply of laborers willing to work long hours for low wages.

While many countries have clearly benefitted from the international trade model, others have actually been damaged by their participation in global competition. Ethiopia's largest export, for example, is coffee and although retail prices for a cup of coffee have increased dramatically, incomes of coffee farmers in that country have not reaped any of those benefits. Instead, prices paid for coffee beans have routinely been set by foreign commodities markets that tend to undervalue the product and harm small producers, leaving only large producers with a meaningful profit. Although just 3 cents more per kilo could bring these farmers out of poverty, global markets don't account for such disparities. One remedy for such a complicate problem is **fair trade**, a system that guarantees a basic living wage for those at the very bottom of the global production cycle.

Another meaningful problem with global trade is that it can be very disruptive and unpredictable. Following years of state controlled markets, Russia began to open itself to foreign competition and free-market reforms after 1989. While Western countries applauded the decision, many people in Russia suffered as they experienced a lower quality of life, loss of jobs, and massive economic and political instability. Social problems, most notably high unemployment and alcohol abuse, began to dominate society as life expectancy actually dropped significantly for Russian men, many of whom could not adapt to the new demands of a fully market-based economy.

In summary, the decisions made by any country about the best path to increase development are not so clear. Leaders and policy-makers face very difficult, often contradictory decisions that vary from place to place, about how to make things better for citizens and residents. Moreover, what might be good in the short-term, can be problematic in the long-term. For example, Jamaica's decision to protect its farmers from outside competition benefitted local farmers in the short-term, but the island government's subsidies to farmers proved too costly for the government in the long term. When the money ran out, the country found itself in crisis. **Collectivist societies** are particularly threatened by international trade models in that many traditional forms of society do not encourage competition, but rather they value cooperation for the good of the group over the desires of any individual. Finally, any development strategy that values short-term benefits over long-term costs are clearly problematic. As India and China have embraced international trade, for example, unprecedented levels of air and water pollution have become the norm in Shanghai, Bombay, Delhi, and Beijing (among others).

Sustainable Development is a strategy that balances both current and future benefits and costs and attempts to balance them with any possible environmental damage (e.g. oil spills, loss of habitat, erosion, increased pollution, dangerous waste) that might occur.

9.4 ROSTOW'S STAGES OF GROWTH AND POLITICAL POLICY

Do all countries follow a similar pattern of development over time? This is a simple question that does not have a simple answer. Just as different people each follow different career paths, depending upon their own unique talents, strengths, and weaknesses, the same is true of countries. While the Asian Tigers have successfully demonstrated the power of embracing an export-oriented economy, countries like Brazil, Jamaica, and South Africa have also embraced global trade but failed to achieve similar results in spite of grand efforts and reforms. Nonetheless, the dominant and consistent advice from Western countries (in the Global North) is for poorer countries to reject protectionism in favor of trade and globalization. Much of the debate is informed by economic theory that argues that the more we trade with one another, the more stuff there is for all of us – and who doesn't want more stuff, right? A leading proponent of trade was an American government official named Walter Rostow, who developed a model known as Rostow's Stages of Growth in 1960. He argued that countries historically follow a similar and predictable pathway to wealth and stability and that each country is in one of the 5 stages of growth described below:

1. **Traditional Society**: characterized by subsistence agriculture with intensive labor and low levels of trade in which most live on small farms, are mostly focused on local concerns, and remain largely disconnected from the rest of the world.

2. **Preconditions for Take-off**: initial manufacturing stage, beginning of trade, new ideas, emergence of banks, but society still dominated by tradition.

3. **Take-off**: a short period of massive disruptive societal change, in which industrialization accelerates, high profits are reinvested in new technologies, often brought on from external forces.

4. **Drive to Maturity**: takes place over a long period of time, as standards of living rise, economic growth outpaces population growth, use of technology and education increases, and the national economy grows and diversifies.

5. **Age of High Mass Consumption**: final stage, in which most enjoy the luxury of consuming far more than they need. Advanced economies have a surplus that can be used do increase social welfare and reduce

risks to society. (Rostow believed the U.S. had reached this stage in the 1920's, Western Europe and Japan in the 1950's and the Soviet Union could potentially reach it if it changed its political system).

Rostow's ideas represent the concept of **modernization theory**, which is the widely held belief that, with the proper intervention each country will pass through a similar pathway of development. Modernization geography recognizes the large disparities that exist from one place to another across the globe and represents an attempt to bring prosperity to places that have been left out or left behind. The goal to address such differences worldwide represents a shift in scale from one in which each country looks after itself to a new paradigm in which wealthy countries felt a keen responsibility to 'bring' development to those across the world, in what's been dubbed the 'development project'. The best evidence of this global effort can be seen in the Millennium Development Goals, which were a series of 8 measurable outcomes decided upon by 189 world leaders in 2000 that were to be achieved by 2015. Although not all of the goals were achieved, significant progress was demonstrated across most of the 8 goals. Much of the progress, however, has been found in Asia – most notably in China, where nearly 200 million people have left farms and moved to cities to work in manufacturing jobs, which pay far more than what they could earn as farmers. Critics of Rostow's model point out that it doesn't account for geographic differences, historical variations, or the long-standing effects of colonialism and they also argue that the model is more of a political document (critical of communism and triumphant of capitalism) than it is a serious economic blueprint for success. Nonetheless, efforts have been spearheaded by Europe and North America to jumpstart all economies of the world into action, so that all can eventually reach the age of high mass consumption. Can you identify any problems with this model? Is it possible for all countries of the world to live in a state of high consumption? Why or why not?

What do you think might be the biggest obstacle to development by poor countries today? It's the same obstacle that keeps you from going out to dinner every night – MONEY, of course! But the kind of money we're talking about in this case is not available from a regular bank. For that reason, the **World Bank** was formed in 1944 to provide funds to countries for the purpose of large development projects. For example, a World Bank loan financed the construction of the Tarbela Dam (1977) – see **Figure 9.10**. One of the largest in the world, the structure facilitates irrigation, helps to control flooding, and (most importantly) provides significant amounts of hydroelectricity to the region in Pakistan along the Indus River. Such projects are often controversial, however. During its construction, the dam displaced 120 villages and many claimed to have never received the compensation they were promised. Similar projects can be found all over the world, and most of them are designed to improve people's daily lives while also greasing the wheels of global commerce.

Figure 9.10 | World Bank Financed Tarbela Dam in Pakistan
Author | Paul Duncan, U.S. Marine Corps
Source | Wikimedia Commons
License | Public Domain

Another source of funding to governments comes in the form of emergency loans, usually from the **International Monetary Fund (IMF)**. The IMF is funded primarily from wealthy private sector banks and offers loans during times of financial crisis. The loans are smaller than those from the World Bank and are not designated for national infrastructure projects, rather they are intended to maintain stability when a country may otherwise be vulnerable to political revolution, dictatorship, or collapse. Such loans, however, come with strings attached. The borrowing country must agree to a series of **structural adjustment programs**. Namely, countries must:

1. Privatize government-owned assets (such as oil companies, utilities companies, or diamond mines)

2. Reduce spending on government programs (such as social services, food programs, or military spending)

3. Open themselves up to foreign investment and foreign competition (such as the case with Mexico and Pemex)

The end result of accepting an IMF loan, then, nudges that country away from the protectionist model of development and towards one that focuses on international trade. Most often, a country following the adjustments will experience a devaluation of their own currency, making imports more expensive, but exports more viable, which can be very beneficial to manufacturers searching for new, lower-cost manufacturing locations around the world. As with any major change to an economic system, there will always be big winners and big losers. The same can be said for structural adjustment programs and the IMF.

9.5 NEW MODELS OF DEVELOPMENT

9.5.1 Is There More to Life Than Money?

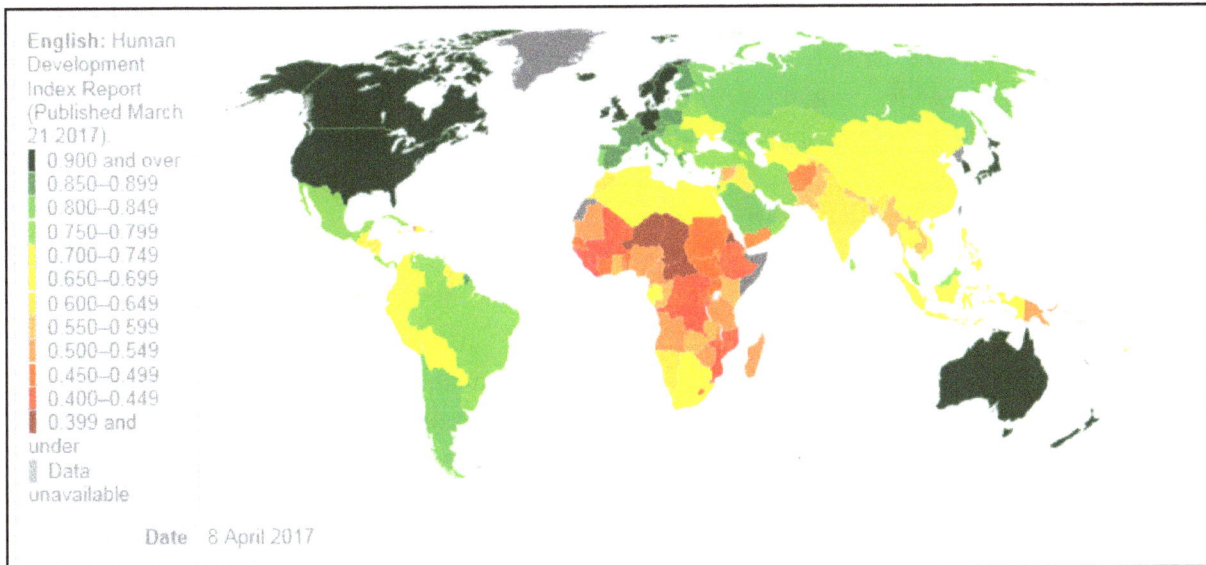

Figure 9.11 | Human Development Index (Based upon 2016 Human Development Report)
Author | User "Happenstance"
Source | Wikimedia Commons
License | CC BY SA 4.0

This may seem like more of a philosophical question than one you'd find in a geography textbook, but it's an important one to consider. In the previous section, we learned that the end-point of development is to arrive to a state of high mass consumption relying primarily on indicators of wealth, income, and production to evaluate one place compared to another. Do you think this a fair assessment or are there other ways to gauge development? Think about some of the things that make you feel good or bad about any particular place. Does it have good parks, safe roads, low levels of traffic, or lots of fun things to do? While some of those things are connected to income and wealth, some markers of development do not necessarily have to be connected to wealth.

Indian economist and Nobel Prize winner Amartya Sen argued that our understanding of development needed to be dramatically altered to include a capabilities approach to measuring, encountering, and improving people's lives. In his book 'Development as Freedom', Sen argues that abject poverty comes from a lack of rights and the absence of freedom that are very different from simplistic notions wealth and poverty. For example, a country that produces a lot of oil like

Saudi Arabia may demonstrate high GDP growth, but if it doesn't positively affect the lives of most people in the country, then the GDP is not improving people's lives in a meaningful way. As obvious as it may sound today, this transformative thinking drove the United Nations to redefine thinking on development as the organization unveiled its first annual **Human Development Report** (http:// www.hdr.undp.org) in 1990, which incorporated a brand new index, called the **Human Development Index (HDI) (Figure 9.9)**. Instead of relying solely on income, the index is derived from indicators of 1) **life expectancy**, 2) **adult literacy**, 3) school enrollment, and 4) income to provide a meaningful comparison for almost all countries (a few countries like North Korea choose not to participate). The index provides reason for optimism in that 105 countries are now categorized as 'very high' (51 countries) or 'high' (54) human development with an HDI of .70 or higher.

You can view the entire list of countries here: http://hdr.undp.org/en/ composite/trends, and the regional trends over time are included in **Figure 9.10**, which shows that the areas with the greatest improvement since 1990 have been East Asia and South Asia. Most notably, China moved from low human development to high and India moved from low to medium during the short period from 1990-2015. Sub-Saharan Africa contains the majority of those with the lowest levels of human development, but nearly every country experienced persistent improvement since 1990. Moreover, developing countries everywhere have seen improvements in educational access and life expectancy during the past 50 years. HDI helps us to quantify those improvements over time. One final point about this index is that some countries do a very good job of providing human development, in spite of low income. Find Mexico and Cuba on the map, for example. In both countries, incomes are low, but life expectancy is high. In the case of Cuba, life expectancy is higher than it is in the U.S. Cuba has followed a protectionist strategy for several years, which has limited financial growth, but its provision of healthcare and education gives it a higher HDI than would otherwise be predicted. Meanwhile, countries like Kuwait and others in the Middle East enjoy high incomes but lower levels of development when measured by the index. HDI also has its limitations since it uses only 4 indicators. Can you think of some other measure that might be meaningful? Some examples might be gender empowerment, levels of inequality, or simply quality of life. Other indexes measure such things, and access to the Internet or to modern sanitation systems. Among those living in rural areas of India, for example, only one quarter of the population has access to a working toilet. The Population Reference Bureau (http://www.prb.org/DataFinder) collects data on these and other indicators by country that are readily accessible. You are encouraged to check out the website to find out more of these!

It's also important to keep in mind that differences within countries can be very stark. People living in Shanghai, China, for example enjoy a standard of living very similar to the average Italian. Meanwhile, rural dwellers in that country are not much better off than an average person from Sub-Saharan Africa. Just as many rural residents in Tennessee, Mississippi, and West Virginia struggle with access to basic health care and have unemployment rates much higher than in other parts of the U.S. The small country of Bhutan has taken a novel approach to measuring its own development by considering one factor above all others – happiness. Rather than measuring GDP, it closely monitors **gross national happiness** as a barometer of the spiritual, physical, social and environmental health of its citizens and natural environment. As stated by the Minister of Education, "It's easy to mine the land and fish the seas and get rich. Yet we believe you cannot have a prosperous nation in the long run that does not conserve its natural environment or take care of the wellbeing of its people." As you may have noticed, none of the indicators of development presented in this chapter take into account the well-being of the environment or the long-term cost of development. As people become wealthier, they consume more of the earth's resources. The Bhutan model might seem a little strange to the average American student, but it is one of the few models that takes seriously the notion of environmental sustainability as a component of development.

9.5.2 What Is Microfinance?

Inherent in all forms of development strategies is the importance of money. Previous sections described how the World Bank and the IMF offer development assistance to developing countries. Both of those entities are limited to relatively large-scale projects, however. In recent decades a new form of development assistance has emerged that has proven much more impactful at the local level. The term **micro-finance** refers to all of financial services normally available through a bank to the wealthy around the world. These include credit cards, short-term loans, interest bearing accounts, and financial insurance. Something as simple as a credit card that allows for a college student to buy books (not this book, however!) can be the difference between staying and leaving college. Small-scale farmers in less developed countries are at constant risk when they do not have access to such credit. With even a very small loan, or **micro-loan** (e.g. $500), hundreds of millions of farmers can increase their income significantly. If you sell milk, but you only own 3 cows, imagine what you could do with 6 cows! Large-scale development programs don't offer such things to individual farmers, but new programs and websites are doing exactly that. Perhaps most well-known among these is Kiva.org (http://kiva. org). What started as a small endeavor to match lenders and borrowers by 2 people has morphed into a major provider of loans to more than 100,000 people in 2016, while raising nearly $150 million. Most incredibly, the repayment rate of loans given to the poorest of the poor stands at 97%, an astounding accomplishment.

9.5.3 What Does the Future Hold For Development Around The World?

Predicting the future is always risky. Climate change, emergent untreatable diseases, and massive political upheavals have historically altered the course of history beyond anyone's imagination. Those caveats aside, the immediate future is expected to be a good one when measured by human development index. As total fertility rates decline across all of the developing world, women's empowerment will also increase. As more women enter the workforce globally, family incomes are expected to continue to rise. The overarching story of the 20th century was that life expectancies skyrocketed, **infant mortality rates** declined, and the spread of many communicable diseases was curtailed. Economists argue that, as more countries increase their connectivity to the global system via trade, that the wealth of everyone increases. As hundreds of millions of farmers left the countryside in China and India to seek manufacturing jobs, levels of development increased dramatically. Many predict that the same spatial pattern will occur in African and in other parts of the world in the coming decades as manufacturers seek new markets and new, lower cost labor deep into the 21st century. Many of the improvements worldwide are directly connected to technological improvements. Anyone with an Internet connection in 2018 has access to more information, imagery, and knowledge than average people had throughout their entire lifetime even just one generation ago. As globalization allows for the free flow of information, commerce, currency, and ideas, there is also an increased risk that dangerous things travel along those same routes. The dark web makes for easy illicit exchanges of drugs, weapons, stolen credit cards, human trafficking victims, and all sorts of other horrible things. The challenge for development in the future clearly lies in balancing the risks and benefits of a world that has experienced such a massive time-space compression in such a short period of time.

Regions	Human Development Index (HDI) Value				Average annual HDI growth (%)			
	1990	2000	2010	2015	1990-2000	2000-2010	2010-2015	1990-2015
Arab States	0.556	0.611	0.672	0.687	0.96	0.95	0.45	0.85
East Asia and the Pacific	0.516	0.595	0.688	0.720	1.45	1.45	0.92	1.35
Europe and Central Asia	0.652	0.667	0.732	0.756	0.23	0.95	0.63	0.59
Latin America and the Caribbean	0.626	0.685	0.730	0.751	0.92	0.63	0.58	0.74
South Asia	0.438	0.502	0.583	0.621	1.38	1.51	1.25	1.40
Sub-Saharan Africa	0.399	0.421	0.497	0.523	0.54	1.67	1.04	1.09
Least developed countries	0.347	0.399	0.481	0.508	1.40	1.90	1.08	1.54
World	0.597	0.641	0.696	0.717	0.71	0.82	0.61	0.74

Figure 9.12 | Human Development Index by Region and Time Period
Author | United Nations Development Programme
Source | United Nations Development Programme
License | CC BY SA 3.0

9.6 CONCLUSION

In some ways, trying to describe and understand all facets of development around the world is a ridiculous task. The previous sections of this chapter barely scratch the surface, and yet the major questions pertaining to development are actually quite simple. Is it better for a country to be intricately connected to the global economy and to willingly import and export lots of things? Or is it better to close oneself off, protecting your own from the potential dangers of the outside world? History seems to be on the side of openness, but that also carries serious risk. Governments can spend huge amounts of money trying to make people's lives better only to find themselves in massive debt, followed by massive political upheaval (as recently happened in Brazil). There is no single magical formula, yet the general principles laid out in this chapter do offer some valuable insight from what has occurred in the past 70 years. On average, human beings in the 21st century are healthier, living longer, earning more money, having fewer children, consuming more, and knowing more about their world than at any other time in the history of our species. Given these realities, there is reason for cautious optimism as we look forward to the next hundred years. In spite of all of that, it's imperative that we realize the fragility of our planet and the environmental cost that comes with the overconsumption of resources. As large countries like China become wealthier, hundreds of millions of new consumers create more plastic waste, more toxic runoff, more carbon emissions, more copper mines, and more demand for energy than ever before. The rush to development has often hastened ecological catastrophe, making it essential that future development efforts consider long-term sustainability as an anchor for decision-making in the 21st century and beyond.

9.7 KEY TERMS DEFINED

Absolute Advantage – In economic terms, a country that is blessed with abundant natural resources or geographic advantages that are rare or in short supply and in high demand elsewhere.

Adult Literacy Rate – The proportion of the adult population aged 15 years and over that is literate. This indicator provides a measure of the stock of literate persons within the adult population who are capable of using written words in daily life and to continue to learn.

Asian Dragons (Asian Tigers) – The high-growth economies of Hong Kong, Singapore, South Korea and Taiwan: all of which focus on exports, an educated populace and high savings rates as pathways to development.

BRICS Countries – The countries of Brazil, Russia, India, China, and South Africa – countries that collectively account for 40% of the world's population and 25% of the world's land. From 1990-2014 these countries share of the global economy rose from 11% to almost 30%. In recent years, economic growth in the BRICS has been slowed by corruption, crisis, and dropping commodity prices.

Capitalism – The historically contingent economic system of trade in which parties are categorized into laborers and capitalists, both of whom seek to maximize their profit/wages. Within capitalism prices, production, and wages are determined by market conditions including, but not limited to supply and demand. Contemporary capitalism is intimately tied to the industrial revolution and ensuing societal transformations in England that diffused across Europe initially and then to many other parts of the world. Capitalism assumes that rational consumers seek to maximize their own utility.

Collectivism (Collective Societies) – A socio-political-economic system that prioritizes the well-being of the group over the individual. Collectivist modes of development, for example tend to include progressive tax regimes, affordable access to healthcare and higher education for all, and protections for marginalized groups. In LDC's this may also include communal ownership of land or other assets.

Comparative Advantage – The principle whereby individuals (or territories) produce those goods or services for which they have the greatest cost or efficiency advantage over others and the lowest **opportunity cost**. The outcome tends to be specialization across places.

Dependency Theory – A theory of development positing that the global economic system disadvantages certain regions and countries. It is argued that prior colonial relationships created systems of trade that benefitted the colonizer much more than the colonized. Countries that were made producers of pineapple, sugar, rubber, and other products during colonial times continue to be dependent upon the production of very low-profit items, limiting opportunities to produce other more valuable products and services long after independence, because the trade systems remains the same.

Developing Country – A term that includes all countries, other than those in the wealthiest category, that continue to improve their levels of development in the late 20th and early 21st centuries. The label has come to replace the less preferred term, LDC, to account for the fact that LDC is a static identifies while 'Developing' is a dynamic one.

Development – Processes related to improving people's lives through improved access to resources, technology, education, wealth, opportunity, and choice. Governments, individuals, non-profit organizations, and inter-government agencies all work towards similar goals with a variety of different approaches.

Domestic subsidy – A government-sponsored financial incentive that provides a production advantage to a company or entity. This may take a variety of forms. Examples include no-interest loans or cash payouts to farmers that meet certain criteria. Such programs are designed to lower risks and increase productivity of particular industries, services, or products and to protect them from competition that comes from outside of that country.

Fair Trade – One of a variety of different global trading systems that seek to guarantee fair (higher) payment for producers; often with other social and environmental considerations.

Free Trade – A system of trade that removes (or attempts to remove) all 'artificial' barriers that otherwise limit exports and exports between countries. A major component of free trade is the elimination of tariffs, duties, domestic subsidies, or laws that favor one country or company over another.

Gig Economy – A labor market characterized by the prevalence of short-term contracts or freelance work as opposed to permanent jobs.

Gross Domestic Product (GDP) – All of the goods and services produced within a country within a given year. The formula for GDP is Consumption + Investment + Government Spending + Net Exports. Global GDP was approximately $76 trillion in 2016.

Gross National Happiness – a holistic and sustainable approach to development, which balances material and non-material values with the conviction that humans want to search for happiness. The objective of GNH is to achieve a balanced development in all the facets of life that are essential for collective and individual happiness. The 4 pillars of happiness are: 1) sustainable & equitable socio-economic development; 2) environmental conservation; 3) preservation & promotion of culture; 4) good governance.

Gross National Income – All of the goods and services produced within a country in addition to all of the net income its companies and citizens receives from overseas.

GDP per capita – GDP divided by total population.

Global North – Those countries generally considered to be 'more developed', which also fall primarily north of the Brandt line as drawn in 1980. See Figure 9.1.

Global South – Those countries generally considered to be 'less developed', which also fall primarily south of the Brandt line as drawn in 1980. See Figure 9.1.

Human Agency – The concept that human beings take an active role in their own situation to invoke change. The concept is of critical importance in understanding how and why models and theories of development are so complex.

Human Development Index (HDI) – A measure developed by the United Nations in 1990 to consider and compare levels of development by all countries of the world using life expectancy, literacy, school enrollment, and income as the indicators. This provides a more meaningful way to compare countries than looking only at income/GDP. Those with the highest HDI tend to be in Australia and Northern Europe. Those with the lowest tend to be in Sub-Saharan Africa.

Human Development Report – An annual comprehensive analysis, assessment, and ranking of every country in the world based upon the Human Development Index. The report has been compiled and released every year since 1990.

Infant Mortality Rate – A measure of how many children die in any given year compared to 1,000 live births. Countries with low levels of development tend to have high infant mortality.

Informal Economy – Those activities within any economic system that are unregulated, untaxed, and/or unquantified. Includes, but not limited to: selling anything illegally, unreported paid work, consuming unlicensed/artificial products (movies, music, watches, etc.). Women, children, and the poor are those most likely engaged in the informal economy, but globalization and technology also plays an important in driving new forms of informalization. This sector of the economy is also related to the rise of the **gig economy**, in which workers increasingly are contractors rather than employees – an important distinction.

International Trade Model of Development – A strategy of development in which a country embraces **free trade** and elects seek specialization of certain products and services that are valuable as exports in the global marketplace. Following such a strategy necessitates the embrace of increasing imports and removing barriers to trade.

International Monetary Fund (IMF) – An intergovernmental organization that provides short-term loans to governments that are in economic crisis.

Least Developed Countries (LDC's) – Defined by the United Nations, those countries with the lowest levels of combined income, human assets, and economic vulnerability. In 2015, LDC criteria was given to 48 countries, where 950 million resided and more than half earned less than $1.25/day.

Life Expectancy – The average predicted number of years of life for any given person beginning at birth. In 2015, global life expectancy was about 72 years.

Macroeconomic Theory – The branch of economics concerned with large-scale or general economic factors, such as interest rates and national productivity.

Market liberalization – A process of removing barriers to foreign companies from operating and competing with domestic ones

Modernization theory –Belief that, with the proper intervention each country will pass through a similar pathway of development

Micro-finance – Financial and banking services designed for those who otherwise would be excluded due to their socioeconomic situation.

Micro-loans – Very small loans designed for those who otherwise would be excluded due to their socioeconomic situation.

More Developed Country (MDC's) – A category for those countries that have the highest levels of development as measured by income, education, and industrialization. MDC's tend to derive most of their GDP from services rather than from agriculture or manufacturing in the 21st century. The term MDC is used less commonly than the term 'Developed Country' but serves as a convenient way to refer to the wealthiest countries collectively.

Newly Industrialized Country (NIC) – Newly industrialized country. Examples include (but not limited to): India, China, Singapore, Taiwan, Turkey, Brazil, Mexico, South Africa and Thailand

Nontariff barrier to trade – Any impediment to trade placed by governments, including regulations based upon environmental, cultural, or political concerns. Examples include political trade embargoes or the prohibition of trade in eagle feathers, 'blood' diamonds, and human organs.

Opportunity Cost – The activity that has to be given up (forgone) in order to conduct the current activity. A country, for example, may choose to specialize in the production of coffee or cocoa. If it chooses coffee, then the opportunity cost is represented by the cocoa production that it could have, but did not produce. In classical economic terms, the opportunity cost is the difference between the two outcomes.

Purchase Power Parity (PPP) – A formula that accounts for cost of living variability from one place to another. PPP adjusted income allows for a meaningful comparison between two places with different cost structures.

Quota – A control on trade that limits amounts of particular items that may be imported or exported from/to a particular country.

Remittance – Money sent to the country of origin by overseas workers. This has become a significant driver of development in many countries in recent years.

Resource Curse/Dutch Disease – The idea that those places blessed with valuable natural resources often are negatively affected by the activities associated with cultivation, mining, and/or extracting those resources.

Self Sufficiency Model of Development – A centralized strategy of development for a country that seeks to develop all sectors of an economy within one's own borders and reduce dependency on outside entities. Such an approach requires tight controls on imports and exports as well as considerable protections on domestic producers against outside competitors.

Specialization – A focus upon skills, experience, and resources that improve capacity to produce one or more types of goods or services that are in high demand in the global economy. Specialization is directly related to the international trade model of development.

Stages of Growth – A concept that aims to categorize any national economy according to its stage. Stages are deterministic. The 5 stages are: traditional, transitional, take-off, drive to maturity, and high mass consumption.

Structural Adjustment Program – A series of requirements (adjustments) placed upon any country that accepts a loan from the IMF.

Sustainable development – A mode of development theory and strategy that considers and accounts for the impacts of economic growth upon society, culture, and environment.

Tariff – A tax placed upon imports.

Universalism – idea that phenomena, conceptual definitions or moral, aesthetic or epistemological truths hold for all times and places, transcending their immediate local circumstances. Its significance for development lies in the belief that human development is not for the few, not even for the most, but for everyone.

World Bank – a financial institution established near the end of WWII with the purpose of providing capital in the form of loans to developing countries and to those in need of reconstruction at the end of the war. The bank offers loans to countries for large-scale projects.

World Systems Theory – an approach to world history and social change that suggests there is a world economic system in which some countries benefit while others are exploited

WTO – World Trade Organization. An intergovernmental organization created in 1994 that promotes international trade between countries. It seeks to reduce trade restrictions, enforce existing agreements, and protect intellectual property.

9.8 WORKS CONSULTED AND FURTHER READING

Binns, Tony, et al. Geographies of Development: An Introduction to Development Studies, 3rd ed. Harlow: Pearson Education, 2008.

Bremmer, Ian. "The Mixed Fortunes in the BRICS countries in 5 Facts." *Time Magazine Online*, September 1, 2017. http://time.com/4923837/brics-summit-xiamen-mixed-fortunes.

Budkin, Yanina. "Understanding Poverty." *World Bank*, April 11, 2018. http://www.worldbank.org/en/topic/poverty/overview.

"Country Comparison of Real GDP (PPP)." In *CIA World Factbook*, 2017. https://www.cia.gov/library/publications/the-world-factbook/rankorder/2001rank.html.

Geary, Kate. "Learning Hard Lessons: AIIB and the Tarbela Dam in Pakistan." *The Wire*, May 9, 2017. https://thewire.in/133795/learning-hard-lessons-aiib-tarbela-dam-pakistan.

Gregory, Derek, ed. *The Dictionary of Human Geography*, 5th ed. Malden, MA: Blackwell, 2009.

Good, Keith. "House Ag Committee Examines NAFTA – Trade Issues." *Farm Policy News, July 27, 2017.* http://farmpolicynews.illinois.edu/2017/07/house-ag-committee-examines-nafta-trade-issues

"Human Development Reports Table 2: Trends in the Human Development Index, 1990-2015." United Nations Development Programme, accessed January 17, 2018, http://hdr.undp.org/en/composite/trends.

Kelly, Annie. "Gross national happiness in Bhutan: the big idea from a tiny state that could change the world." *The Guardian: US Edition*, Dec 1, 2012. https://www.theguardian.com/world/2012/dec/01/bhutan-wealth-happiness-counts.

Kiev, C.W. "What Dutch disease is, and why it's bad." *The Economis*, November 5, 2014. https://www.economist.com/blogs/economist-explains/2014/11/economist-explains-2.

"LDC's in Facts and Figures." United Nations, accessed December 1, 2017, http://unohrlls.org/about-ldcs/facts-and-figures-2.

Long, Heather. "Reality check: U.S. manufacturing jobs at 1940s levels." *CNN Money Online. U.S. Version*, April 7, 2017. http://money.cnn.com/2017/04/07/news/economy/us-manufacturing-jobs/index.html.

Lopez, Ditas and Karl M Yap. "Donald Trump Has Call Centers in the Philippines Worried." *Bloomberg News,* March 9, 2017. https://www.bloomberg.com/news/articles/2017-03-10/trump-risk-prompts-philippine-call-centers-to-seek-lobbyist.

McMichael, Philip. *Development and Social Change: A Global Perspective*, 3rd ed. Thousand Oaks, CA: Pine Forge Press, 2005.

Neate, Rupert. "Richest 1% own half the world's wealth, study finds." *The Guardian (US Edition)*, November 14, 2017. https://www.theguardian.com/inequality/2017/nov/14/worlds-richest-wealth-credit-suisse.

Peet, Robert, and Ellen Hartwick. *Theories of Development : Contentions, arguments, alternatives,* 2nd ed. New York: Guilford Press, 2009.

"Population Reference Bureau Data Finder." United Nations Population Reference Bureau, accessed February 11, 2018. http://www.prb.org/DataFinder.

Reuters Staff. "Mexican agency finds irregularities in Pemex, Odebrecht contract." *Reuters Online*, September 11, 2017. https://www.reuters.com/article/us-mexico-brazil-corruption/mexican-agency-finds-irregularities-in-pemex-odebrecht-contract-idUSKCN1BN08D

Simoes, Alexander. "South Korea." *The Observatory of Economic Complexity*, Accessed January 17, 2018. https://atlas.media.mit.edu/en/profile/country/kor/.

Stanton, Elizabeth. "The Human Development Index: A History." *Working Paper Series Number 127. Political Economy Research Institute*. University of Massachusetts Amherst. February 2007. https://scholarworks.umass.edu/cgi/viewcontent.cgi?article=1101&context=peri_workingpapers

Stephenson, Wesley. "Have 1,200 World Cup workers really died in Qatar?" *BBC News Magazine*, June 6, 2017. http://www.bbc.com/news/magazine-33019838.

Yueh, Linda. "Can de-industrialisation be reversed?" *BBC News*, February 16, 2015. http://www.bbc.com/news/business-31495851.

10

Agriculture and Food

Georgeta Connor

STUDENT LEARNING OUTCOMES

By the end of this section, the student will be able to:

1. Understand: the *origin* and *evolution* of agriculture across the globe

2. Explain: the environment-agriculture relationship, market forces, institutions, agricultural industrialization, and biorevolution versus sustainable agriculture

3. Describe: agricultural regions, comparing and contrasting subsistence and commercial agriculture

4. Connect: the factors of global changes in food production and consumption

CHAPTER OUTLINE

10.1 INTRODUCTION

Before the invention of agriculture, people obtained food from hunting wild animals, fishing, and gathering fruits, nuts and roots. Having to travel in small groups to obtain food, people led a nomadic existence. This remained the only mode of subsistence until the end of the Mesolithic period, some 12,000-10,000 years ago. Then, agriculture gradually replaced the **hunting and gathering** system, constituting the spread of the Neolithic revolution. Even today, some isolated groups survive as they did before agriculture developed. They can be found in some remote areas such as in Amazonia, Congo, Namibia, Botswana, Tanzania, New Guinea, and the Arctic latitude, where hunting dominates life (**Figures 10.1, 10.2,** and **10.3**).

The term **agriculture** refers to the *cultivation of crops* and the *raising of livestock* for both sustenance and economic gain. The origin of agriculture goes back to prehistoric time, starting when humans domesticated plants and animals. The domestication of plants and animals as the *origin of agriculture* was a pivotal transition in human history, which occurred several times independently. Agriculture originated and spread in different regions (*hearths*) of the world, including the Middle East, Southwest Asia, Mesoamerica and the Andes, Northeastern India, North China, and East Africa, beginning as early as 12,000 – 10,000 years ago. People became sedentary, living in their villages, where new types of social, cultural, political, and economic relationships were created. This period of momentous innovations is known as the *First Agricultural Revolution*.

Figure 10.1 | Hadza Hunters
Author | User "Idobi"
Source | Wikimedia Commons
License | CC BY SA 3.0

Figure 10.2 | Pumé Hunter and Gatherers
Author | User "Ajimai"
Source | Wikimedia Commons
License | CC BY SA 4.0

Figure 10.3 | Inuit Hunters
Author | User "Wiki-profile"
Source | Wikimedia Commons
License | CC BY SA 3.0

10.2 AGRICULTURAL PRACTICES

Agriculture is a science, a business, and an art (**Figures 10.4** and **10.5**). Spatially, agriculture is the world's most widely distributed industry. It occupies more area than all other industries combined, changing the surface of the Earth more than any other. Farming, with its multiple methods, has significantly transformed the landscape (small or large fields, terraces, polders, livestock grazing), being an important reflection of the two-way relationship between people and their environments. The world's agricultural societies today are very diverse and complex, with agricultural practices ranging from the most rudimentary, such as using the ox-pulled plow, to the most complex, such as using machines, tractors, satellite navigation, and genetic engineering methods. Customarily, scholars divide agricultural societies into categories such as *subsistence, intermediate,* and *developed*, words that express the same ideas as *primitive, traditional,* and *modern*, respectively. For the purpose of simplification, farming practices described in this chapter are classified into two categories, *subsistence* and *commercial*, with fundamental differences between their practice in developed and developing countries.

Figure 10.4 | Trading Floor at the Chicago Board of Trade
Author | Jeremy Kemp
Source | Wikimedia Commons
License | Public Domain

Figure 10.5 | Tulip Fields in The Netherlands
Author | Alf van Beem
Source | Wikimedia Commons
License | CC 0

10.2.1 Subsistence Agriculture

Subsistence agriculture replaced hunting and gathering in many parts of the globe. The term **subsistence**, when it relates to farming, refers to growing food only to sustain the farmers themselves and their families, consuming most of what they produce, without entering into the cash economy of the country. The farm size is small, 2-5 acres (1-2 hectares), but the agriculture is less mechanized; therefore, the percentage of workers engaged directly in farming is very high, reaching 50 percent or more in some developing countries (**Figure 10.6**). Climate regions play an important role in determining agricultural regions. Farming activities range from *shifting cultivation* to *pastoralism*, both extensive forms that still prevail over large regions, to *intensive subsistence*.

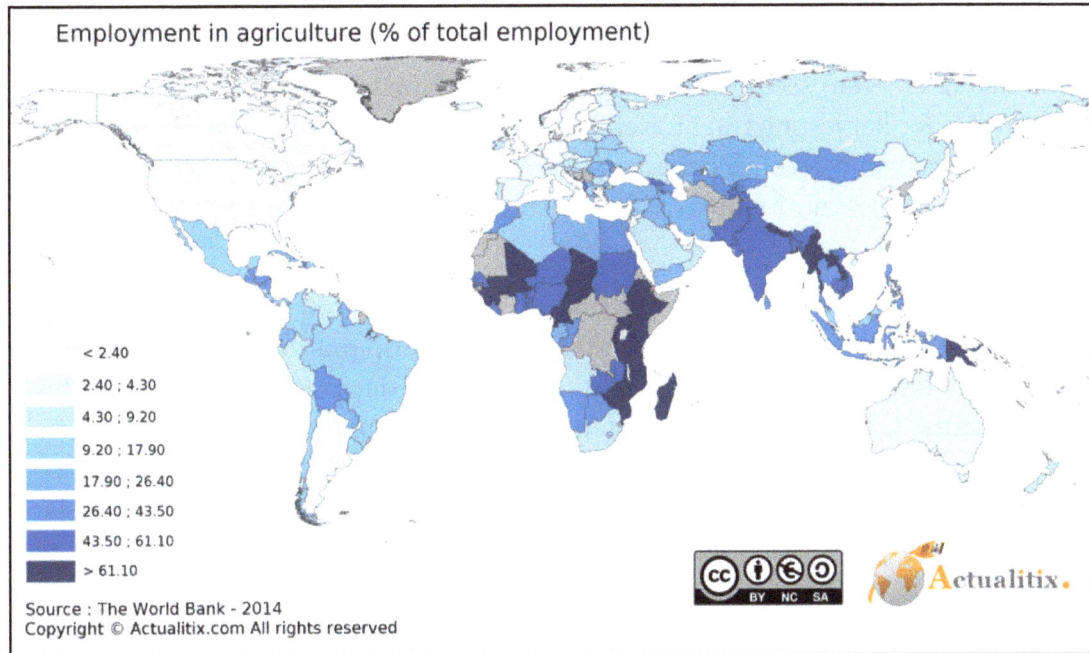

Employment in agriculture (% of total employment)

< 2.40
2.40 ; 4.30
4.30 ; 9.20
9.20 ; 17.90
17.90 ; 26.40
26.40 ; 43.50
43.50 ; 61.10
> 61.10

Source : The World Bank - 2014
Copyright © Actualitix.com All rights reserved

Figure 10.6 | Employment in Agriculture, 2014
Author | Actualitix
Source | Actualitix.com
License | CC BY NC SA 4.0

Shifting Cultivation

Shifting cultivation, also known as **slash-and-burn agriculture**, is a form of subsistence agriculture that involves a kind of natural rotation system. Shifting cultivation is a way of life for 150-200 million people, globally distributed in the tropical areas, especially in the rainforests of South America, Central and West Africa, and Southeast Asia. The practices involve removing dense vegetation, burning the debris, clearing the area, known as **swidden,** and preparing it for cultivation **(Figures 10.7)**. Shifting cultivation can successfully support only low population densities and, as a result of rapid depletion of soil fertility, the fields are actively cultivated usually for three years. As a result, the infertile land has to be abandoned and another site has to be identified, starting again the process of clearing and planting. The

Figure 10.7 | Slash-and-Burn Farming in Thailand
Author | User "mattmangum"
Source | Wikimedia Commons
License | CC BY SA 2.0

slash-and-burn technique thus requires extensive acreage for new lots, as well as a great deal of human labor, involving at the same time a frequent gender division of labor. The kinds of crops grown can be different from region to region, dominated by tubers, sweet potatoes especially, and grains such as rice and corn. The practice of mixing different seeds in the same swidden in the warm and humid tropics is favorable for harvesting two or even three times per year. Yet, the slash-and-burn practice has some negative impacts on the environment, being seen as ecologically destructive especially for areas with vulnerable and endangered species.

Pastoralism

Involving the breeding and herding of animals, **pastoralism** is another extensive form of subsistence agriculture. It is adapted to cold and/or dry climates of savannas (grasslands), deserts, steppes, high plateaus, and Arctic zones where planting crops is impracticable. Specifically, the practice is characteristic in Africa [north, central (Sahel) and south], the Middle East, central and southwest Asia, the Mediterranean basin, and Scandinavia. The species of animals vary with the region of the world including especially sheep, goats, cattle, reindeer, and camels. Pastoralism is a successful strategy to support a population on less productive land, and adapts well to the environment.

Three categories of pastoralism can be individualized: sedentary, nomadic, and transhumance.

Sedentary pastoralism refers to those farmers who live in their villages and their herd animals in nearby pastures. A number of men usually are hired by the villagers in order to take care of their animals. Equally important is the practice in which the hired men gather the animals (cattle especially) in the morning, feed them during the day in the nearby pasture, and then return them to the village early in the evening. This is the typical pattern for many traditional European pastoralists.

Nomadic pastoralism is a traditional form of subsistence agriculture in which the pastoralists travel with their herds over long distances and with no fixed pattern. This is a continuous movement of groups of herds and people such as the Bedouins of Saudi Arabia, the Bakhtiaris of Iran, the Berbers of North Africa, the Maasai of East Africa, the Zulus of South Africa, the Mongols of Central Asia, and other groups. The settlement landscape of pastoral nomads reflects their need for mobility and flexibility. Usually, they live in a type of tent (known as *yurt* in Central Asia) and move their herds to any available pasture (**Figure 10.8**). Although there are approximately 10-15 million nomadic pastoralists in the world, they occupy about 20 percent of Earth's land area. Today, their life is in decline, the victim of more constricting political borders, competing land uses, selective overgrazing, and government resettlement programs.

Transhumance is a seasonal vertical movement by herding the livestock (cows, sheep, goats, and horses) to cooler, greener high-country pastures in the summer and then returning them to lowland settings for fall and winter grazing.

Figure 10.8 | Mongolian Nomads Moving to Autumn Encampment
Author | User "Yaan"
Source | Wikimedia Commons
License | CC BY SA 3.0

Figure 10.9 | Transhumance in the Pyrenees Mountains
Author | User "Clicgauche"
Source | Wikimedia Commons
License | CC BY SA 1.0

Herders have a permanent home, typically in the valleys. Generally, the herds travel with a certain number of people necessary to tend them, while the main population stays at the base. This is a traditional practice in the Mediterranean and the Black Sea basins such as southern European countries, the Carpathian Mountains, and the Caucasus countries (**Figures 10.9** and **10.10**). In addition, near highland zones such as the Atlas Mountains (northwest Africa) and the Anatolian Plateau (Turkey), as well as in Sub-Saharan Africa, the Middle East countries, and Central Asia, the pastoralists have to practice another type of transhumance, such as the movement of animals between wet-season and dry-season pasture.

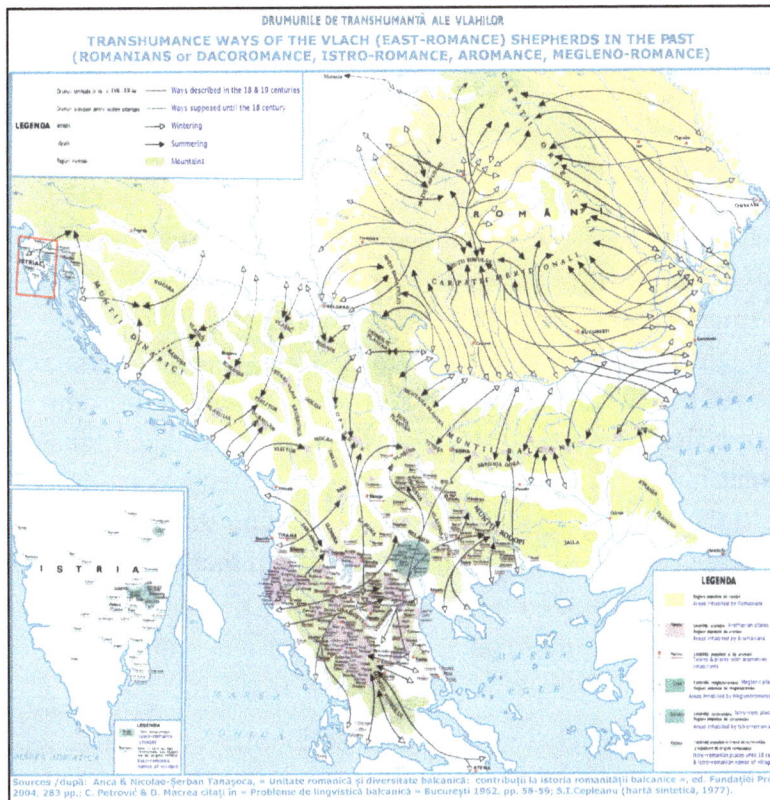

Figure 10.10 | Romanian and Vlachs Transhumance in Balkans
Author | User "Julieta39"
Source | Wikimedia Commons
License | CC BY SA 4.0

Intensive Subsistence Agriculture

Intensive subsistence agriculture, characteristic of densely populated regions especially in southern, southeastern, and eastern Asia, involves the effective and efficient use of small parcels of land in order to maximize crop yield per acre. The practice requires intensive human labor, with most of the work being done by hand and/or with animals. The landscape of intensive subsistence agriculture is significantly transformed, including hillside terraces and raised fields, adding the irrigation systems and fertilizers (**Figures 10.11** and **10.12**). As a result, intensive subsistence agriculture is able to support large rural populations. Rice is the dominant crop in the humid areas of southern, southeastern, and eastern Asia. In the drier areas, other crops are cultivated such as grains (wheat, corn, barley, millet, sorghum, and oats), as well as peanuts, soybeans, tubers, and vegetables. In both situations, the land is intensively used, and the milder climate of those regions allows double cropping (the fields are planted and harvested two times per year).

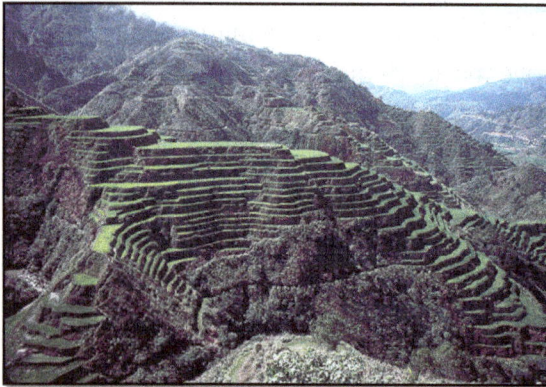

Figure 10.11 | Rice Terraces, the Philippines
Author | Susan McCouch
Source | Wikimedia Commons
License | CC BY 2.5

Figure 10.12 | Raised fields, Vietnam
Author | Dennis Jarvis
Source | Wikimedia Commons
License | CC BY SA 2.0

In recent decades, as the result of the introduction of higher-yielding grain varieties such as wheat, corn, and rice, known as **Green Revolution**, tens of millions of subsistence farmers have been lifted above the survival level. The spread of these new varieties throughout the farmlands of South, Southeast, and East Asia, and Mexico greatly improved the supply of food in these areas. Equally important was the use of fertilizers, pesticides, irrigation, and new machines. Today, China and India are self-sufficient in basic foods, while Thailand and Vietnam are two of the top rice exporters in the world. Although hunger and famine still persist in some regions of the world, especially in Africa, many people accept that they would be much worse without using these innovations.

10.2.2 Commercial Agriculture

Commercial agriculture, generally practiced in core countries outside the tropics, is developed primarily to generate products *for sale* to food processing companies. An exception is plantation farming, a form of commercial agriculture which persists in developing countries side by side with subsistence. Unlike the small subsistence farms (1-2 hectares/2-5 acres), the average of the commercial farm size is over 150 hectares/370 acres (178 ha/193 acres U.S.) and, being mechanized, many of them are family owned and operated. Mechanization also determines the percentage of the labor force in agriculture, with many developed countries being even below two percent of the total employment, such as Israel, the United Kingdom, Germany, the United States, Canada, Norway, Denmark, and Sweden (**Figure 10.6**). Moreover, as the result of industrialization and urbanization, many developed countries continue to lose significant areas of agricultural land. North America, for example, had 28.3 percent agricultural land out of the total land area in 1961 and 26 percent in 2014. The European Union decreased its agricultural land from 54.7 percent to 43.8 percent for the same period, during which some countries recorded outstanding decreases, such as Ireland from 81.9 to 64.8 percent, the United Kingdom from 81.8 to 71.2 percent, and Denmark from 74.6 to 62.2 percent to mention only a few. In addition to the high level of mechanization, in order to increase their productivity, commercial farmers use scientific advances in research and technology such as the Global Positioning System (autonomous precision seed-planting robot, intelligent systems for animal monitoring, savings in field vegetable-growing through the use of a GPS automatic steering system), and satellite imagery (finding efficient routes for selective harvesting based on remote sensing management).

Climate regions also play an important role in determining *agricultural regions*. In developed countries, these regions can be individualized as six types of commercial agriculture: mixed crop and livestock, grain farming, dairy farming, livestock ranching, commercial gardening and fruit farming, and Mediterranean agriculture

Mixed Crop and Livestock

Mixed crop and livestock farming extends over much of the eastern United States, central and western Europe, western Russia, Japan, and smaller areas in South America (Brazil and Uruguay) and South Africa. The rich soils, typically involving **crop rotation**, produce high yields primarily of corn and wheat, adding also soybeans, sugar beets, sunflower, potatoes, fruit orchards, and forage crops for livestock. In practice, there is a wide variation in mixed systems. At a higher level, a region can consist of individual specialized farms (corn, for example) and service systems that together act as a mixed system. Other forms of mixed farming include cultivation of different crops on the same field or several varieties of the same crop with different life cycles, using space more efficiently and spreading risks more uniformly. The same farm may grow cereal crops or orchards, for example, and keep cattle, sheep, pigs, or poultry (**Figure 10.13**).

Figure 10.13 | Crop-livestock Integration
Sheep grazing under tall-stemmed fruit trees (the Netherlands).
Author | FAO
Source | FAO
License | © FAO. Used with permission.

Grain Farming

Commercial grain farming is an extensive and mechanized form of agriculture. This is a development in the continental lands of the mid-latitudes (mostly between 30° and 55° North and South latitudes), in regions that are too dry for mixed crop and livestock farming. The major world regions of commercial grain farming are located in Eurasia (from Kiev, in Ukraine, along southern Russia, to Omsk in western Siberia and Kazakhstan) and North America (the Great Plains). In the southern hemisphere, Argentina, in South America, has a large region of commercial grain farming, and Australia has two such areas, one in the southwest and another in the southeast. Commercial grain farming is highly specialized and, generally, one single crop is grown. The most important crop grown is wheat (winter and spring), used to make flour (**Figure 10.14**). The wheat farms are very large, ranging from 240 to 16,000 hectares (593-40000 acres). The average size of a farm in the U.S. is about 1000 acres (405 hectares). In these areas land is cheap, making it possible for a farmer to own very large holdings.

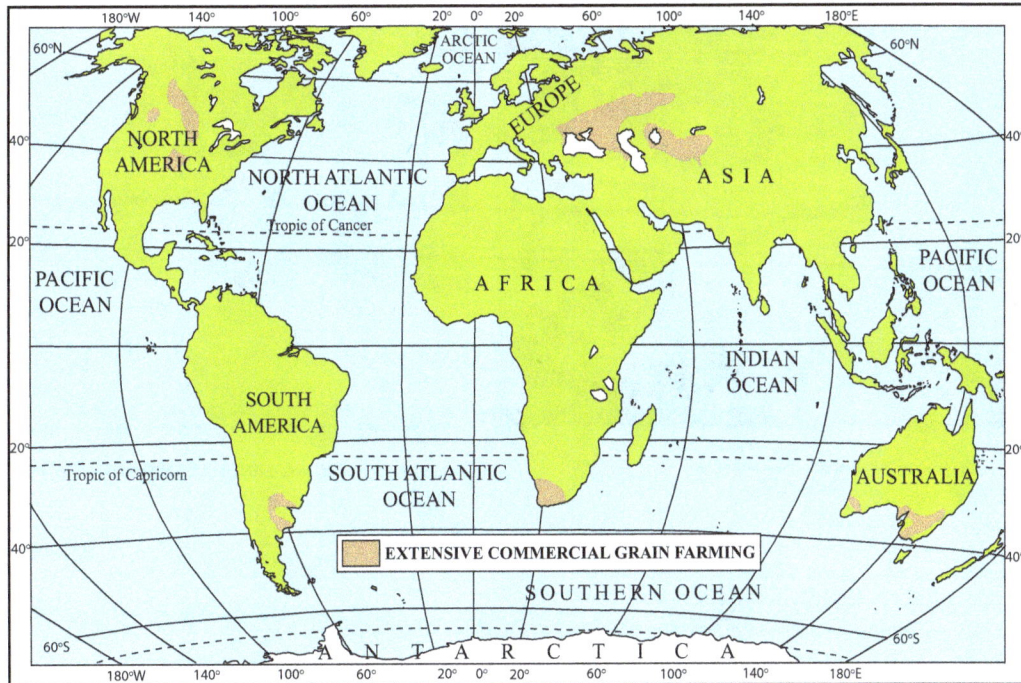

Figure 10.14 | World Commercial Grain Farming
Author | CIET NCERT
Source | NROER
License | CC BY SA 4.0

Dairy Farming

Dairy farming is a branch of agriculture designed for long-term production of milk, processed either on a farm or at a dairy plant, for sale. It is practiced near large urban areas in both developed and developing countries. The location of this type of farm is dictated by the highly perishable milk. The ring surrounding a city where fresh milk is economically viable, supplied without spoiling, is about a 100-mile radius. In the 1980s and 1990s, robotic milking systems were developed and introduced in some developing countries, principally in the EU (**Figure 10.15**). There is an important variation in the pattern of dairy production worldwide. Many countries that are large producers consume most of this internally, while others, in particular New Zealand, export a large percentage of their production, some from the organic farms (**Figure 10.16**).

Figure 10.15 | Rotary Milking Parlor
Author | Gunnar Richter
Source | Wikimedia Commons
License | CC BY SA 3.0

Figure 10.16 | Calves at Organic Dairy Farm
Author | Julia Rubinic
Source | Wikimedia Commons
License | CC BY 2.0

Livestock Ranching

Ranching is the commercial grazing of livestock on large tracts of land. It is an efficient way to raise livestock to provide meat, dairy products, and raw materials for fabrics. Contemporary ranching has become part of the meat-processing industry. Primarily, ranching is practiced on semiarid or arid land where the vegetation is too sparse and the soil too poor to support crops, being a vital part of economies and rural development around the world. In Australia, like in the Americas, ranching is a way of life (**Figure 10.17**). In the United States, near Greeley, Colorado, there is the world's largest cattle feedlot, with over 120,000 head, a subsidiary of the food giant ConAgra (**Figure 10.18**). The largest beef-producing company in the world is the Brazilian multinational corporation JBS-Friboi. Argentina and Uruguay are the world's top per capita consumers of beef. China is the leading producer of pig meat while the United States leads in the production of chicken and beef.

Figure 10.17 | Ranching
Author | William Henry Jackson
Source | Wikimedia Commons
License | Public Domain

Figure 10.18 | Feedlot in the Texas Panhandle
Author | User "H2O"
Source | Wikimedia Commons
License | CC BY SA 3.0

Commercial Gardening and Fruit Farming

A market garden is a relatively small-scale business, growing vegetables, fruits, and flowers (**Figure 10.19**). The farms are small, from under one acre to a few acres (.5-1.5 hectares). The diversity of crops is sometimes cultivated in greenhouses, distinguishing it from other types of farming. Commercial gardening and fruit farming is quite diverse, requiring more manual labor and gardening techniques. In the United States, commercial gardening and fruit farming is the predominant type of agriculture in the Southeast, the region with a warm and humid climate and a long growing season. In addition to the traditional vegetables and fruits (tomatoes, lettuce, onions, peaches, apples, cherries), a new kind of commercial gardening has developed in the Northeast. This is a non-traditional market garden, growing crops that, although limited, are increasingly demanded by consumers, such as

Figure 10.19 | Market Farming
A garden with edible plants for use in a culinary school in Lawrenceville, Georgia.
Author | U.S. Department of Agriculture
Source | Wikimedia Commons
License | Public Domain

asparagus, mushrooms, peppers, and strawberries. Market gardening has become an alternative business, significantly profitable and sustainable especially with the recent popularity of organic and local food.

Mediterranean Agriculture

The term 'Mediterranean agriculture' applies to the agriculture done in those regions which have a Mediterranean type of climate, hot and dry summers and moist and mild winters. Five major regions in the world have a Mediterranean type of agriculture, such as the lands that border the Mediterranean Sea (South Europe, North Africa, and the Middle East), California, central Chile, South Africa's Cape, and in parts of southwestern and southern Australia (**Figure 10.20**). Farming is intensive, highly specialized and varied in the kinds of crops raised. The hilly Mediterranean lands, also known as *'orchard lands of the world,'* are dominated by citrus fruits (oranges, lemons, and grapefruits), olives (primary for cooking oil), figs, dates, and grapes (primarily for wine), which are mainly for export. These and other commodities flow to distant markets, Mediterranean products tending to be popular and commanding high prices. Yet, the warm and sunny Mediterranean climate also allows a wide range of other food crops, such as cereals (wheat, especially) and vegetables, cultivated especially for domestic consumption.

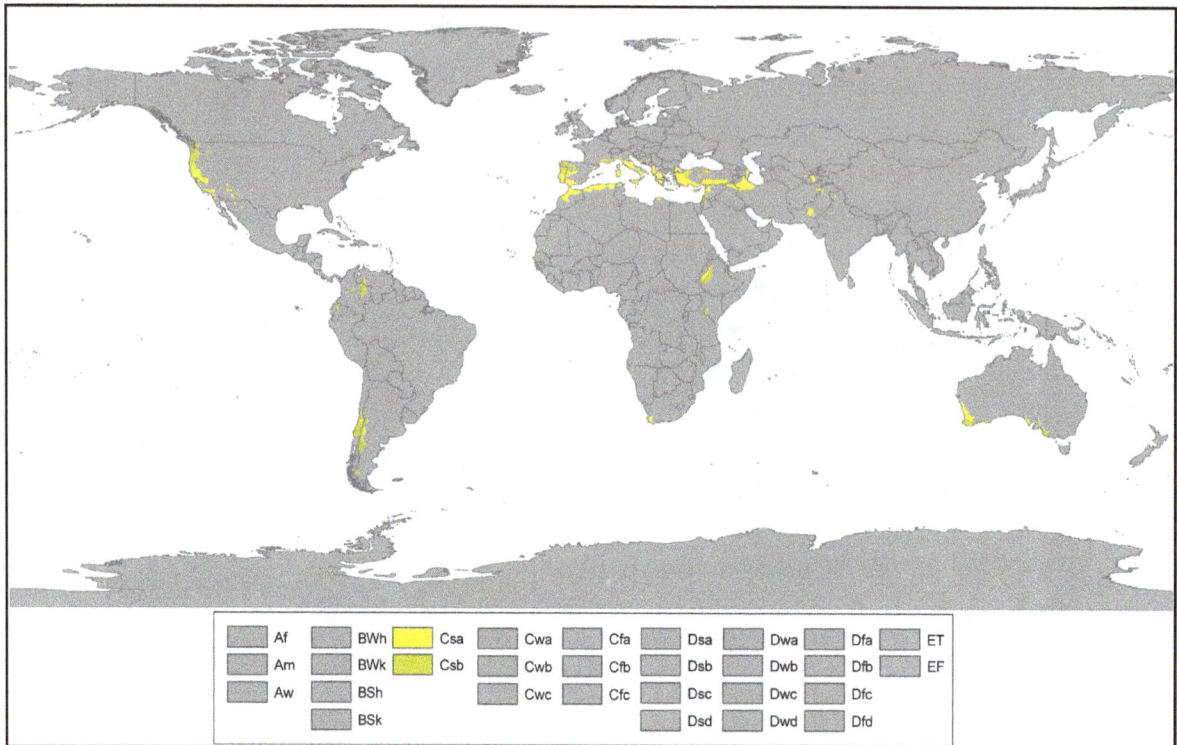

Figure 10.20 | Mediterranean Regions
Author | User "me ne frego"
Source | Wikimedia Commons
License | CC BY SA 4.0

Plantation Farming

Plantations are large landholdings in developing regions designed to produce crops for export. Usually, they specialize in the production of one particular crop for market laid out to produce coffee, cocoa, bananas, or sugar in South and Central America; cocoa, tea, rice, or rubber in West and East Africa; tea in South Asia; rubber in Southeast Asia, and/or other specialized and luxury crops such as palm oil, peanuts, cotton, and tobacco (**Figures 10.21** and **10.22**). Plantations are located in the tropical and subtropical regions of Asia, Africa, and Latin America and, although they are located in the developing countries, many are owned and operated by European or North American individuals or corporations. Even those taken by governments of the newly independent countries continued to be operated by foreigners in order to receive income from foreign sources. These plantations survived during decolonization, continuing to serve the rich markets of the world.

Unlike coffee, sugar, rice, cotton and other traditional crops, exported from large plantations, other crops can be required by the international market such as flowers and specific fruits and vegetables. These represent the **nontraditional agricultural exports,** which have become increasingly important in some countries or regions such as Argentina, Colombia, Chile, Mexico, and Central America, to mention a few. One important reason for sustaining nontraditional exports is that they complement the traditional exports, generating foreign exchange and employment. Thus, plantation agriculture, designed to produce crops for export, is critical to the economies of many developing countries.

Figure 10.21| Coffee Plantation
Author | User "Prince Tigereye"
Source | Wikimedia Commons
License | CC BY 2.0

Figure 10.22 | Tea Plantation
Author | User "Joydeep"
Source | Wikimedia Commons
License | CC BY SA 3.0

10.3 GLOBAL CHANGES IN FOOD PRODUCTION AND CONSUMPTION

10.3.1 Commercial Agriculture and Market Forces

Farming is part of **agribusiness** as a complex political and economic system that organizes food production from the development of seeds to the retailing and consumption of the agricultural product. Although farming is just one stage of the complex economic process, it is incorporated into the world economic system of capitalism (*globalized*). Most farms are owned by individual families, but, in this context, many other aspects of agribusiness are controlled by large corporations. Consequently, this type of farming responds to market forces rather than to feeding the farmer. Using **Von Thünen's isolate state model**, which generated four concentric rings of agricultural activity, geographers explain that the choice of crops on commercial farms is only worthwhile within certain distances from the city. The effect of distance

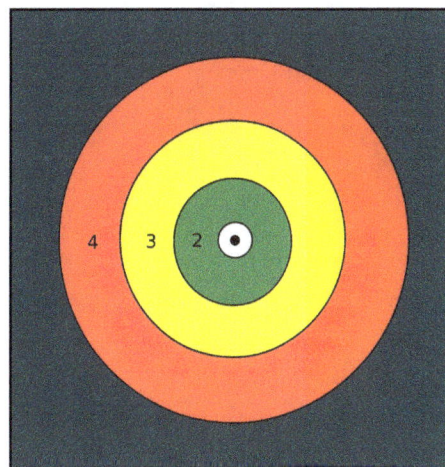

Figure 10.23 | Von Thünen Model
The dot represents a city. The white area around it (1) represents dairy and market gardening; 2) (green) the forest for fuel; 3) (yellow) field crops and grains; 4) (red) Ranching and livestock ;and the outer (dark green) region represents the wilderness where agriculture is not practised.
Author | Erin Silversmith
Source | Wikimedia Commons
License | Public Domain

determines that highly perishable products (milk, fresh fruits, and vegetables) need to be produced near the market, whereas grain farming and livestock ranching can be located on the peripheral rings (**Figure 10.23**).

New Zealand, for example, is a particular case of a country whose agriculture was thrown into a global free market. More specifically, its agriculture has changed in response to the restructuring of the global food system and, at the same time, is responding to a new global food regime. For New Zealand to remain competitive, farmers have to intensify production of high added value or more customized products, also focusing on *nontraditional exports* such as kiwi, Asian pears, vegetables, flowers, and venison (meat produced on deer farms) (**Figure 10.24**). The New Zealand agricultural sector is unique in being the only developed country to be totally exposed to the international markets since the government subsidies were removed.

Figure 10.24 | Deer Farm, New Zealand
Author | User "LBM1948"
Source | Wikimedia Commons
License | CC BY SA 4.0

10.3.2 Biotechnology and Agriculture

Since the 19th century, manipulation and management of biological organisms have been a key to the development of agriculture. In addition to Green Revolution, agriculture has also undergone a **Biorevolution**, involving agricultural **biotechnology** (*agritech*), an area of agricultural science involving the use of scientific tools and genetic engineering techniques to modify living organisms (or part of organisms) of plants and animals with the potential of outstripping the productivity increases of the Green Revolution and, at the same time, reducing agricultural production costs. Within the agricultural biotechnology process, desired traits are exported from a particular species of crop or animal to the different species obtaining *transgenic crops*, which possess desirable characteristics in terms of flavor, color of flowers, growth rate, size of harvested products, and resistance to diseases and pests (BT corn, for example, can produce its own pesticides).

By removing the genetic material from one organism and inserting it into the permanent genetic code of another, the biotech industry has created an astounding number of organisms that are not produced by nature. It has been estimated that upwards of 75 percent of processed foods on supermarket shelves – from soda to soup, crackers to condiments – contain genetically engineered ingredients. So far, little is known about the impacts of genetically modified (GM) foods on human health and the environment. Consequently, it is difficult to sort the benefits from the costs of their increasing incorporation into global food production. The United States is the leader not only for the number of the genetically engineered (GE) food crops but also for the largest areas planted with commercialized biotech crops. Many countries, in Europe, for example, consider that the genetic modification has not been proved safe, the reason for which they require all food to be labeled and refuse to import GM food. Yet, in the United States, genetic modification is permitted, taking into consideration that there is no evidence yet supporting that it is dangerous. Many people instead consider that they have the right to decide what they eat and, consequently, in their opinion, *labeling* of GM products must be mandatory. Protests against **GMO** regulatory structures have been very effective in many counties including the United States (**Figure 10.25**).

Figure 10.25 | Protest against GMOs
Author | Rosalee Yagihara
Source | Wikimedia Commons
License | CC BY SA 2.0

Currently, over 60 countries around the world require labeling of genetically modified foods, including the 28 nations in the European Union, Japan, Australia, Brazil, Russia, India, South Africa, China, and other countries (**Figure 10.26**). The debates regarding labeling certainly will continue. Since no one knows whether GM foods are entirely bad or entirely good, regulatory structures are crucial, protecting human health and the environment.

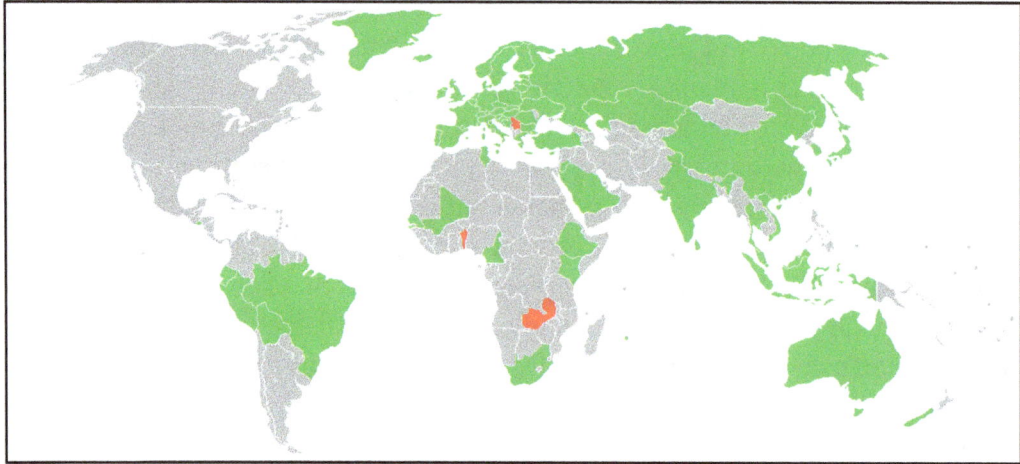

Figure 10.26 | GM Labeling Around the Globe
Author | User "Co9man"
Source | Wikimedia Commons
License | CC 0

10.3.3 Food and Health

Since the end of World War II, the world's technically and economically feasible food production potential has significantly expanded. As a result, today, there is more than enough food to feed all the people on the Earth (**Figure 10.27**) Yet, the major issue is the *access to food*, which is uneven, the reason for which millions of individuals in both the core and the periphery are affected by poverty, preventing them from securing adequate nutrition.

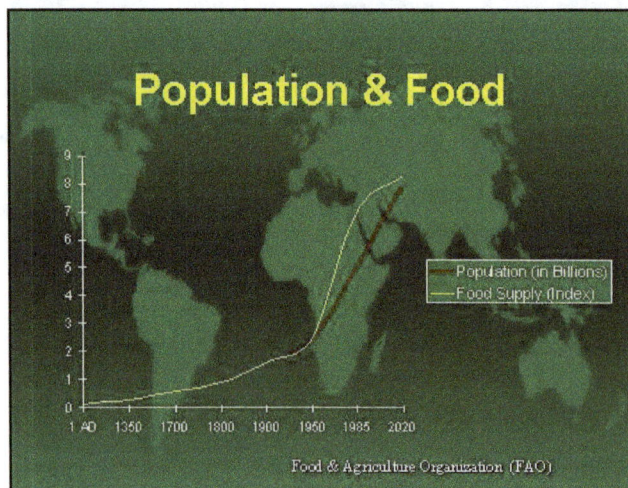

Figure 10.27 | Population & Food supply
Author | FAO
Source | Infogram
License | © FAO. Used with permission.

Hunger, *chronic* (long-term) or *acute* (short-term), therefore, is one of the most pressing issues facing the world today. Chronic hunger, also known as **undernutrition**, is an inadequate consumption of the necessary nutrients and/or calories. The Food and Agriculture Organization of the United Nations (FAO) considers necessary at least 1,800 kcal/day for an individual to consume in order to maintain a healthy life. The world average consumption is 2,780 kcal/day, but there is a significant difference between developed countries, with an average of 3,470 kcal/day (3,800 kcal/day in the U.S.), and developing countries, recording an average of 2,630 kcal/day (even less in sub-Saharan countries). FAO estimates that currently about 800 million people are undernourished globally, significantly less than in the early 1990s, but the majority continue to be counted in southern Asia and sub-Saharan Africa (**Figure 10.28**). One form of hunger is **famine**, an acute starvation caused even by a population's command over food resources, natural disasters (e.g., drought, Ethiopia in 1984-1985), or wars. In contrast, in North America, the United States especially, where the food is abundant and inspected for quality, overeating is a national problem, the reason for which the general condition of the population is reflected more by **obesity** (**Figure 10.29**).

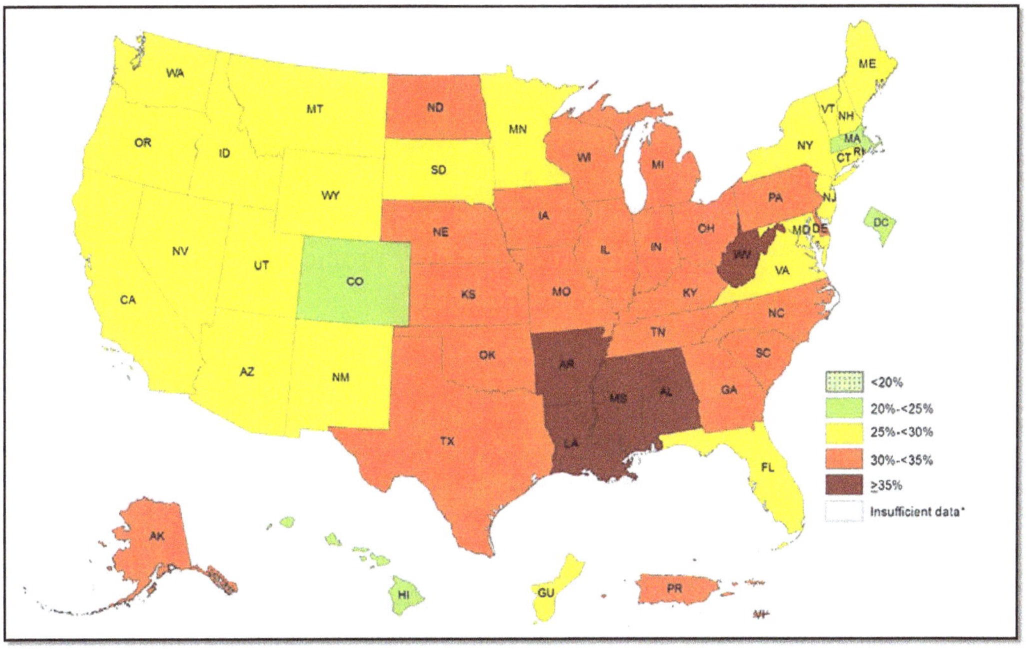

Figure 10.29 | U.S. Adult obesity rates, 2016
Author | CDC
Source | CDC
License | Public Domain

Nutritional vulnerability is conceptualized in terms of the notion of food security. According to FAO, **food security** exists when all people, at all times, have access to food for an active and healthy life. Related to food security is the concept of **food sovereignty**, which is the right of people, communities, and countries to define their own agricultural policies. One factor connected with food in general and food sovereignty especially is the fact that more cropland is redirected to raising **biofuels**, fuels derived from biological materials. They not only have a significant and increasing impact on global food systems but also result in evictions of small farmers and poor communities.

10.3.4 Sustainable Agriculture

Alongside the emergence of a core-oriented **food regime** especially of fresh fruits and vegetables, a new orientation in agriculture is sustainability. **According to the Sustainable Agriculture Initiative (SAI), "sustainable agriculture** *is the efficient production of safe, high quality agricultural products, in a way that protects and improves the natural* **environment**, *the* **social and economic conditions** *of farmers, their employees and local communities, and safeguards the* **health and welfare** *of all farmed species" (SAI Platform 2010-2018)*. More specifically, sustainability in agriculture is the increased commitment to **organic farming**, the principles and practices for sustainable agriculture developed by SAI being articulated around **three main pillars**: *society, economy,* and *environment*.

Although organic food production is not the primary mode of the agricultural practice, it has already become a growing force alongside the dominant conventional farming. Yet, unlike *conventional farming*, which promotes monoculture on large commercial farms and uses chemicals and intensive hormone-practices, *organic farming*, which puts small-scale farmers at the center of food production, does not use genetically modified seeds, synthetic pesticides, herbicides, or fertilizers. Thus, sustainable agricultural practices not only promote diversity and healthy food but also preserve and enhance environmental quality.

10.4 CONCLUSION

Agriculture was the key development in the rise of sedentary human civilization, domesticating species of plants and animals and creating food surpluses that nurtured the development of civilization. It began independently in different parts of the globe, both the Old and New World. Throughout history, agriculture played a dynamic role in expanding food supplies, creating employment, and providing a rapidly growing market for industrial products. Although subsistence, self-sufficient agriculture has largely disappeared in Europe and North America, it continues today in large parts of rural Africa, and parts of Asia and Latin America. While traditional forms of agricultural practices continue to exist, they are overshadowed by the global industrialization of agriculture, which has accelerated in the last few decades. Yet, commercial agriculture differs significantly from subsistence agriculture, as the main objective of commercial agriculture is achieving higher profits.

Farmers in both the core and the periphery have had to adjust to many changes that occurred at all levels, from the local to the global. Although states have become important players in the regulation and support of agriculture, at the global level, the World Trade Organization (WTO) has significant implications in agriculture. Social reactions to genetically engineered foods have repercussions throughout the world food system. Currently, the focus is especially on the option that a balanced, safe, and sustainable approach can be the solution not only to achieve sustainable intensification of crop productivity but also to protect the environment. Therefore, agriculture has become a highly complex, globally integrated system, and achieving the transformation to sustainable agriculture is a major challenge.

10.5 KEY TERMS DEFINED

agribusiness: commercial agriculture engaged in the production, processing, and distribution of food

agriculture: a science, art, and business directed to modify some specific portions of the Earth's surface through the cultivation of crops and the raising of livestock for sustenance and profit

biofuels: fuel derived from biological materials

biorevolution: the genetic engineering of plant and animals with the potential to exceed the production of the Green Revolution

biotechnology: the manipulation through genetic engineering of living organisms or their components to make or modify products or processes for specific use

commercial agriculture: a system in which farmers produce crops and animals primarily for sale

conventional farming: agriculture that uses chemicals (fertilizers, pesticides, and herbicides) and/or hormone-based practices

crop rotation: method in which the field under cultivation remain the same, but the crop is changed in order to avoid exhausting the soil

double cropping: method used in the milder climates in which intensive subsistence fields are planted and harvested twice per year

famine: extreme scarcity of food

food regime: specific set of links, indicating the ways a particular type of food is dominant during a specific time

food security: the situation when all people, at all times, have access to food for an active and healthy life

food sovereignty: the right of people, communities, and countries to define their own agricultural policies

globalized agriculture: agriculture increasingly influenced more at the global or regional levels than at national level

genetically modified organisms (GMOs): organisms that have their DNA modified in a laboratory

Green Revolution: a new agricultural technology characterized by high-yield seeds and fertilizers exported from the core to the periphery in order to increase their agricultural productivity

hunting and gathering: activities through which people obtain food from hunting wild animals, fishing, and gathering fruits, nuts and roots

intensive subsistence agriculture: a form of subsistence agriculture in which farmers involve the effective and efficient use of small parcels of land in order to maximize crop yield per hectare

nontraditional agricultural exports: new export crops that contrast with traditional exports

organic farming: a method of crop and livestock production without commercial fertilizers, pesticides, growth hormones, and genetically modified organisms

pastoralism: subsistence activity that involves the breeding and herding of animals to satisfy the human needs of food, shelter, and clothing

pastoral nomadism: a traditional form of subsistence agriculture in which the pastoralists travel with their herds over long distances and with no fixed pattern

plantation: large landholdings in developing regions specialized in the production of one or two crops usually for export to more developed countries

ranching: a form of commercial agriculture in which the livestock graze over an extensive area

shifting cultivation: a form of subsistence agriculture, which involves a kind of natural rotation system

slash-and-burn agriculture: a method for obtaining more agricultural land in which fields are cleared (*swidden*) by slashing the vegetation and burning the debris

subsistence agriculture: farming designed to grow food only to sustain farmers and their families, consuming most of that they produce without entering into cash economy of the country

sustainable agriculture: *the efficient production of safe, high quality agricultural products, in a way that protects and improves the natural* **environment**, *the* **social and economic conditions** *of farmers, and safeguards the* **health and welfare** *of all farmed species*

swidden: land that is cleared for planting using the slash-and-burn process

transhumance: a seasonal vertical movement by herding the livestock to cooler, greener high country pastures in the summer and returning them to lowland settings for fall and winter grazing

undernourishment/undernutrition: inadequate dietary consumption that is below the minimum requirement for maintaining a healthy life

10.6 WORKS CONSULTED AND FURTHER READING

Alexandratos, N. and Bruinsma, J. 2012. World agriculture towards 2030/2050. The 2012 revision. [On-line]. *www.fao.org/docrep/016e/ap106e.pdf.*

Bernstein, H. 2015. Food regimes and food regime analysis: A selective survey. [On-line]. *https://www.iss.nl/fileadmin/ASSETS/iss/Research_and_projects/Research_networks/LDPI/CMCP_1_Bernstein.pdf.*

____. *https://en.wikipedia.org/wiki/Agricultural_biotechnology.*

____. AGBIOS (Agriculture and Biotechnology Strategies). Database. [On-line]. *http://www.agbios.com.*

____. *https://www.google.com/agricultural hearths.*

____. 2008. Biofuel and food security. International Food Policy Research Institute (IFPRI). [On-line]. *https://www.ifpri.org/publication/biofuels-and-food-security.*

Brandt, K. 1967. Can food supply keep pace with population growth? Foundation for Economic Education (FEE). [On-line]. *https://fee.org/articles.*

Colman, D. and Nixon, F. 1986. Economics of change in less developed countries. 2nd Ed. Totowa, NJ: Barnes and Noble Books.

____. *www.abbreviations.com/commercial agriculture.*

____. *https://www.google.com/corn production.*

____. *https://en.wikipedia.org/wiki/Dairy_farming.*

De Blij, H. J. 1995. *The Earth: An introduction to its physical and human geography.* 4E. New York: John Wiley & Sons, Inc.

De Blij, H. J. and Muller, P. 2010. *Geography: Realms, regions, and concepts.* 14 E. New York: John Wiley & Sons, Inc.

Dickenson, J., Gould, B., Clarke, C., Marther, S., Siddle, D., Smith, C., and Thomas-Hope, E. 1996. *A geography of the Third World.* Second edition. New York: Routledge.

Domosh, M., Neumann, R., and Price, P. Contemporary Human Geography: Culture, Globalization, Landscape. New York: W. H. Freeman and Company.

____. *https://www.google.com/search?q=maps+world+agric+employment.*

____. *https://en.actualitix.com/country/wld/employment-in-agriculture.php/.*

Evenson, R. 2008. Environmental planning for a sustainable food supply. In *Toward a vision of land in 2015: International perspectives*, ed. G. C. Cornia and J. Riddell, 285-305. Cambridge, MA: Lincoln Institute of Land Policy.

____. *https://typesoffarming.wordpress.com/2016/01/21/what-are-the-main-types-of-farming.*

____. The state of food and agriculture: Climate change, agriculture, and food security. [On-line]. *http://www.fao.org/3/a-i6132e.pdf.*

____. The state of food insecurity in the world 2015. [On-line]. *http://www.fao.org/3/a-i4646e.pdf.*

____. Protecting our food, farm, and environment. Center of Food Safety. [On-line]. *http://www.centerforfoodsafety.org/*.

____. Food security statistics. [On-line]. *www.fao.org/economic/ess/ess-fs/en*.

Friedmann, H. Food regimes and their transformation. Food Systems Academy. [On-line]. *www.foodsystemsacademy.org.uk/audio/docs/HF1-Food-regime-Transcript.pdf*.

____. Tulip fields, Lisse, Netherlands. [On-line]. https://in.pinterest.com/pin/.

____. About genetically engineered food. Center for Food Safety. [On-line]. *www.centerforfoodsafety.org/issues/311/ge-food/about-ge-foods*.

____. Genetically modified foods. Learn.Genetics. Genetic Science Learning Center. [On-line]. *http://learn.genetics.utah.edu/content/science/gmfoods/*.

____. Genetically modified food. [On-line]. *https://en.wikipedia.org/wiki/Genetically_modified_food*.

____. GM food: Viewpoints. [On-line]. *http://www.pbs.org/wgbh/harvest/viewpoints*.

____. 2016. Global status of commercialized biotech/GM crops: Brief 52. ISAAA. [On-line]. *www.isaaa.org/resources/publications/Brief/52/doawnload/isaaa-brief-52-2016.pdf*.

____. How to avoid GMOs in your food. [On-line]. *https://www.foodandwaterwatch.org/live-healthy/how-avoid-gmos-your-food*.

____. GMO: Labeling around the world. [On-line]. *http://www.justlabelit.org/a-gmo-labeling-around-the-world/*.

____. Just label it: We have a right to know. [On-line]. *http://www.justlabelit.org/right-to-know-center/right-to-know/*.

Gimenez, E. H. and Shattuck, A. 2011. Food crises, food regimes and food movements: Rumblings of reform or tides of transformations? Taylor & Francis Online. *http://www.tandfonline.com/doi/abc/10.1080/03066150.2010.538578*

____. Commercial grain farming: Location and characteristics. [On-line]. *www.yourarticlelibrary.com/family/commercial-grain-farming-location-and-characteristics-with maps/25446*.

Gruber, Carl. Hunter-Gatherer Men Not the Selfless Providers We Thought. March 31, 2016. *http://www.australiangeographic.com.au/topics/history-culture/2016/03/hunter-gatherer-men-not-the-selfless-providers-we-thought/*.

____. *Harvest of Fear*. NOVA Frontline. DVD.

____. Herding: Pastoralism, mustering, droving. [On-line]. *https://www.nationalgeographic.org/encyclopedia/herding*.

____. *https://en.wikipedia.org/wiki/History-_of_agriculture*

Hueston, W. and McLeod, A. Overview of the global food system: Changes over time/space and lessons for future food safety. [On-line]. *https://www.ncbi.nlm.nih.gov/books/NBK114491/*.

____. Hunger. [On-line]. *http://www.fao.org/hunger.en/*.

____. Hunger Map. 2015. [On-line]. *http://www.fao.org/3/a-i4674e.pdf*.

____. *https://wikipedia.org/wiki/hunter-gatherer*.

____. Survival International. Accessed May 5, 2018. *https://www.survivalinternational. org/tribes/congobasintribes/*.

____. The Ifugao rice terraces. June 1, 2014. *https://arlenemay.wordpress.com/2014/06/01/ the-banaue-rice-terraces/*.

Knox, P. and Marston, S. 2013. *Human Geography: Places and regions in global context*. Sixth Edition. Upper Saddle River, NJ: Pearson.

____. ISAAA (International Service for the Acquisition of Agri-Biotech Applications). [On-line]. *http://www.isaaa.org*.

Lernoud, J. and Willer, H. 2017. Organic agriculture worldwide: Key results from the FIBL survey on organic agriculture worldwide. Research Institute of Organic Agriculture (FIBL), Frick, Switzerland. [On-line]. *http://orgprints.org/3424/7/fibl-2017- global-data-2015.pdf*.

Magnan, A. 2012. Food regime. In The Oxford handbook of food history, ed. J. M. Pilcher. Oxford Handbooks Online. *www.oxfordhandbooks.com/view/10.1093/ oxfordhb/9780199729937.001.0001/oxf*.

____. *https://en.wikipedia.org/wiki/Market_garden*.

____. *www.yourarticlelibrary.com/agriculture/mediterranean-agriculture-location-and- characteristics-with-diagrams/25443*.

____. *https://en.wikipedia.org/wiki/Mixed-farming*.

____. Characterization of mixed farms. [On-line]. *www.fao.org/docrep/004/Y0501E/ y0501e03.htm*.

Nap, J.-P., Metz, P., Escaler, M., and Conner, A. 2003. The release of genetically modified crops into the environment. [On-line]. *http://onlinelibrary.wiley.com/ doi/10.1046/j.0960-7412.2003.01602.x/full*.

____. *https://en.wikipedia.org/wiki/Agriculture_in_New_Zealand*.

____. *https://en.wikipedia.org/wiki/wiki/Economy_of_New_Zealand*.

____. Interactive map tracks obesity in the United States: 2013 U.S. adult obesity rate. [On-line]. *https://www.sciencenews.org/article/interactive-map-tracks-obesity-united- states*.

____. Organic area. [On-line]. *http://faostat.fao.org/static/syb/syb_5000.pdf*.

____. Organic farming by country. [On-line]. *http://en.wikipedia.org/wiki/Organic_ farming_by_country*.

____. Organic farming statistics. [On-line]. *http://www.fibl.en/themes/organic-farming- statistics.html*.

____. Organic universe. Down to Earth. Fortnightly of politics of development, environment and health [On-line]. *www.downtoearth.org.in/coverage/organic-universe-38665.*

____. Organic world: Global organic farming statistics and news. 2015. [On-line]. *http://www.organic-world.net/statistics/statistics-data-tables.html.*

____. *https://en.wikipedia.org/wiki/Pastoralism.*

____. *https://www.videoblocks.com/video/cows-sheep-and-goats-in-front-of-a-yurt-ger-from-a-mongolian-nomads-family-xlzkuzf/.*

Pulsipher, L. M. 2000. *World Regional Geography.* New York: W. H. Freeman and Company

____. Rancing. [On-line]. *https://www.nationalgeographic.org/encyclopedia/rancing.*

____. *https://en.wikipedia.org/wiki/Ranch.*

Roser, M. and Ritchie, H. Land use in agriculture. [On-line]. *https://ourworldindata/land-use-in-agriculture.*

Rowntree, L., Lewis, M., Price, M., and Wyckoff, W. 2006. *Diversity amid globalization: World regions, environment, development.* 3rd Ed. Upper Saddle River, NJ: Pearson.

Rubenstein, J. 2013. *Contemporary Human Geography.* 2e. Upper Saddle River, NJ: Pearson.

Rubenstein, J. 2016. *Contemporary Human Geography.* 3e. Upper Saddle River, NJ: Pearson.

____. Slash-and-burn farming in Congo. [On-line]. *https://www.youtube.com/channel/UCxM5DOEmlPN4rd9_Egj12pA/.*

____. Intensive subsistence agriculture. [On-line]. *www.yourarticlelibrary.com/agriculture-intensive-subsistence-agriculture/44620.*

____. *http://en.wikipedia.org/wiki/Subsistence_agriculture.*

____. *www.anthro.palomar.edu/subsistence/sub_3.htm.*

____. Types of subsistence farming: Primitive and intensive subsistence farming. [On-line]. *www.yourarticlelibrary.com/fg/types-of-subsistence-farming-primitive-and-intensive-subsistence-farming/25457.*

____. Sustainable agriculture initiative platform: The global food value chain initiative for sustainable agriculture. [On-line]. *http://www.saiplatform.org/sustainable-agriculture/definition.*

Than, Ker. Climate change linked to waterborne diseases in Inuit communities. April 7, 2012. *https://news.nationalgeographic.com/news/2012/04/120405-climate-change-waterborne-diseases-inuit/.*

____. Chicago Board of Trade. [On-line]. *https://en.wikipedia.org/wiki/Chicago_Board_of_Trade/.*

____. Transhumance. [On-line]. *https://www.britannica.com/topic/transhumance.*

____. *www.medconsortium.org/projects/tranhumance.*

____. *https://en.wikipedia.org/wiki/Transhumance.*

True, J. 2005. Country before money? Economic globalization and national identity in New Zealand. In Economic nationalism in a globalizing world, ed. E. Helleiner and A. Pickel, Chapter 9. Ithaca: Cornell University Press. [On-line]. *https://books. google.com/*

Tyson, P. Should we grow GM crops? [On-line]. *http://www.pbs.org/wgbh/harvest/ exist/www.fao.org/fileadmin/templates/FCIT/PDF/UPA-WBpaper-Final_ October_2008.pdf.*

____. *https://en.wikipedia.org/wiki/Johann_Heinrich_von_Thunen.*

____. *https://www.google.com.wheat production.*

World Bank. World Development Indicators: Agricultural inputs. [On-line]. *wdi. worldbank.org/table/3.2.*

World Bank. 2015. Data Bank: World Development Indicators. Agricultural land. [On-line]. *www://data.worldbank.org/indicator/AG.LND.AGRI.ZS?view=chart.*

World Bank. 2016. Data Bank: World Development Indicators. Employment in agriculture.

[On-line]. *www://databank.worldbank.org/data/reports.aspx?source=SL.AGR. EMPL.ZS&country.*

11

Industry
David Dorrell and Todd Lindley

Figure 11.1 | Factory in Katerini, Greece
Author | Jason Blackeye
Source | Wikimedia Commons
License | CC 0

STUDENT LEARNING OUTCOMES

By the end of this section, the student will be able to:

1. Understand: the origins and diffusion of industrial production

2. Explain: the impact of industry on places

3. Describe: the industrial basis of modern cultures

4. Connect: industrialization, technology, the service sector and globalization

CHAPTER OUTLINE

11.1 INTRODUCTION

We live in a globalized world. Products are designed in one place, assembled in another from parts produced in multiple other places. These products are marketed nearly everywhere. Until a few decades ago, such a process would have been impossible. Two hundred years ago, such an idea would have been beyond comprehension. What happened to change the world in such a way. What eventually tied all the economies of the world into a global economy? Industry did. The Industrial Revolution changed the world as much as the Agricultural Revolution. Industry has made the modern lifestyle possible.

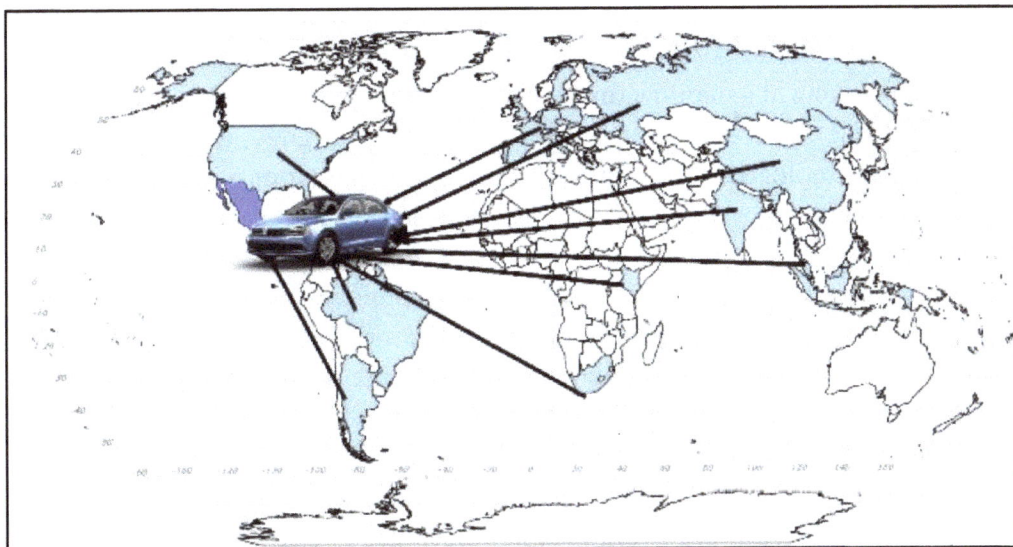

Figure 11.2 | The Volkswagen Jetta
The Volkswagen Jetta is designed in Germany and assembled in Mexico from parts from the represented countries. It is marketed as "German Engineered." Does it matter if the car wasn't made in Germany from German parts?
Author | David Dorrell
Source | Original Work
License | CC BY SA 4.0

During the 2016 U.S. presidential election, something very peculiar happened. Candidates from both major parties (Donald Trump and Hillary Clinton) agreed over and over again on one thing (and only one thing). The U.S. needed to create and/or bring back manufacturing jobs. Both candidates promised, if elected, to create new, well-paid manufacturing jobs. This was an odd shift, because for the previous 40+ years Republicans typically embraced free trade that allows manufacturers to choose where and what to produce (and many chose to move operations outside of the U.S.), while Democrats claimed to be working for the interests of blue-collar, working class people, whose jobs and wages had diminished since the 1980s period of **de-industrialization** both in the U.S. and throughout the developed/industrialized world. In the U.S. manufacturing provided jobs to 13 million workers in 1950, rising to 20 million in 1980 but by 2017 that number was back to 12 million – similar to levels last seen in 1941. A similar story can be found in Great Britain where jobs in manufacturing in 2017 were half of what they were in 1978 and output that once was 30% of GDP accounts for only 10% in 2017. Similar stories can be found in Germany, Japan, and other 'industrialized' economies. You may ask yourself, "Where did all of those jobs go?" But if you think about it, you can probably come up with your own answers.

It's important to note that even as jobs declined, manufacturing output in most industrialized countries continued to increase, so fewer people were producing more things. The first and simplest explanation for this is automation. For years, science fiction writers have warned us that the robots are coming. In the case of manufacturing technology...they're already here! Workers today are aided by software, robots, and sophisticated tools that have simply replaced millions of workers. Working at a manufacturing facility is no longer simply a labor-intensive effort, but one that requires extensive training, knowledge, and willingness to learn new technologies all the time. The second explanation is the relocation of manufacturing from wealthy countries to poorer ones because of lower wages in the latter.

As discussed earlier in this chapter, most countries have moved away from the protectionist model of development, allowing corporations to choose for themselves the location of production. No better example of this shift can be found than Walmart, which in the 1970's advertised that the majority of all of the products it sold were made in the USA. Thirty years later, it would be difficult for to find ANY manufactured product that was still 'made in the USA'. A third reason for the decline in manufacturing jobs is a decrease in demand for certain types of items. Steel production in the U.S. and England dropped precipitously during the period of deindustrialization (since the 1980's) not just due to automation or cheaper wages elsewhere, but also because demand for steel also declined. During the 20th century the U.S., Europe, and Japan required enormous amounts of steel in the construction of bridges, dams, railroad, skyscrapers, and even automobiles. Building of such items in the 21st century has slowed down, not because those countries are in decline, but because there is a limit as to how many bridges and skyscrapers are needed in any country! Demand for steel in a country diminishes as GDP per capita reaches about $20,000. Meanwhile, demand for steel will continue to rise in Japan and India for several years as they (and other industrializing countries) continue to expand cities, rail lines, and other large-scale construction projects. Such a decline in steel production does NOT mean that a country is in decline, but rather that there has been a shift in the type of manufacturing that occurs. The U.S., Germany, and Japan all continue to increase manufacturing output (**Figure 11.3**), even as their share of global output continues to decline.

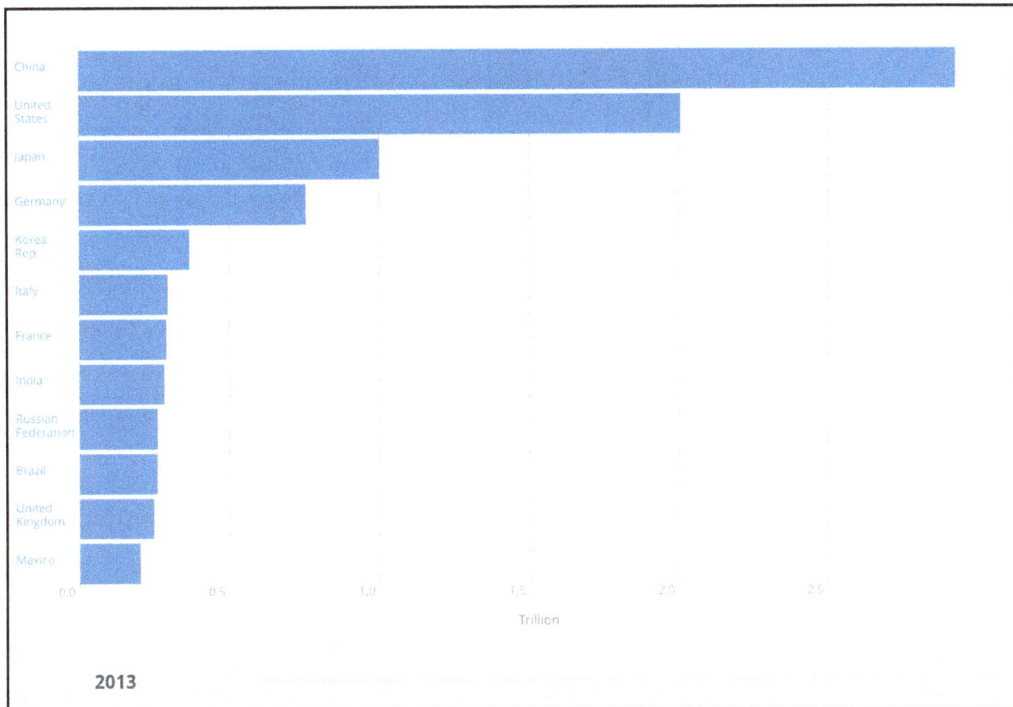

Figure 11.3 | Leading Countries in Manufacturing Output, 2013
Author | The World Bank
Source | The World Bank
License | CC BY 4.0

Another significant shift in manufacturing relates specifically to the geography of production and is best understood in the consideration of 2 different modes of production: 1) **Fordism** 2) **Post-Fordism**. Fordism is associated with the assembly line style of production credited to Henry Ford, who dramatically improved efficiency by instituting assembly line techniques to specialize/simplify jobs, standardize parts, reduce production errors, and keep wages high. Those techniques drove massive growth in manufacturing output throughout most of the 20th century and brought the cost of goods down to levels affordable by the masses. Nearly all of the automobile assembly plants located in and around the Great Lakes region of North America adopted the same strategies, which also provided healthy amounts of competition and new innovation for decades, as North America became the world's leading producer of automobiles. Post-Fordism begins to take hold in the 1980's as a new, global mode of production that seeks to relocate various components of production across multiple places, regions, and countries. Under Fordism, the entire unit would be produced locally, while Post-Fordism seeks the lowest cost location for every different component, no matter where that might be. Consider an optical, wireless mouse for a moment. The optical component may come from Korea, the rubber cord from Thailand, the plastic from Taiwan, and the patent from the U.S. Meanwhile, all of those items are most likely transported to China, where low-wage workers manually assemble the finished project and an automated packing system boxes and wraps it for shipping to all corners of the world. Global trade has been occurring for hundreds of years, dating back to the days of the Silk Road, Marco Polo, and the Dutch East India Company, but Post-Fordism, in which a single item is comprised of multiple layers of manufacturing from multiple places around the world, is a very recent innovation. The system has reconfigured the globe, such that manufacturers are constantly searching for new locations of cheap production. Consumers tend to benefit greatly from the system in that even poor middle school students in the U.S. can somehow afford to own a pocket computer (smart phone) that is more powerful than the most advanced computer system in the world from the previous generation. This is kind of a miracle. On the other hand, manufacturing jobs that once were a pathway to upward economic mobility, no longer assure people of such a decent standard of living as they once did.

11.1.1 History of Industrialization

Industrialization was not a process that emerged, fully-formed in England in the eighteenth century. It was the result of centuries of incremental developments that were assembled and deployed in the 18th century. Early industrialization involved using water power to run giant looms that produced cloth at a very low cost. This early manufacturing didn't use coal and belch smoke into the sky, but it initiated an industrial mindset. Costs could be reduced by relying on inanimate power (first water, then steam, then electricity), converting production to simple steps that cheap low-skilled laborers could do (**Taylorism**), getting larger and concentrated in an area (economy of scale), and cranking out large numbers of the same thing (**Fordism**). This is industry in a nutshell. The advantage of industry was that a company could sell a cheaper product, but at a greater profit.

As this mindset was applied to other goods, and then services, the world was changed forever. Places which had been producing goods for millennia suddenly (really suddenly) found themselves competing with a product that was far cheaper. Economist Joseph Schumpeter coined the term "creative destruction" to describe the process in which new industries destroy old ones. Hand production of goods for the masses began to decline precipitously. They quickly became too expensive in comparison to manufactured goods. Today, hand produced goods are often reserved for the wealthy.

A contemporary example of the industrial mode of production is fast food. Looking inside the kitchen of a fast food restaurant will reveal industrially prepared ingredients prepared just in time for sale to a customer. It is not the same process that you would use at home.

In the abstract, companies do not exist to provide jobs or even to make things. Companies exist to produce a profit. If changing the method of making a profit is necessary, then the company will do that in order to survive. If it cannot, then it will go away. For example, many companies today are highly diversified, for example Mitsubishi produces such unrelated products as cars and tuna fish. What is the connection between the two. They both produce profit.

11.1.2 Industrial Geography

How is industry related to geography? For one thing, industrial societies have more goods in them. Since the goods are cheaper, people just have more things. For another, the means of production, the factories, shipping terminals, and distribution centers are visible for anyone to see. The lifestyles of industrialized people are different. Pre-industrialized societies are not regulated by clocks, for example, people wear lass-produced similar clothing. They listen to globally marketed music. If it seems like you already read this in the chapter on pop culture, you have. Pop culture is a function of industry. Geography is concerned with places and industry changed the way the world operates. It changed the relationships between places. Places that industrialized early gained the ability to economically and politically dominate other parts of the world that had not industrialized. Something as simple as having access to cheap, mass produced guns had impacts far beyond mere trade relationships.

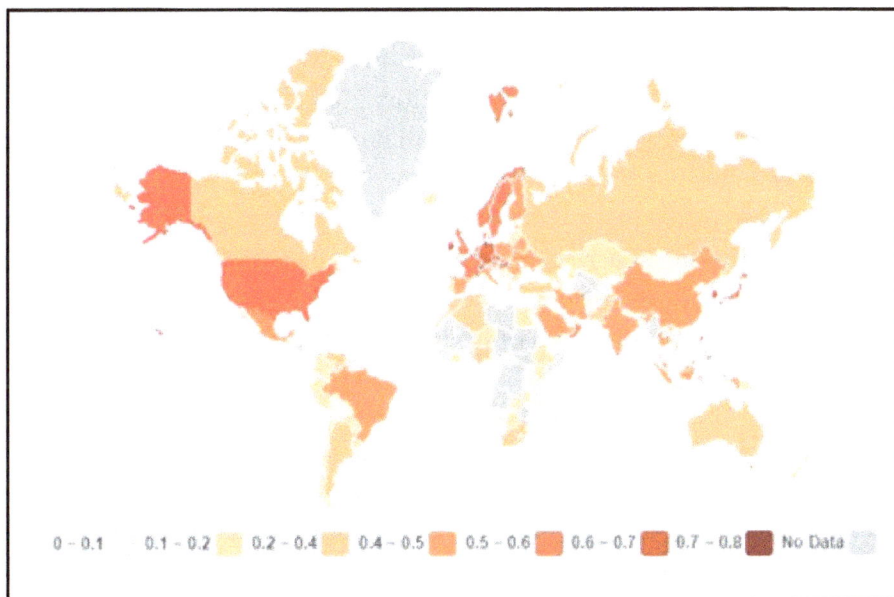

Figure 11.4 | Map of Medium- And High-Tech Manufacturing Value Added Share in Total Manufacturing[1]
Author | David Dorrell
Source | Original Work
License | CC BY SA 4.0

Over time industrial production changed from one that disrupted local economies to one that completely changed the relationship that most human beings had with their material culture, their environment, and with one another. Industry has improved standards of living and increased food production on one hand and it has despoiled environments and promoted massive inequality on the other.

Industrialization is about applying rational thought to production of goods. Specifically, this form of rational thought refers to discovering ways to reduce unnecessary labor, materials, capital (money), and time. In the same way that factories changed how things were produced, it also changed where things were produced. Locational criteria are used to determine where a factory even gets built.

11.2 MARX'S TENDENCY OF THE RATE OF PROFIT TO FALL

Karl Marx spent his working life trying to understand the nature of production. One of the things he noticed was that products have lifespans. When they are invented, they are new on the market and the producer has a monopoly. As soon as a product is released, competitors will quickly begin to provide alternative products at a lower price. The race begins to produce the product at an ever cheaper price, but also with an acceptable level of profit. This process occurs across both time and space as any mechanism possible to reduce the cost to manufacture the product is discovered and used.

Advancements in materials will often occur. Instead of using a metal case, plastic may be good enough. Capital infusion may allow automation of the production. Eventually, after every other possibility to reduce costs is exhausted, the only way to maintain production is to cut labor costs. Few workers will accept a dramatic reduction in pay. It's time to move the factory to a place with lower labor costs. Marx called this footloose capitalism. We call it offshoring. It's the same thing and it has always been part of capitalism. One aspect of this is that we who have grown up in a capitalistic world just naturally expect the price of goods to fall over time. In the United States we experienced a shift in manufacturing from the Northeast to the Midwest, then to the South and West. People in the United States have long moved to follow employment, and only stopped doing so when it left the confines of the country.

11.3 FACTORS FOR LOCATION

Industrial location is a balance between capital, material, and labor and markets. The goal is overall lowest cost. Sometimes pushing down one category, like labor, can increase other costs, like transportation. Substitution is possible across categories. For example, additional capital can replace labor through automation.

Earlier factories were built in cities in order to use the labor that was available there. Building in the middle of nowhere could have created an immediate labor shortage. Of course, labor will also migrate to places with available employment.

Some industrial activities are determined by site, the physical characteristic of a location. If you want to have a coal mine, it is probably a good idea to locate in a place that has coal. This is the most extreme form of restriction, but bear in mind that many places have mineral wealth, but not all of them are extracting that wealth right now. Many places that would otherwise be candidates for resource extraction are not currently being used because the resources cannot be extracted and sold for a profit. If the resource cannot make money, it will not be used. Remember that government subsidies can make some resources more profitable than they normally would be.

Other industrial locations are products of their situation. The maquiladoras that line the southern side of the Rio Grande (Rio Bravo in Mexico) border would not be there if the United States were not on the other side of the river. The proximity to the United States is the determining characteristic in the choice to locate there. Industries moved to Mexico because it appeared to make economic sense to do so. They were able to lower labor costs while remaining within the US/Canadian market. Transportations costs increased initially, but transportation costs overall have fallen due to more efficient methods of moving goods. Were the North American Free Trade Agreement (NAFTA) to be revoked the situation of Mexico would change, their access to the markets of the United States and Canada would diminish, and the factories that are there would have a much harder time selling their goods.

11.3.1 Land (or Materials/Energy) Costs

Classic economics lumped raw materials and energy under the category of land. Few companies bother to buy the land that produces a particular material, nor do they generally invest in power generation, or their own oil fields. These inputs are subject to the same price pressures as everything other part of the manufacturing process.

The land that a factory sits on has become less important over time. In earlier factories, most workers walked to work and factories were compelled to locate on expensive land in cities. That is no longer the case. Due to the widespread diffusion of the automobile, factories can be built in more suburban or even rural areas and workers will commute to the factory. This shifts part of the cost of securing labor onto the workers themselves. It lowers costs for the company. It is still necessary for the company to have access to the transport network, so new factories are usually built to take advantage of existing road networks, often next to interstates.

Sourcing materials has become global. Commodified inputs like steel are bought wherever they are relatively cheap and are shipped to where they are needed. Whichever source provider has the lowest cost, including shipping, will likely be chosen by the company. If you have ever bough something online, generally you want to know which company has the product with the lowest overall price. Companies operate the same way.

There are differences in energy costs across space. For energy intensive industrial activities, cheap energy is essential. China rose from producing very little steel a few decades ago to leading the world today. It this by leveraging low labor costs and a very cheap energy source- vast supplies of (highly-polluting) coal.

11.3.2 Labor

Labor costs can be reduced in a number of ways. One way is simply to pay the workers that you have less money. Workers, however do not like having their pay cut. Sending the work to a place with lower wages has been a common response to higher wages. A large portion of industrial labor requires minimal education. A company can choose to hire high school graduates in wealthy countries for a high wage or equivalently educated workers in poorer countries for far less. The labor pool has thus become globalized. Instead of competing for jobs with the other workers in a particular city, today's workers are competing for jobs with much of the human population. Labor-intensive industries are particularly sensitive to labor costs. Clothing production has shifted to places with cheap labor as wages in developed countries have increased.

Besides the option of simply shipping jobs offshore, there is also the option of replacing workers outright. Most of the industrial jobs that have been lost in the United States in the last 30 years have not been to overseas production, they have been due to automation. Replacing people with machines is as old as industry itself. The difference now is that machines are much more capable than they were in the past and they are much cheaper. Referencing the earlier example of clothing manufacture, production of athletic shirts in the United States got a large boost with the introduction of a factory in Little Rock, Arkansas that relies on a robot called Sewbot. The factory will produce millions of shirts at very low costs ($.33 in labor per shirt). This low cost of labor renders in nearly negligible in importance compared to other factors, such as transporting the materials to the factory and transporting the finished clothes to market. The factory was built by a Chinese company in the center of the United States to minimize transport costs. More factories will likely migrate nearer their markets as the cost of labor becomes less important. Developed countries are already seeing an expansion of manufacturing, but not an expansion of manufacturing employment, due to the influence of automation.

11.3.3 Transportation

An important contributor to geographic thinking regarding industrial locations was Alfred Weber (1868-1958). Weber took the concept of using lowest overall cost for the locations of industry and expanded it. He developed models that held many inputs to manufacturing constant in order to demonstrate the importance of transportation in determining least cost.

Weber believed that transportation costs were the most relevant factor in determining the location of an industry. The best way to minimize the cost of transport depends on the raw materials being transported. There are two kinds of raw material, things that are more-or-less everywhere (ubiquitous) and local raw materials. For the ubiquitous material (e.g. water), you have more freedom to locate, because it's commonly available. In this case, you should build your facility near your market, then you won't really need to transport much. We see examples of this in the locations of breweries and soft-drink manufacture.

On the other hand, if the material is only in a particular place (like bauxite), then you have some calculating to do. First you need to figure out if your manufacturing condenses, distills or otherwise shrinks your material. We call these processes bulk-reducing. If you aren't doing that, but instead took small pieces and assembled them into something that was harder to transport (like a tractor) then it is called bulk-gaining. In order to minimize transport costs for bulk-reducing activities, you want to process them as close as possible to their extraction site. Examples of this are metal smelters and lumber mills.

Bulk-gaining processes are a little more complicated. You need to find the least cost point between your source materials (which may be numerous) and your market. Remember that the goal is overall lowest cost, so that involves many calculations to determine which is the cheapest location, meaning your location might be in between several your inputs and/or market. A general rule of this is that these sorts of businesses tend to be fairly close to their final markets.

There were two last considerations that Weber discussed, agglomeration and deagglomeration. Weber called them secondary factors, because they were less important than the previously mentioned characteristics. Agglomeration is related to the idea of economy of scale. Sometimes there are advantages to having similar industries near one another. Consider the manufacture of computers. Computer manufacturers don't make their own components, they buy them, and they use similar tools to put them together, and they use similar labor, and so forth. When industries concentrate in a place they can share some resources and split the cost. This is not to say that this is a conscious process, in many instances it just develops on its own. Degglomerative factors produce diseconomies of scale and are responsible for redistribution of industry. Examples of this could be escalating prices for land or labor which drive production away from an area.

Figure 11.5 | Mills and Mines
This is a photo of a lead mine in South Dakota in the 19th Century. Notice the large piles of waste material left behind.
Author | John C. H. Grabill
Source | Library of Congress
License | Public Domain

11.3.4 Reducing Transportation Costs

Containerization has greatly changed the nature and cost of transportation. In the past, transporting goods required large numbers of people to move goods around. People loaded and unloaded goods at break-of-bulk points. Break-of-bulk points were places like railroad terminals, where goods were loaded onto trains, or ports, where goods were loaded into ships. Break-of-bulk points had large numbers of people loading and unloading items. Containerization changed that process tremendously. Goods are now packed into large metal boxes, then the bokes are moved from point to point. Cranes move the boxes from trucks to trains to ships, then reverse the process the process at the shipping destination. Intermodal transportation assumes that a container will seamlessly be transported by any number of different shipping methods. The number of people necessary to ship goods plummeted. The speed at which goods moved increased tremendously, since the bottlenecks were removed from the system. Containerization is a good example of an innovation that did not require a large technological shift, it simply required rationalizing a system that was labor intensive. Logistics, the commercial activity of transporting goods, is the glue that holds the global production network together. Without relatively inexpensive shipping, many offshored industries would not be able to make goods in distant factories, ship them to other places, and still make a profit.

Figure 11.6 | Shipping Containers
Uniform size and shape allow for rapid transport and sorting at minimal cost.
Author | Frank McKenna
Source | Unsplash
License | CC 0

11.3.5 Reducing Capital Costs

Operating an industry has more costs than materials, labor, and shipping. Other factors such as taxes, regulatory compliance, and financial incentive packages can either attract of repel manufacturing. These factors increased tremendously in importance. It has become possible for many industries to ignite a "bidding war" in order to secure increasingly advantageous incentives to locate in a particular place. Tax breaks, construction allowances and other benefits will be paid by the places that desperately struggle to attract industries. This has triggered what has been described as a race for the bottom as places promise more than they can afford in the hopes of securing outside investment.

Another consideration is access to capital. Businesses often need investment funds, or short-term credit. Being unable to secure capital prevents many businesses in developing countries from starting or continuing. Countries without banking infrastructure find it nearly impossible to raise sufficient money to develop their own industries. Companies in countries like this are forced to hunt for financial backing in other countries.

11.3.6 Risk

Industries have large sunk costs. Corruption can take advantage of this to demand bribes, protection money, or other expenses that siphon off profit. High levels of corruption inhibit investment. Industries are risk averse. Places that are politically unstable will also have difficulty industrializing, since companies will be reluctant to build in places where any investment could be lost in a revolution or other political violence. This doesn't mean that industry requires representative government. Many places have experienced tremendous economic development without democracy. It just means that companies have to feel that any money that they invest in a place is safe.

11.3.7 Accessing a Market

One of the interesting examples of this concept is the number of foreign companies that establish themselves in the United States. Companies such as Foxconn (Taiwan), Hyundai (Korea), and BMW (Germany) build products in the United States. A good question to ask yourself would be "Why do they do that?" It turns out that locating in the United States is advantageous for them. First, they can reduce shipping costs, we have already addressed that. Second, although labor in the United States is relatively expensive, it's not appreciably different than in their home markets. Third, they may be able to avoid tariffs on imports. Finally, being in the United States places these products near a large market. The North American market is very large, being embedded in it can be advantageous for some companies.

11.4 THE EXPANSION (AND EVENTUAL DECLINE) OF INDUSTRY

11.4.1 Transnational Corporations

Modern industry has become dominated by Transnational Corporations. Transnational Corporations (TNCs) have the ability to coordinate and control various processes and transactions within production networks, both within and between different countries. They also have the ability to take advantage of geographical differences of factors of production and state policies. Potential geographical flexibility for sales is a final benefit.

Foreign Direct Investment (FDI) is exactly what is sounds like. Companies invest in other countries. The usual narrative of this in the United States is simply offshoring for cheaper labor. The reality is more complicated. Just consider the examples above of the manufacturers who build facilities in the United States. They aren't seeking cheap labor. America companies moving production may be seeking to tap into a large pool of labor, but that labor is not as cheap as it was 20 years ago. Another reason for investing in China is the same reason that other companies invest here in North America, to gain entry into a large market. Companies and individuals investing in other countries have numerous motivations. Some of these motivations are altruistic. There is no shortage of entities that seek to use FDI as a mechanism to alleviate poverty. The earlier critique of financial incentives applies

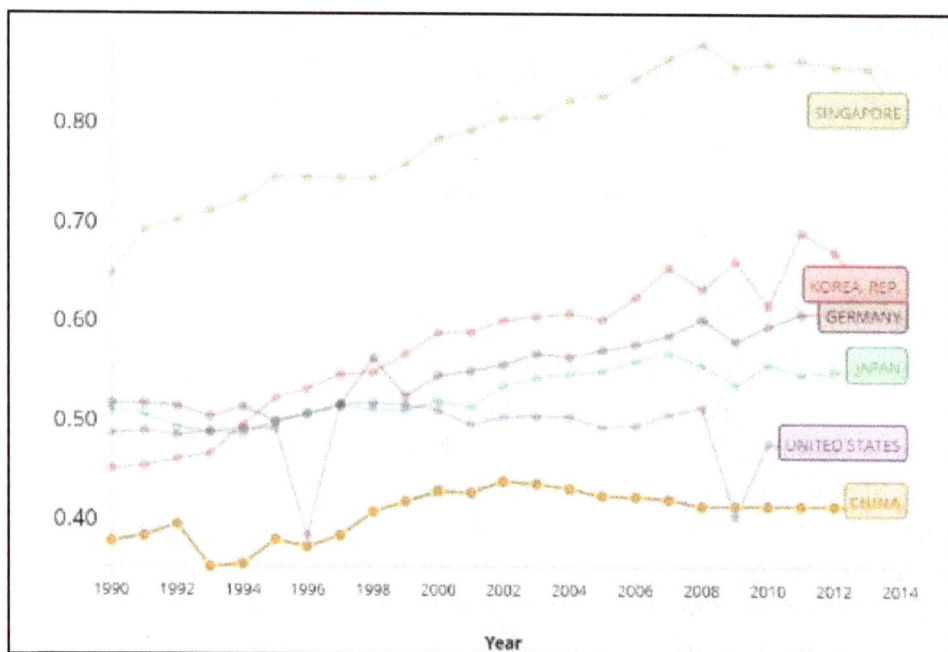

Figure 11.7 | Medium-Tech And High-Tech Manufacturing Value Added Share In Total Manufacturing
Author | The World Bank
Source | The World Bank
License | CC BY 4.0

directly of foreign direct investment, for obvious reasons. Poor and desperate places will sometimes make economic decisions that make little economic sense.

FDI has an established history. Initially companies in wealthy countries were mostly interested in extracting raw materials from other, usually poorer, countries. This continues to this day. Countries like Niger, The Democratic Republic of Congo, and Venezuela function almost solely as a source of inputs for other companies based in other countries. Other places have seen investment in factories. Sometimes this is due to very low-cost labor, but in many instances, it is due to the fact that their wages are relatively low and they are relatively close to their market. An example of this was seen in Eastern Europe when many of these countries entered the European Union.

The next graph is slightly different. It shows the different trajectories of manufacturing in general as part of a countries economy. As you can see, manufacturing as a share of GPD in the United States has been quite low for some time. Manufacturing in China and Korea is much more important, but in both these places, the relative value of manufacturing is falling. In Germany, Japan and Singapore, the value of manufacturing is rising, although at very moderate levels. Note that neither of these graphs account for wages produced from manufacturing employment.

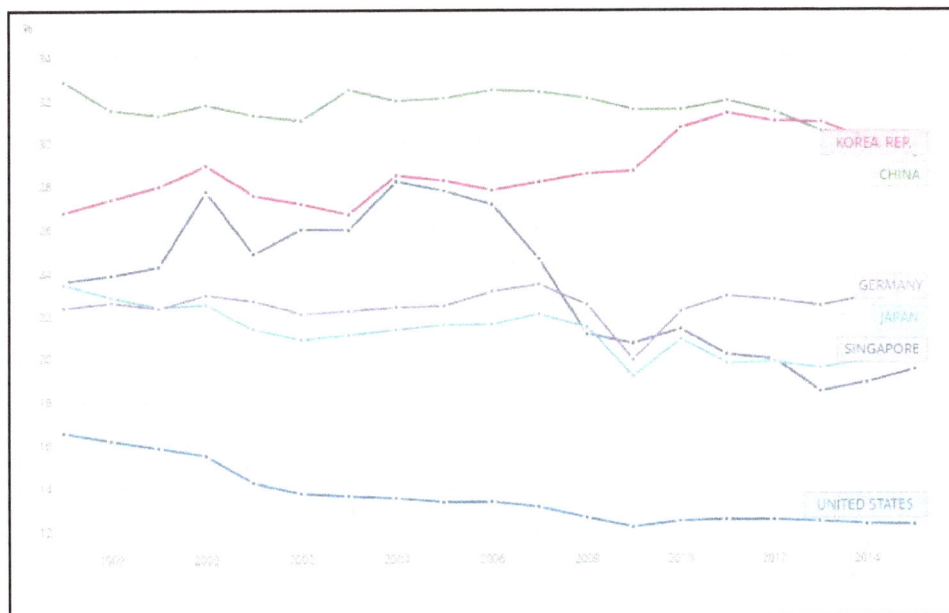

Figure 11.8 | Manufacturing Value Added Share In GDP
Author | The World Bank
Source | The World Bank
License | CC BY 4.0

11.5 GLOBAL PRODUCTION

11.5.1 Hegemony and Economic Ascendency

At times industrialization has propelled countries to great economic heights. Britain, The United States and Japan all rode an industrial wave to international prominence. In those countries and others, a (largely mythical) golden age centers around a time when low-skilled workers could earn a sufficient wage to secure economic security. This is more-or-less what the "American Dream" was. Deindustrialization has changed the economic trajectories of these countries and the people living in these countries. However, it must be noted that post-industrial countries that have not seen rapid increases in poverty. Wages have been largely stagnant for decades, but they have not generally gone down. The largest difference has to do with relative prosperity for industrial and post-industrial countries. Countries such as Japan, the UK or the US are no longer far wealthier than their neighbors. In the same way that flooding a market with a particular product reduces the value of that product, flooding the world with industrial capacity lowers the relative value of that activity. Developing countries function as appendages to the larger economies in the world. The poor serve the needs of the wealthy. Unindustrialized countries buy goods from developed countries, or they license or copy technology and make the products themselves.

11.5.2 Space and Production

In the context of a globalized market, a factory built in one market may not built in another. This is not to say that producing goods is a zero-sum game, but there are limits to the amount of any good that can be sold. It's a valid question to ask why transnational corporations (TNCs) have bought into China at rates far greater than in Cuba, Russia, or other Communist or formerly Communist (to varying degrees) countries? There is only so much spare capacity for production in the world. If one giant country (China) is taking all the extra capacity, then there will be none left for others. FDI is simply easier in China, since there is more bang for the buck. This is largely a function of population. The population of China is roughly two times the population of Sub Saharan Africa. And China has a single political/economic running class, as opposed to 55 different sets of often fractious political classes. If the industrialization of Africa happens at all, it will occur after China and its immediate neighbors who have been drawn into its larger economic functioning have largely finished their own industrializing. An example of this proximate effect is seen in the shift of some industries from China to Vietnam and Indonesia.

China's industrialization had to do with promoting itself as a huge cheap labor pool, and as a gigantic market for goods. It successfully leveraged both of these characteristics to attract foreign investment, and to gain foreign technology from the companies that have invested in producing goods there. Industrialization overall seems to have slowed. The speed at which China industrialized has not been matched by other countries following China. One current idea is that the world is in a race between industrial expansion and rapid over capacity of production. In other words, the reason that industrialization isn't expanding as rapidly as before is that we are already making enough goods to satisfy demand. Remember that goods require demand. Unsold goods don't produce any income. If the factories in the world are already producing enough, or even too much, then new factories are much less likely to be built. Technological advances and the massive industrialization of China might have ended the expansion of industry.

It also appears that the highest levels of manufacturing income are well in the past. According to economist Dani Rodrik, the highest per capita incomes from manufacturing occurred between 1965 and 1975, and has fallen dramatically since then. This is even considering inflation. Many countries industrializing now only see modest improvements in income. This is related to supply and demand. When there are fewer factories, they make comparatively more money. When factories are everywhere, they are competing with everyone.

11.5.3 Trade

Even more than expansion of industrial production, the world has seen an expansion of trade. Global trade has produced an intricate web of exchanges as products are now designed in one country, parts are produced in 10 others, assembled in yet another country, and then marketed to the world. Consider something as complex as an automobile. The parts of a car can be sourced from any of dozens of countries, but they all have to be brought to one spot for assembly. Such coordination would have been impossible in the past.

Individuals can buy directly from another country on the internet, but most international trade is business-to-business. TNCs are able to conduct an internal form of international trade in goods that can be moved and produced in a way that is most advantageous for the company. Tax breaks, easy credit and banking privacy laws exist to siphon investment from one place to another.

Because of global trade, improvements in communication and transportation have enabled some companies to enact just in time delivery, in which the parts need for a product only arrive right before they are needed. The advantage of this is that a company has less money trapped in components in a storage facility, and it becomes easier to adjust production. Once again, such coordination at a global scale was not possible even in the relatively recent past.

11.5.4 Deindustrialization

Historically industrialized countries were the wealthy countries of the world. Industrialization, however, is now two centuries old. In the last decades of the twentieth century, deindustrialization began in earnest in the United Kingdom, the United States, and many other places. Factories left and the old jobs left with them. Classical economics holds that such jobs had become less valuable, and that moving them offshore was a good deal for everyone. Offshored goods were cheaper for consumers, and the lost jobs were replaced by better jobs. The problem with this idea is that it separates the condition of being a consumer from the condition of being a worker. Most people in any economy are workers. They can only consume as long as they have an income, and that is tied to their ability to work. Many workers whose jobs went elsewhere found that their new jobs paid less than their old jobs.

11.5.5 What Happens After Deindustrialization?

The simple answer to the above question is this. The service economy happens! As manufacturing provides fewer jobs, service industries tend to create new jobs. This is a very delicate balance, however. If you are a 50-year old coal miner whose job has been eliminated by automation, it is very difficult for you to simply change jobs and enter the 'service sector'. This transition is very damaging to those without the right skills, training, education, or geographic location. Many parts of the American Midwest, for example, have become known as the "rust belt" as industrial facilities closed, decayed, and literally rusted to the horror of those residents who once had good jobs there. The city of Detroit, for example, lost nearly half of its urban population from 1970-2010. Meanwhile the state of Illinois loses one resident every 15 minutes as job growth has weakened in the post-industrial age. However, job growth and productivity in the service economy have strengthened and provide more job opportunities today than the industrial era ever did in the U.S.

There are 3 sectors to every economy:

1. Primary (agriculture, fishing, and mining)
2. Secondary (manufacturing and construction)
3. Tertiary (service related jobs)

The vast majority of economic growth in the post-industrialized world comes in the tertiary sector. This doesn't mean that all tertiary jobs pay well. Just ask any fast-food worker if the service sector is making them rich! However, service sector jobs are very dynamic and offer tangible opportunities to millions of people around the world to earn a living providing services to somebody else. We can further break down the service sector into 1) public (the post office, public utilities, working for the government) 2) business (businesses providing services to other businesses) 3) consumer (anything that provides a service to a private consumer e.g. hotels, restaurants, barber shops, mechanics, financial services). Traditionally, service sector jobs worked very much like manufacturing jobs in that employees worked regular hours, earned benefits from the employer, gained raises through increased performance, and went to work somewhere outside of their home. Many service jobs in the 21st century, however, have been categorized as the **gig economy**, in which workers serve as contractors (rather than employees), have no regular work schedules, don't earn benefits, and often work in isolation from other workers rather than as a part of a team. Examples of 'jobs' in the gig economy include private tutor, Uber/Lyft driver, AirBNB host, blogger, and YouTuber. Work, in this economy, is not necessarily bound by particular places and spaces in the way that it did in manufacturing. Imagine a steel worker calling in to tell his/her boss that they're just going to work from home today! Even public schools have adapted to this model in the following manner. As schools cancel class due to weather, the new norm is to hold class online, whereby students do independent work submitted to the teacher even though nobody is at school. As such, some workers are freed up from the traditional constraints of time and place and can choose to live anywhere as long as they maintain access to a computer and the Internet. Services like fiverr. com facilitate a marketplace for freelance writers compose essays for others or for graphic artists to sell their design ideas directly to customers without every meeting one another.

The global marketplace continues to be defined as a place where the traditional relationships between employer and employee are changing dramatically. A word of caution is necessary here, however. As many choose to celebrate the freedom that accompanies flexible work schedules, there is also a darker side in that the traditional 'contract' and social cohesive element between workers and owners is very much at risk. One defining factor of the 20th century was the development of civil society that fought for and won a host of protective measures for workers, who otherwise could face abusive work conditions. Child labor laws, minimum wage, environmental safety measures, overtime pay, and guards against discrimination were all based upon an employer-employee relationship that seems increasingly threatened by the gig economy. Uber drivers can work themselves to exhaustion since they are not employees. AirBNB hosts can skirt environmental safety precautions since they do not face the same safety inspections required at hotels. These are just a few examples, but they are very worth consideration. Regardless of the positives and negatives, the new service economy is having a transformative effect upon all facets of society. Although, the authors of this textbook are all geography professors with PhD's from a variety of universities, perhaps the next version of this textbook will simply draw upon the gig economy to seek the lowest cost authors who are willing to write about all things geographical. Will you be able to tell the difference? (We hope so!!!)

11.6 SUMMARY

Industrial production changed the relationship of people to their environments. Folk (pre-industrial) cultures used local resources and knowledge to hand-produce goods. Now the productions of goods and the provisioning of services can be split into innumerable spatially discrete pieces. Competition drives the costs of goods and services downward providing relentless pressure to cut costs. This process has pushed industrialization into most corners of the world as companies have looked further and further afield to find cheaper labor and materials and to find more customers. Industrialization has fueled a change in lifestyle, as goods have become cheaper, they have become more accessible to more people. Our lives have changed. We now live according to a schedule dictated by international production.

11.7 KEY TERMS DEFINED

Back office services – interoffice services involving personnel who do not interact directly with clients.

Break-of-bulk point – point of transfer from one form of transport to another.

Bulk reducing – industrial activity that produces a product that weighs less than the inputs.

Commodification – The process of transforming a cultural activity into a saleable product.

Containerization – transport system using standardized shipping containers.

Deindustrialization – process of shifting from a manufacturing based economy to one based on other economic activities.

Economies of scale – efficiencies in production gained from operating at a larger scale.

Footloose capitalism – spatial flexibility of production.

Fordism – rational form of mass production for standardizing and simplifying production.

Gig economy – a labor market characterized by freelance work.

Globalization – the state in which economic and cultural systems have become global in scale.

Intermodal – transportation system using more than one of transport.

Just in time delivery – manufacturing system in which components are delivered just before they are need in order to reduce inventory and storage costs.

Locational criteria – factors determining whether an economic activity will occur in a place.

Logistics – the coordination of complex operations.

Outsourcing – shifting the production of a good or the provision of a service from within a company to an externals source.

Offshoring – shifting the production of a good or the provision of a service to another country.

Supply chain – all products and process involved in the production of goods.

Taylorism – the scientific management of production.

11.8 WORKS CONSULTED AND FURTHER READING

Berkhout, Esmé.. 2016. "Tax Battles: The Dangerous Global Race to the Bottom on Corporate Tax," 46.

Dicken, Peter. 2014. Global Shift: Mapping the Changing Contours of the World Economy. SAGE.

Dorrell, David. 2018. "Using International Content in an Introductory Human Geography Course." In Curriculum Internationalization and the Future of Education.

Goodwin, Michael, David Bach, and Joel Bakan. 2012. Economix: How and Why Our Economy Works (and Doesn't Work) in Words and Pictures. New York: Harry N. Abrams.

Grabill, John C. H. 1889. "'Mills and Mines.' Part of the Great Homestake Works, Lead City, Dak." Still image. 1889. //www.loc.gov/pictures/resource/ppmsc.02674.

Gregory, Derek, ed. 2009. The Dictionary of Human Geography. 5th ed. Malden, MA: Blackwell.

Griswold, Daniel. n.d. "Globalization Isn't Killing Factory Jobs. Trade Is Actually Why Manufacturing Is up 40%." Latimes.Com. Accessed April 21, 2018. http://www.latimes.com/opinion/op-ed/la-oe-griswold-globalization-and-trade-help-manufacturing-20160801-snap-story.html.

Howe, Jeff. 2006. "The Rise of Crowdsourcing." WIRED. 2006. https://www.wired.com/2006/06/crowds/.

Massey, Doreen B. 1995. Spatial Divisions of Labor: Social Structures and the Geography of Production. Psychology Press.

Rendall, Matthew. 2016. "Industrial Robots Will Replace Manufacturing Jobs — and That's a Good Thing." TechCrunch (blog). October 9, 2016. http://social.techcrunch.com/2016/10/09/industrial-robots-will-replace-manufacturing-jobs-and-thats-a-good-thing/.

Rodrik, Dani. 2015. "Premature Deindustrialization." Working Paper 20935. National Bureau of Economic Research. https://doi.org/10.3386/w20935.

Sherman, Len. 2017. "Why Can't Uber Make Money?" December 14, 2017. https://www.forbes.com/sites/lensherman/2017/12/14/why-cant-uber-make-money/#2fb5abc10ec1.

Sumner, Andrew. 2005. "Is Foreign Direct Investment Good for the Poor? A Review and Stocktake." Development in Practice 15 (3–4): 269–85. https://doi.org/10.1080/09614520500076183.

Zhou, May, and Zhang Yuan. 2017. "Textile Companies Go High Tech in Arkansas - USA - Chinadaily.Com.Cn." July 25, 2017. http://usa.chinadaily.com.cn/world/2017-07/25/content_30244657.htm?utm_campaign=T-shirt%20line&utm_source=hs_email&utm_medium=email&utm_content=54911122&hsenc=p2ANqtz-9WyxMiFliVTrpO35Quk5KNoXpHHHj2bYn9-7WKp3Tt_iF8LUsO9Q6m6OEH892iW9QcXJ4kvAk8C1Ooiy5TffzH6URrPVnKTrvZ3TEFQ_zyt6rIjp0&_hsmi=54911122.

11.9 ENDNOTES

1. Data source: World Bank. https://tcdata360.worldbank.org

12

Human Settlements

Georgeta Connor

STUDENT LEARNING OUTCOMES

By the end of this section, the student will be able to:

1. Understand the similarities and differences between rural and urban

2. Explain urban origins and how the earliest settlements developed independently in the various hearth areas

3. Describe the models of rural and urban structure, comparing and contrasting urban patterns in different regions of the world

4. Connect the nature and causes of the problems associated with overurbanization in developing countries

CHAPTER OUTLINE

12.1 INTRODUCTION

Rural areas cover a multitude of natural and cultural landscapes, activities, and functions, including not only villages and agricultural areas, ranging from traditional to intensive monoculture systems, forests, various parks, and wilderness, but also services and commercial sites, as well as educational and research centers. Specifically, rural areas provide living space for their inhabitants and for flora and fauna and, as buffer zones, fulfill significant balance functions between unpopulated wilderness zones and overloaded centers of dense development. Because of this complex diversity, our understanding of rural areas must consider more than how land is used by nature and humans. That is, our understanding must also encompass the economic and social structures in rural areas in which farming and forestry, handicraft, and small, middle, or large companies produce and trade, where services, from the most local to the most international (such as tourism), are provided. In addition, some rural areas represent valuable ecological balance zones through preservation and/or conservation establishments. All these factors create and evolve into a tight interdependence, interconnection, and competition.

Yet, today, over 54 percent of the world population (7,536 million)[1] lives in urban areas and the proportion of the urban population is growing at a rapid rate. Thus, urbanization is one of the most important geographic phenomena in today's world. Towns and cities are in constant flux. Historically, cities have been influenced by technological developments such as the steam engine, railroads, the internal combustion engine, air transport, electronics, telecommunications, robotics, and the internet. As the result of the global shift to technological-, industrial-, and service-based economies, the growth of cities and urbanization of rural areas are now irreversible. Moreover, another phase of transformation is under way, involving global processes of economic, cultural, and political changes.

Within the cities of the developed world, the economic reorganization has determined a selective recentralization of residential and commercial land use connected especially with a selective industrial decentralization. In contrast to the core regions, where urbanization has largely resulted from economic growth, the urbanization of peripheral regions has been a consequence of demographic growth, generating large increases in population (overurbanization) well in advance of any significant levels of urban or rural economic development. Luxury homes and apartment complexes, corresponding with a dynamic formal sector of the economy, contrast sharply with the slums and squatter settlements of people, working in the informal (not regulated by the state) sector.

12.2 RURAL SETTLEMENT PATTERNS

There are many types of rural settlements. Using as classification criteria the shape, internal structure, and streets texture, settlements can be classified into two broad categories: *clustered* and *dispersed.*

12.2.1 Clustered Rural Settlements

A clustered rural settlement is a rural settlement where a number of families live in close proximity to each other, with fields surrounding the collection of houses and farm buildings. The layout of this type of village reflects historical circumstances, the nature of the land, economic conditions, and local cultural characteristics. The rural settlement patterns range from *compact* to *linear*, to *circular*, and *grid*.

Compact Rural Settlements

This model has a center where several public buildings are located such as the community hall, bank, commercial complex, school, and church. This center is surrounded by houses and farmland. Small garden plots are located in the first ring surrounding the houses, continued with large cultivated land areas, pastures, and woodlands in successive rings. The compact villages are located either in the plain areas with important water resources or in some hilly and mountainous depressions. In some cases, the compact villages are designed to conserve land for farming, standing in sharp contrast to the often isolated farms of the American Great Plains or Australia (**Figure 12.1**).

Figure 12.1 | A Compact Village in India
Author | User "Parthan"
Source | Wikimedia Commons
License | CC BY SA 2.0

Linear Rural Settlements

The linear form is comprised of buildings along a road, river, dike, or seacoast. Excluding the mountainous zones, the agricultural land is extended behind the buildings. The river can supply the people with a water source and the availability to travel and communicate. Roads were constructed in parallel to the river for access to inland farms. In this way, a new linear settlement can emerge along each road, parallel to the original riverfront settlement (**Figure 12.2**).

Figure 12.2 | Linear Village of Outlane
Author | Mark Mercer
Source | Wikimedia Commons
License | CC BY SA 2.0

Circular Rural Settlements

This form consists of a central open space surrounded by structures. Such settlements are variously referred to as a *Rundling, Runddorf, Rundlingsdorf, Rundplatzdorf* or *Platzdorf* (Germany), *Circulades* and *Bastides* (France), or *Kraal* (Africa). There are no contemporary historical records of the founding of these circular villages, but a consensus has arisen in recent decades. The current leading theory is that *Rundlinge* were developed at more or less the same time in the 12th century, to a model developed by the Germanic nobility as suitable for small groups of mainly Slavic farm-settlers. Also, in the medieval times, villages in the Languedoc, France, were often situated on hilltops and built in a circular fashion for defensive purpose (**Figures 12.3** and **12.4**).

Although far from the German territory, Romania has a unique, circular German village. Located southwestern Romania, Charlottenburg is the only round village in the country. The village was established around 1770 by Swabians who came to the region as part of the second wave of German colonization. In the middle of the village is a covered well surrounded by a perfect circle of mulberry trees behind which are houses with stables, barns, and their gardens in the external ring. Due to its uniqueness, the beautiful village plan from the baroque era has been preserved as a historical monument (**Figure 12.5**).

Figure 12.3 | Bastide in France
Author | User "Chensiyuan"
Source | Wikimedia Commons
License | CC BY SA 4.0

Figure 12.4 | Kraal - A circular village in Africa
Author | User "Hp.Baumeler"
Source | Wikimedia Commons
License | CC BY SA 4.0

Figure 12.5 | Charlottenburg, Romania
Author | German Wikipedia user "Eddiebw"
Source | Wikimedia Commons
License | CC BY SA 3.0

12.2.2 Dispersed Rural Settlements

Dispersed Rural Settlements

A dispersed settlement is one of the main types of settlement patterns used to classify rural settlements. Typically, in stark contrast to a nucleated settlement, dispersed settlements range from a *scattered* to an *isolated* pattern (**Figure 12.6**). In addition to Western Europe, dispersed patterns of settlements are found in many other world regions, including North America.

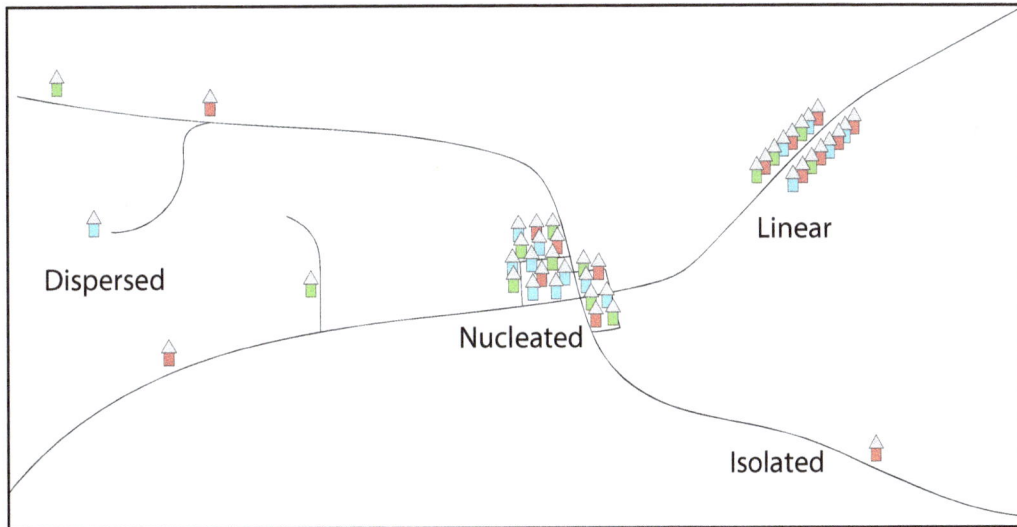

Figure 12.6 | Settlement Patterns²
Author | Corey Parson
Source | Origina Work
License | CC BY SA 4.0

Scattered Rural Settlements

A scattered dispersed type of rural settlement is generally found in a variety of landforms, such as the foothill, tableland, and upland regions. Yet, the proper scattered village is found at the highest elevations and reflects the rugged terrain and pastoral economic life. The population maintains many traditional features in architecture, dress, and social customs, and the old market centers are still important. Small plots and dwellings are carved out of the forests and on the upland pastures wherever physical conditions permit. Mining, livestock raising, and agriculture are the main economic activities, the latter characterized by terrace cultivation on the mountain slopes. The sub-mountain regions, with hills and valleys covered by plowed fields, vineyards, orchards, and pastures, typically have this type of settlement.

Isolated Rural Settlements

This form consists of separate farmsteads scattered throughout the area in which farmers live on individual farms isolated from neighbors rather than alongside other farmers in settlements. The isolated settlement pattern is dominant in rural areas of the United States, but it is also an important characteristic for Canada, Australia, Europe, and other regions. In the United States, the dispersed settlement pattern was developed first in the Middle Atlantic colonies as a result of the individual immigrants' arrivals. As people started to move westward, where land was plentiful, the isolated type of settlements became dominant in the American Midwest. These farms are located in the large plains and plateaus agricultural areas, but some isolated farms, including *hamlets*, can also be found in different mountainous areas (**Figures 12.7** and **12.8**).

Figure 12.7 | Isolated Horse Farm
Author | Randy Fath
Source | Unsplash
License | CC 0

Figure 12.8 | Undredal, Norway
Author | Micha L. Rieser
Source | Wikimedia Commons
License | © Micha L. Rieser. Used with permission.

12.3 URBANIZATION

12.3.1 Urban Origins

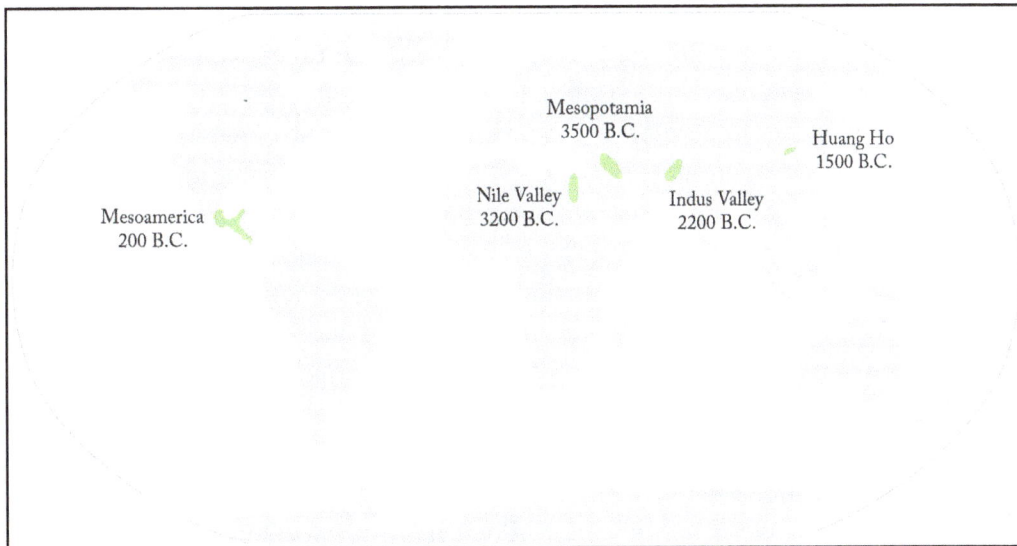

Figure 12.9 | Five Hearths of Urbanization
Author | User "Canuckguy" and Corey Parson
Source | Wikimedia Commons
License | CC BY SA 4.0

The earliest towns and cities developed independently in the various regions of the world. These *hearth areas* have experienced their *first agricultural revolution*, characterized by the transition from hunting and gathering to agricultural food production. Five world regions are considered as hearth areas, providing the earliest evidence for urbanization: Mesopotamia and Egypt (both parts of the Fertile Crescent of Southwest Asia), the Indus Valley, Northern China, and Mesoamerica (**Figure 12.9**). Over time, these five hearths produced successive generations of urbanized world-empires, followed by the diffusion of urbanization to the rest of the world.

The first regions of independent urbanism were in Mesopotamia and Egypt from around 3500 B.C. Mesopotamia, the land between the Tigris and Euphrates rivers, was the eastern part of the so-called **Fertile Crescent (Figure 12.10)**. From the Mesopotamian Basin,

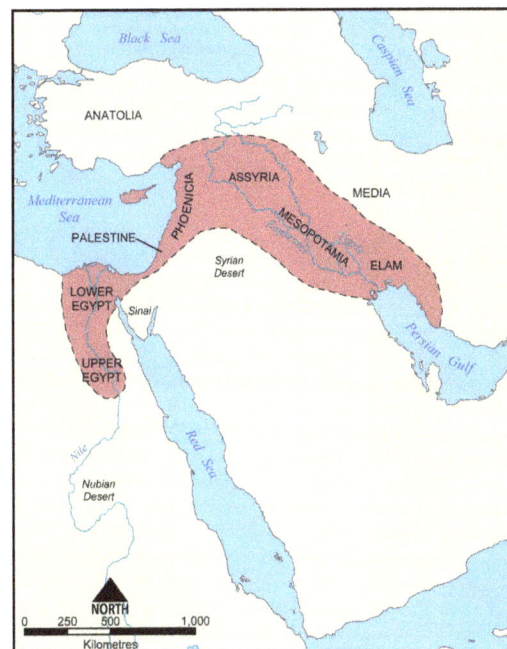

Figure 12.10 | Fertile Crescent
Author | User "NormanEinstein"
Source | Wikimedia Commons
License | CC BY SA 3.0

the Fertile Crescent stretched in an arc across the northern part of the Syrian Desert as far west as Egypt, in the Nile Valley. In Mesopotamia, the significant growth in size of some of the agricultural villages formed the basis for the large fortified **city-states** of the Sumerian Empire such as Ur, Uruk, Eridu, and Erbil, in present-day Iraq. By 1885 B.C., the Sumerian city-states had been taken over by the Babylonians, who governed the region from Babylon, their capital city. Unlike in Mesopotamia, internal peace in Egypt determined no need for any defensive fortification. Around 3000 B.C. the largest Egyptian city was probably Memphis (over 30,000 inhabitants). Yet, between 2000 and 1400 B.C., urbanization continued with the founding of several capital cities such as Thebes and Tanis.

About 2500 B.C, large urban settlements were developed in the Indus Valley (Mohenjo-Daro, especially), in modern Pakistan, and later, about 1800 B.C., in the fertile plains of the Huang He River (or Yellow River) in Northern China, supported by the fertile soils and extensive irrigation systems. Other areas of independent urbanism include Mesoamerica (Zapotec and Mayan civilizations, in Mexico) from around 100 B.C. and, later, Andean America from around A.D. 800 (Inca Empire, from northern Ecuador to central Chile). Teotihuacan (**Figure 12.11**), near modern Mexico City, reached its height with about 200,000 inhabitants between A.D. 300 and 700.

Figure 12.11 | Teotihuacan, Mexico
Author | User "BrCG2007"
Source | Wikimedia Commons
License | Public Domain

The city-building ideas eventually spread into the Mediterranean area from the Fertile Crescent. In Europe, the urban system was introduced by the Greeks, who, by 800 B.C., founded famous cities such as Athens, Sparta, and Corinth. The city's center, the "acropolis," (**Figure 12.12**), was the defensive stronghold, surrounded by the "*agora*" suburbs, all surrounded by a defensive wall. Except for Athens, with approximately 150,000 inhabitants, the other Greek cities were quite small

Figure 12.12 | Athens, Greece, Acropolis
Author | Uswer "Jebulon"
Source | Wikimedia Commons
License | CC 0

by today's standards (10,000-15,000 inhabitants). The Greek urban system, through overseas colonization, stretched from the Aegean Sea to the Black Sea, around the Adriatic Sea, and continued to the west until Spain (**Figure 12.13**). Although the Macedonians conquered Greece during the 4th century BC, Alexander the Great extended the Greek urban system eastward toward Central Asia. The location of the cities along Mediterranean coastlines reflects the importance of long-distance sea trade for this urban civilization.

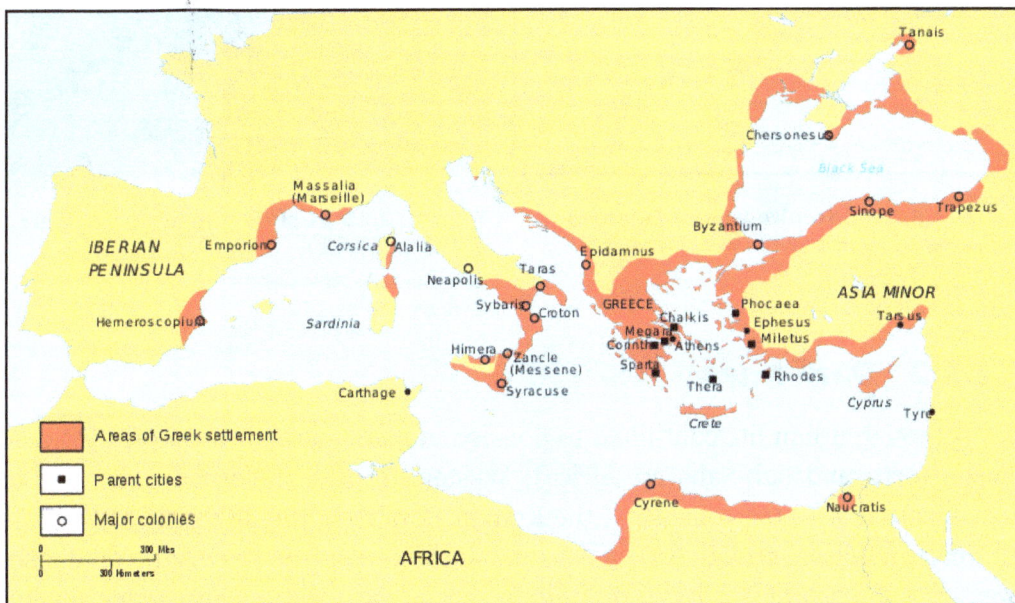

Figure 12.13 | Greek city-states in the Mediterranean
Author | User "Dipa1965"
Source | Wikimedia Commons
License | CC BY SA 4.0

With the impressive feats of civil engineering, the Romans had extended towns across southern Europe (**Figure 12.14**), connected with a magnificent system of roads. Roman cities, many of them located inland, were based on the grid system. The center of the city, "forum," surrounded by a defensive wall, was designated for political and commercial activities. By A.D. 100, Rome reached approximately one million inhabitants, while most towns were small (2,000-5,000 inhabitants). Unlike Greek cities, Roman cities were not independent, functioning within a well-organized system centered on Rome. Moreover, the Romans had developed very sophisticated urban systems, containing paved streets, piped water and sewage systems and adding massive monuments, grand public buildings (**Figure 12.15**), and impressive city walls. In the 5th century, when Rome declined, the urban system, stretching from England to Babylon, was a well-integrated urban system and transportation network, laying the foundation for the Western European urban system.

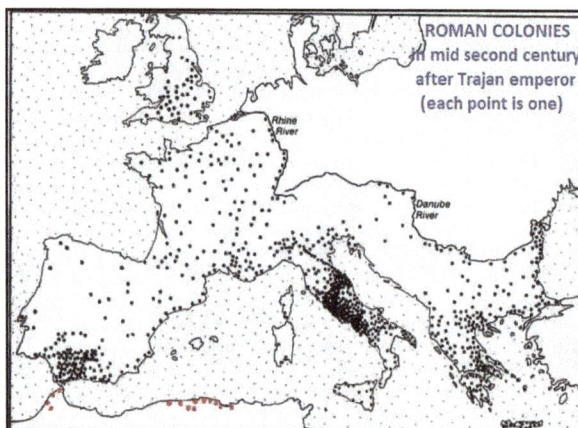

Figure 12.14 | Roman Empire and its Colonies
Author | User "Cresthaven"
Source | Wikimedia Commons
License | CC BY SA 4.0

Figure 12.15 | Colosseum in Rome, Italy
Author | User "Diliff"
Source | Wikimedia Commons
License | CC BY SA 2.5

12.3.2 Dark Ages

Although urban life continued to flourish in some parts of the world (Middle East, North and sub-Saharan Africa), Western Europe recorded a decline in urbanization after the collapse of the Roman Empire in the fifth century. During this early medieval period, A.D. 476-1000, also known as the Dark Ages, feudalism was a rurally oriented form of economic and social organization. Yet, under Muslim influence in Spain or under Byzantine control, urban life was still flourishing. As Rome was falling into decline, Constantine the Great moved the capital of the Roman Empire to Byzantium, renaming the city Constantinople (current Istanbul, Turkey). With its strategic location for trade, between Europe and Asia, Constantinople became the world's largest city, maintaining this status for most of the next 1000 years (**Figure 12.16**).

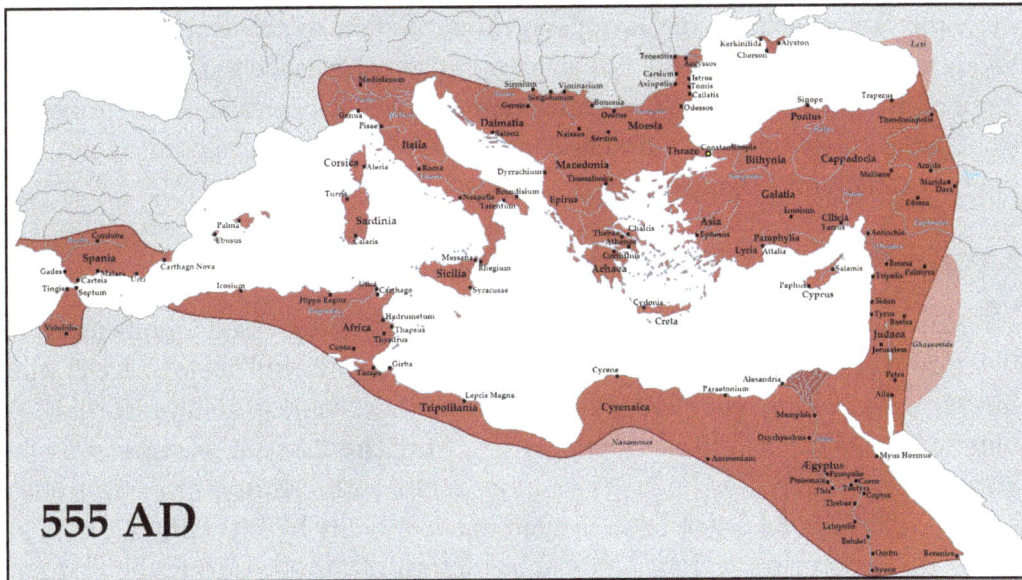

Figure 12.16 | Byzantine Empire
Author | User "Tataryn"
Source | Wikimedia Commons
License | CC BY SA 3.0

Most European regions, however, did have some small towns, most of which were either ecclesiastical or university centers (Cambridge, England and Chartres, France), defensive strongholds (Rasnov, Romania), gateway towns (Bellinzona, Switzerland), or administrative centers (Cologne, Germany). The most important cities at the end of the first millennium were the seats of the world-empires/ kingdoms: the Islamic caliphates, the Byzantine Empire, the Chinese Empire, and Indian kingdoms.

12.3.3 Urban Revival in Europe

From the 11th century onward, a more extensive money economy developed. The emerging regional specializations and trading patterns provided the foundations for a new phase of urbanization based on **merchant capitalism**. By 1400, long distance trading was well established based not only on luxury goods but also on metals, timber, and a variety of agricultural goods. At that time, Europe had about 3000 cities, most of them very small. Paris, with about 275,000 people was the dominant European city. Besides Constantinople (Byzantium) and Cordoba (Spain), only cities from northern Italy (Milan, Florence), and Bruges (Belgium) had more than 50,000 inhabitants.

Between the 14th and 18th centuries, fundamental changes occurred that transformed not only the cities and urban systems of Europe but also the entire world economy. Merchant capitalism increased in scale and the **Protestant Reformation** and **scientific revolution** of the **Renaissance** stimulated economic and social reorganization. Overseas **colonization** allowed Europeans to shape the world's economies and societies. Spanish and Portuguese colonists were the first to extend the European urban system to the world's peripheral regions. Between 1520 and 1580 Spanish colonists established the basis of a **Latin American urban system**. Centralization of political power and the formation of national states during the Renaissance determined the development of more integrated national urban systems.

12.3.4 Industrial Revolution and Urbanization

Although the urbanization process had already progressed significantly by the 18th century, the Industrial Revolution was a powerful factor accelerating further urbanization, generating new kinds of cities, some of them recording an unprecedented concentration of population. **Manchester**, for example, was the **shock city** of European industrialization in the 19th century, growing from a small town of 15,000 inhabitants in 1750 to a world city by 1911 with 2.3 million. Industrialization spread from England to the rest of Europe during the first half of the 19th century and then to different parts of the world. Moreover, followed by successive innovations in transport technology, urbanization increased at a faster pace. The canals, railroads, and steam-powered transportation reoriented urbanization toward the interior of the countries. In North America, the shock city was **Chicago,** which grew from 4,200 inhabitants in 1837 to 3.3 million in 1930. Its growth as an *industrial* city followed especially the arrival of the railroads, which made the city a major transportation hub. In the 20th century, the new innovations significantly helped urban development.

By the 19th century, urbanization had become an important dimension of the **world-system.** The higher wages and greater opportunities in the cities (industry, services) transformed them into a significant pull factor, attracting a massive influx of rural workers. Consequently, *the percentage of people living in cities* increased from 3 percent in 1800 to 54 percent in 2017.[3] In developed countries, 78 percent live in urban areas, compared with 49 percent in developing countries, reflecting the region's and/or the country's level of development.

12.3.5 City-Size Distribution

Most developed countries have a higher percentage of urban people, but developing countries have more of the very large urban settlements (**Table 1, Figure 12.17**). In 1950, out of the world's 30 largest metropolitan areas, the first three metropolitan areas were in developed countries: New York (U.S.), Tokyo (Japan), and London (UK), two of which (New York and Tokyo) had more than 10 million inhabitants. After 30 years, in 1980, a significant change was recorded. Although metro New York increased from 12.3 million to 15.6 million, Tokyo, with 28.5 million inhabitants, became the largest metropolitan area in the world, a position which the city still maintains. In addition, except for Osaka, the second metropolitan area from Japan, two large metropolitan areas in developing countries were added, Mexico City (Mexico) and Sao Paulo (Brazil). The number of large metropolitan areas continued to increase after 2010, adding more developing countries such as India (Delhi and Mumbai/Bombay), China (Shanghai and Beijing), Bangladesh (Dhaka), and Pakistan (Karachi) from Asia; and Egypt (Cairo), Nigeria (Lagos), and Democratic Republic of the Congo (Kinshasa) from Africa. Each of these metropolitan areas is expected to have over 20 million inhabitants after 2020, adding Delhi and Shanghai to the largest metropolitan areas with over 30 million inhabitants. In the United States, New York-Newark is the largest metropolitan area, in which the population was constantly increasing from 12.3 million in 1950 to 15.6 million in 1980 and 18.3 million in 2010, having the potential to reach 20 million in 2030. Yet, unlike the developing countries, characterized by a very fast urban growth rate, the developed countries had recorded a moderate urban growth rate.

Table 12.1 | The World's 30 Largest Metropolitan Areas, Ranked by Population Size[4]
1950, 1980, 2010, 2020, 2025, 2030 (in millions)

	1950	**1980**	**2010**	**2020**	**2025**	**2030**
World	New York-12.3 Tokyo-11.2 London-8.3	Tokyo-28.5 Osaka-17.0 New York-15.6 Mexico City-13.3 Sao Paulo-12.0	Tokyo-36.8 Delhi-21.9 Mexico City-20.1 Shanghai-19.9 Sao Paulo-19.6 Osaka-19.4 Bombay-19.4 New York-Newark-18.3 Cairo-16.9 Beijing-16.1	Tokyo-38.3 Delhi-29.3 Shanghai-27.1 Beijing-24.2 Bombay-22.8 Sao Paulo-22.1 Mexico City-21.8 Dhaka-20.9 Cairo-20.5 Osaka-20.5	Tokyo-37.8 Delhi-32.7 Shanghai-29.4 Beijing-26.4 Bombay-25.2 Dhaka-24.3 Mexico City-22.92 Sao Paulo-22.90 Cairo-22.4 Karachi-22.01 Osaka-20.3 Lagos-20.03	Tokyo-37.19 Delhi-36.06 Shanghai-30.7 Bombay-27.8 Beijing-27.7 Dhaka-27.3 Karachi-24.8 Cairo-24.5 Lagos-24.2 Mexico City-23.8 Sao Paulo-23.4 Kinshasa-20.00
United States	New York-Newark-12.3 Chicago-5.0 Los Angeles-Long Beach-Santa Ana 4.0 Philadelphia 3.1 Detroit 2.7 Boston 2.5 San Francisco-Oakland-1.8	New York-Newark-15.6 Los Angeles-Long Beach-Santa Ana-9.5 Chicago-7.2 Philadelphia-4.5	New York-Newark-18.3 Los Angeles-Long Beach-Santa Ana-12.1	New York-Newark-18.7 Los Angeles-Long Beach-Santa Ana-12.4	New York-Newark-19.3 Los Angeles-Long Beach-Santa Ana-12.8	New York-Newark-19.89 Los Angeles-Long Beach-Santa Ana-13.26

Summary:

Year	1 - 5 mil	5 - 10 mil	10 - 15 mil	15 - 20 mil	20 - 25 mil	25 - 30 mil	> 30 mil
1950	22	6	2	-	-	-	-
1980	6	19	2	2	-	1	-
2010	-	7	13	7	2	-	1
2020	-	-	12	8	7	2	1
2025	-	-	10	8	7	3	2
2030	-	-	10	8	6	3	3

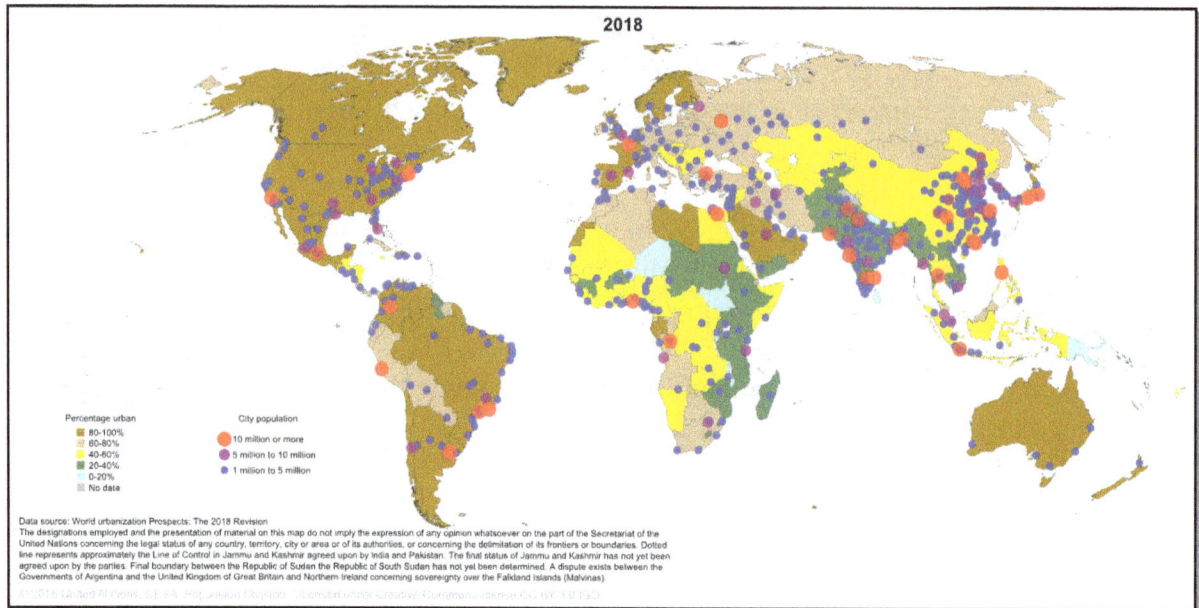

Figure 12.17 | Percentage Urban and Urban Agglomerations
Author | United Nations, DESA, Population Division
Source | United Nations Population Division
License | CC BY 3.0 IGO

The relationship between the size of cities and their rank within an urban system is known as **rank-size rule**, describing a regular pattern in which the nth-largest city in a country or region is 1/n the size of the largest city. According to the rank-size rule, the second-largest city is one-half the size of the largest, the third-largest city is one-third the size of the largest city, and so on. In the United States, the distribution of settlements closely follows the rank-size rule, but in other countries, the population of the largest city is disproportionately large in relation to the second- and third-largest city in that urban system. These cities are called **primate cities** (Buenos Aires, for example, is 10 times larger than Rosario, the second largest city in Argentina). Yet, cities do not necessarily have to be primate in order to be functionally dominant (economic, political, cultural) within their urban system. The functional dominance of the city is called **centrality**. Moreover, some of the largest metropolises, closely integrated within the global economic system and playing key roles beyond their own national boundaries, qualify as **world cities.** Today, because of the importance of their financial markets and associated business services, London, New York, and Tokyo dominate Europe, the Americas, and Asia, respectively.

12.4 URBAN PATTERNS

12.4.1 North American Cities

The contemporary North American scene dramatically displays how its population has refashioned the settlement landscape to meet the needs of a modern postindustrial society. In North American cities, a city's center, commonly called *downtown*, has historically been the nucleus of commercial and services land use.

It is also known as the **central business district (CBD)**, usually being one of the oldest districts in a city and the nodal point of transportation routes. The CBD gives visual expression to the growth and dynamic of the industrial city, becoming a symbol of progress, modernity, and affluence. It contains the densest and tallest nonresidential buildings and its accessibility attracts a diversity of services.

Urban decentralization also reconfigured *land-use patterns* in the city, producing today's *metropolitan areas*. If during the first half of the 20th century the **concentric zone model** was idealized, in which urban land is organized in rings around the CBD, today's urban model highlights new suburban growth characterized by a mix of peripheral retailing, industrial areas, office complexes, and entertainment facilities called "**edge cities.**" In addition, the **sector** and **multiple nuclei** (*multiple growth points*) models of urban structure were developed to help explain where different socio-economic classes tend to live in an urban area (**Figures 12.19, 12.20, and 12.21**). All models show new residential districts added beyond the CBD as the city expanded, with higher-income groups seeking more desirable, peripheral locations.

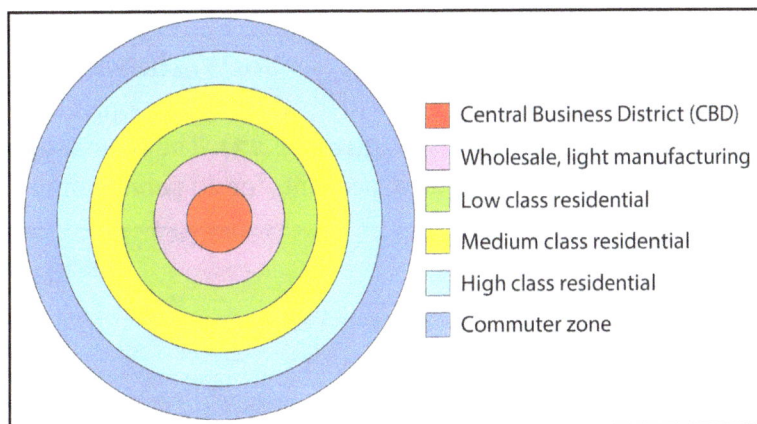

Figure 12.19 | Concentric-Ring Theory[5]
Author | Corey Parson
Source | Original Work
License | CC BY SA 4.0

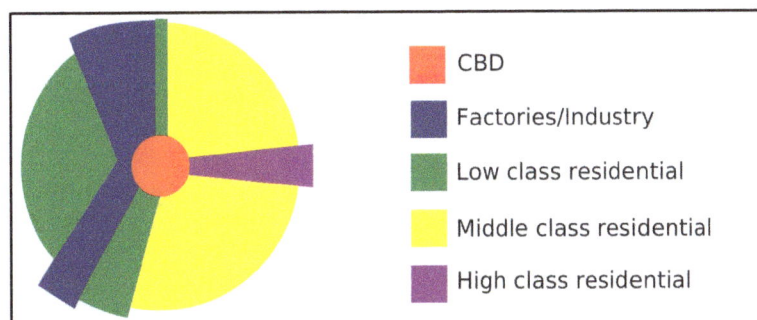

Figure 12.20 | Sector Theory
Author | User "Cieran 91"
Source | Wikimedia Commons
License | Public Domain

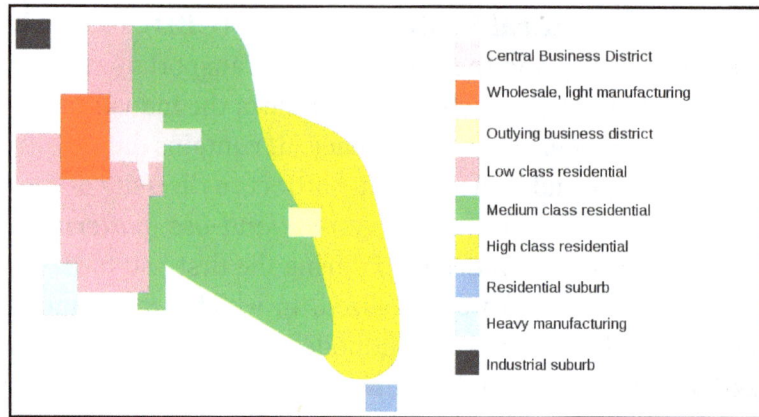

Figure 12.21 | Multiple-Nuclei Theory
Author | User "Cieran 91"
Source | Wikimedia Commons
License | Public Domain

Metropolitan clusters produce uneven patterns of settlement across North America. Eleven urban agglomerations, also known as **megalopolises**, exist in the United States and Canada, 10 of them being located in the United States (**Figure 12.22**). **Boston–Washington Corridor,** or **BosWash**, with its roughly 50 million inhabitants, representing 15 percent of the U.S. population, is the most heavily urbanized region of the United States (**Figure 12.23**). The region, located on less than two percent of the nation's land area, accounts for 20 percent of the U.S. GDP.

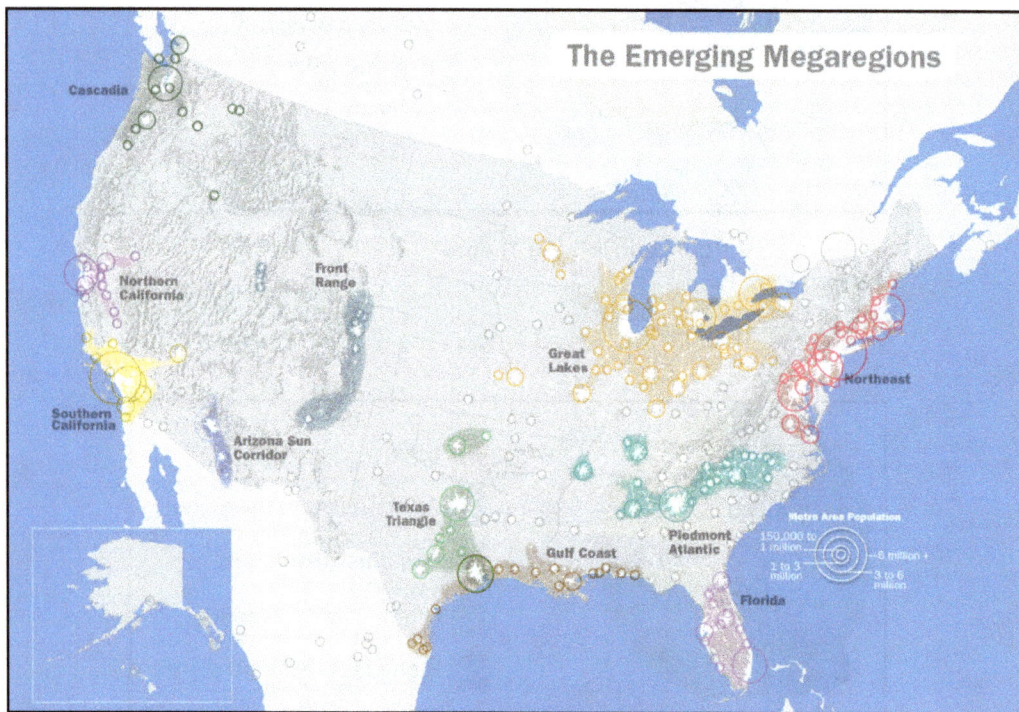

Figure 12.22 | Map of Emerging US Megaregions
Author | User "IrvingPINYC"
Source | Wikimedia Commons
License | CC BY SA 3.0

Like many other cities in the core regions, North American cities are also recognized for their prosperity. Yet, some problems still exist such as **fiscal squeeze** (less money from taxes)**, poverty, home-lessness, neighborhood decay,** and **infrastructure** needs. By contrast, some inner cities experience the process of **gentrification**, invaded by higher-income people who work downtown and who are seeking the convenience of less expensive and centrally located houses, larger and with attractive architectural details.

12.4.2 European Cities

One characteristic of Europe is its high level of urbanization. Even in sparsely settled Northern Europe, over 80 percent of the people live in urban area (Iceland 94 percent)[6]. Southern and Eastern Europe are the least urbanized, with an average of 69-70 percent. While widespread urbanization is relatively recent in Europe, dating back only a century or two, the spread of cities into Europe can be associated with the classical Greek and Roman Empires, making many cities in Europe more than 2,000 years old. Among the largest metropolitan areas are Paris (10.9 mill.), London (10.4 mill.), and Madrid (6.1 mill.)[7]. European cities, like North American cities, reflect the operation of competitive land markets and they also suffer from similar problems of urban management, infrastructure maintenance, and poverty. Yet, what makes most European cities distinctive in comparison with North American cities is their *long history*, being the product of several major epochs of urban development.

The three zone models (concentric, sector, and multiple nuclei) characterizing North American urban areas are also valid in Europe, but there are significant differences regarding the spatial distribution of social groups, who may not have the same reasons for selecting particular neighborhoods within their cities. The European Central Business Districts (CBDs), for example, are inhabited by more residents than CBDs of the United States, most of them being higher-income people attracted by the opportunities to have access to commercial and cultural facilities, as well as to occupy beautiful old buildings located in some elegant residential districts. As a result, European CBDs are less dominated by business services than American CBDs. CBDs are, in general, very expensive urban areas. Consequently, poor people are more likely to live in outer rings in European cities.

Because many of today's most important European cities were founded in the Roman and medieval periods, there are strict rules for preservation of their historic

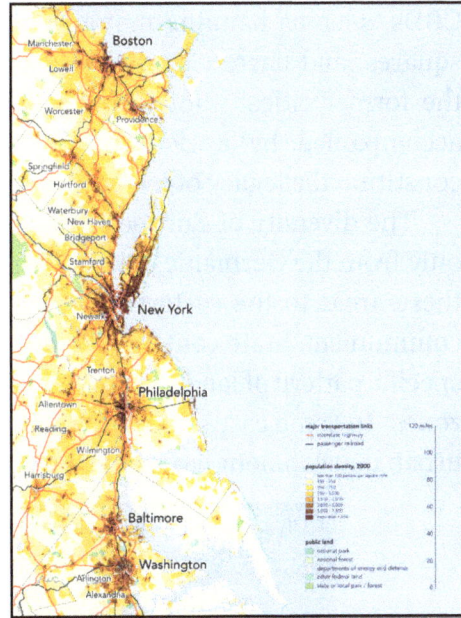

Figure 12.23 | Megalopolis BosWash
Author | Bill Rankin
Source | Wikimedia Commons
License | CC BY SA 3.0

CBDs, such as banning motor vehicles, maintaining low-rise structures, plazas, squares, and narrow streets, and preserving the original architecture, including the former cities' walls. Impressive palaces, cathedrals, churches, monasteries, accompanied by a rich variety of symbolism—memorials and statues—also constitute the legacy of a long and varied history (**Figure 12.24**).

The diversity of Europe's geography means that are important variations not only from the Germanic cities to the cities of Mediterranean Europe but also from these areas to the eastern European cities, which experienced over 40 years of communism. State control of land and housing determined the development of a specific pattern of land use, expressed by huge public housing estates and industrial zones. In some cases, the structure of the old cities has been altered, and the new urban development has extended over rural areas (**Figure 12.25**).

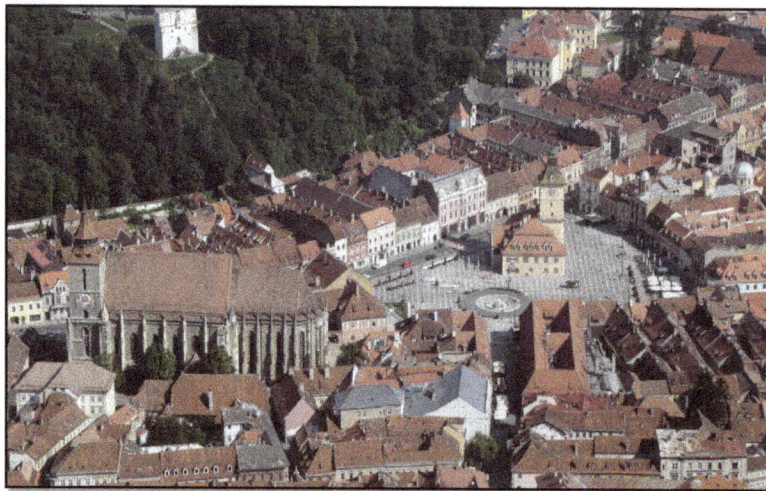

Figure 12.24 | Braşov, Romania: CBD square and city hall
Author | David Stanley
Source | Wikimedia Commons
License | CC BY 2.0

Figure 12.25 | Onesti, Romania: Urban-rural interaction
Author | Paul Istoan
Source | Wikimedia Commons
License | CC BY SA 4.0

Europe has overlapping metropolitan areas, the most extensive **megalopolis, Blue Banana**, also known as the **Manchester-Milan Axis**, being a discontinuous urban corridor in Western Europe, with a population of around 111 million inhabitants. It stretches approximately from Northwest England across Greater London to the Benelux states and along the German Rhineland, Southern Germany, Alsace in France in the west, and Switzerland to Northern Italy in the south (**Figure 12.26**). New regions that have been compared to the Blue Banana can be found on the Mediterranean coast between Valencia and Genoa, as part of the *Golden Banana* or *European Sunbelt*, and in the north of Germany, where another conurbation lies on the North Sea coast, stretching into Denmark, and from there into southern Scandinavia.

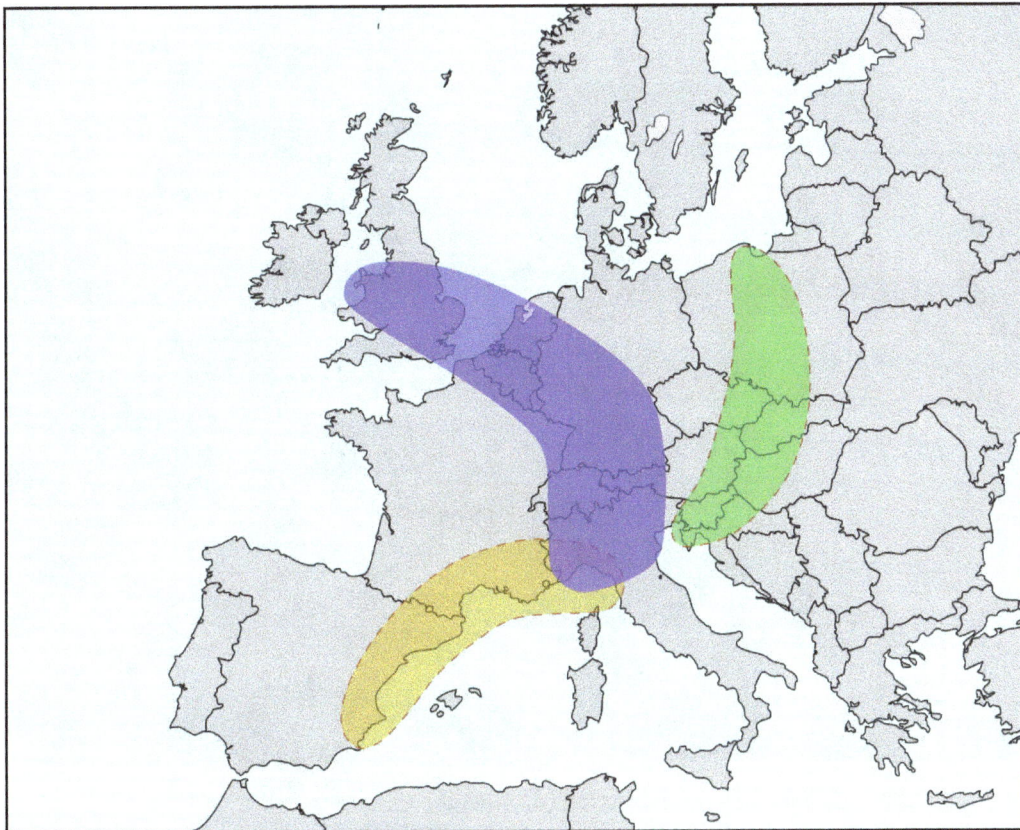

Figure 12.26 | The Bananas of Europe
Author | User "Luan"
Source | Wikimedia Commons
License | CC BY SA 3.0

12.4.3 Sustainable Cities versus Suburban Sprawl

Suburban Sprawl

The growth in automobile ownership in the U.S., the new infrastructure systems, and the long-term home financing, as well as the tremendous amount of **annexations** of territory from surrounding counties, resulted in a dramatic spurt in suburban growth (**Figure 12.27**). Thus, **sprawl** is endemic to North American urbanization. Low density and single-family suburban development is a positive aspect of suburbanization. Today, about 50 percent of Americans live in suburbs. Sprawl is less common outside European cities. Yet, the personal and environmental costs of this development are also significant. Noteworthy among these are automobile dependence, increased commute time and cost of gasoline, as well as air pollution and health problems. Even worse, more and more agricultural land is lost in favor of residential developments. Equally important, local governments must spend more money than is collected through taxes to provide services to the suburban areas.

Figure 12.27 | Suburban sprawl outside of Chicago, Illinois
Author | User "Wjmummert"
Source | Wikimedia Commons
License | Public Domain

Sustainable Cities

The compromise solution is '*smart growth*,' known as '*compact city*' in Europe, particularly in the United Kingdom. Smart growth is urban planning that concentrates growth in compact areas, strategically designed with adequate infrastructure and walkable, transit-oriented, bicycle-friendly urban centers. Characteristics of smart development include higher density living spaces, preserving open space, farmland, and natural and cultural resources, providing a variety of transportation choices, making development decisions predictable, fair, and cost-effective, equitably distributing the costs and benefits of development, and encouraging community collaboration in developing decisions. Smart growth and related concepts are not necessarily new, but are a response to car culture and sprawl. Smart growth values *long-range*, regional considerations of **sustainability** over a short-term focus. Yet, in practice, the process experiences challenges even from citizens, in some cases expressing their opposition to the local smart growth projects.

12.5 CONCLUSION

In summary, definitions of the terms *rural* and/or *rurality* and delineation of rural from urban areas have been long debated topics in many academic works. Rural space comes into existence in certain areas, typified by a series of factors such as land use (especially for agriculture), population density, agricultural employment, and built areas. Generally, rural areas are considered to be synonymous with more extensive land use activities in agriculture and forestry, low population density, small settlements, and an agrarian way of life. Rural space is divided into territorial entities, with variable scales, covering the local or regional economy, and each unit includes both agricultural and non-agricultural activities. Different countries have varying definitions of rural for statistical and administrative purposes. Although urbanization is a global phenomenon, today about 45.5 percent of the world's population still lives in rural areas.[8]

There are many types of rural settlements. Early villages had to be near a reliable water supply, be defensible, and have sufficient land nearby for cultivation, to name but a few concerns. They also had to adapt to local physical and environmental conditions, conditions that can be identified with a practiced eye. Villages in the Netherlands are linear, crowded on the dikes surrounding land reclaimed from the sea. Circular villages in parts of Africa indicate a need for a safe haven for livestock at night. A careful examination of the rural settlement of a region reveals much about its culture, history and traditions.

Most people can agree that cities are places where large numbers of people live and work, and are hubs of government, commerce, and transportation. But how best to define the geographical limits of a city is a matter of some debate. So far, no standardized international criteria exist for determining the boundaries of a city. Often, different boundary definitions are available for any given city. The "city proper," for example, describes a city according to an administrative boundary, while "urban agglomeration," considers the extent of the built-up area, to delineate the city's boundaries. A third concept of the city, the "metropolitan area," defines its boundaries according to the degree of economic and social interconnectedness of nearby areas.

The United Nations, the most comprehensive source of statistics regarding urbanization, emphasizes the fact that more than one half of the world population now lives in urban areas (54.5 percent in 2016)[9], and virtually all countries of the world are becoming increasingly urbanized. Yet, given the fact that different countries use different definitions, it is difficult to know exactly how urbanized the world has become. Canada and Australia consider urban any settlements of 1,000 inhabitants or more, for example, while a settlement of 2,000 is a significant urban center in Peru. Other countries consider other limits for urban settlements such as 5,000 inhabitants in Romania and 50,000 in Japan. Consequently, in 2016, the percentage of urban population by continent was as follows: North America, 81 percent, the most urbanized continent in the world; Latin America, 80 percent; Europe, 74 percent; Oceania, 70 percent; Asia, 48 percent; and Africa, 41 percent.[10]

12.6 KEY TERMS DEFINED

Annexation – legally adding land area to a city

Blue Banana – a discontinuous corridor of urbanization in Western Europe, from North West England to Northern Italy

Boswash – the United States megalopolis, extending from Boston to Washington D.C.

central business district (CBD) – the central nucleus of commercial land uses in a city

Centrality – the functional dominance of cities within an urban system

City-state – a sovereign state that consists of a city and its dependent territories

Clustered rural settlement – an agricultural based community in which a number of families live in close proximity to each other, with fields surrounding the collection of houses and farm buildings

Concentric zone model – a model of the internal structure of cities in which social groups are spatially arranged in a series of rings

City – an urban settlement that has been legally incorporated into an independent, self-government unit

Dark Ages – early medieval period, A.D. 476-1000

Dispersed rural settlement – a rural settlement pattern in which farmers live on individual farms isolated from neighbors

Dualism – the juxtaposition in geographic space of the formal and informal sectors of the economy

Edge city – a nodal concentration of shopping and office space situated on the outer fringes of metropolitan areas, typically near major highway intersections

Fiscal squeeze – increasing limitations on city revenues, combined with increasing demands for expenditure

Fordism – principles for mass production based on assembly-line techniques, scientific management, mass consumption based on higher wages, and sophisticated advertising techniques

Gateway city – serves as a link between one country or region and others because of its physical situation

Gentrification – invasion of older, centrally located, working-class neighborhoods by higher-income households seeking the character and convenience of less expensive and well-located residences

Hearth areas – the locations of the five earliest urban civilizations

Informal sector – economic activities that take place beyond official record, not subject to formalized systems of regulation or remuneration

Iraal – a traditional African village of huts, typically enclosed

Megacity – very large city characterized by both primacy and high centrality within its national economy

Megalopolis (megapolitan region) – a continuous urban complex (the chain of metropolitan areas) along a specific area (a clustered network of cities)

Merchant capitalism – the earliest phase in the development of capitalism as an economic and social system

Multiple-nuclei model – a model of the internal structure of cities in which social groups are arranged around a collection of nodes of activities

Neo-Fordism – economic principles in which the logic of mass production coupled with mass consumption is modified by the addition of more flexible production, distribution, and marketing systems

Primacy – condition in which the population of the largest city in an urban system is disproportionately large in relation to the second- and third-largest cities

Primate city – the largest settlement in a country, if it has more than twice as many people as the second-ranking settlement

Protestant Reformation – a schism from the Roman Catholic Church initiated by Martin Luther

Rank-size rule – statistical regularity in size distribution of cities and regions

Renaissance – a period in European history, from the 14th to the 17th century, regarded as the cultural bridge between the Middle Ages and modern history

Reurbanization – growth of population in metropolitan central cores, following a period of absolute or relative decline in population

Scientific Revolution – a concept used by historians to describe the emergence of modern *science* during the early modern period

Sector model – a model of the internal structure of cities in which social groups are arranged around a series of sectors, radiating out from the central business district

Shock city – a city recording surprising and disturbing changes in economic, social, and cultural life in a short period of time

Sprawl – development of new housing sites at relatively low density and at locations that are not contiguous to the existing built-up area

Suburbanization – growth of population along the fringes of large metropolitan areas

Underemployment – situation in which people work less than full-time even though they would prefer to work more hours

Urban area – a dense core of census tracts, densely settled suburbs, and low-density land that links the dense suburbs with the core

Urban forms – physical structure and organization of cities

Urban system – interdependent set of urban settlements within a specified region

urbanism – way of life, attitudes, values, and patterns of behavior fostered by urban settings

Urbanization – increasing concentration of population into growing metropolitan areas

World city – city in which a disproportionate part of the world's most important business is conducted

WorLd-empire – minisystems that have been absorbed into a common political system while retaining their fundamental cultural differences

Zone in transition – area of mixed commercial and residential land uses surrounding the CBD

12.6 WORKS CONSULTED AND FURTHER READING

Berkovitz, P. and Schulz-Greve, W. 2001. Defining the concept of rural development: A European perspective. In *The challenge of rural development in the EU accession countries*. Third World Bank/FAO EU Accession Workshop, Sofia, Bulgaria, June 17-20, 2000, ed. C. Csaki and Z. Lerman, 3-9. Washington, D. C.: The World Bank

____. Blue Banana. [On-line]. *https://en.wikipedia.org/wiki/Blue_Banana*

____. *https://en.wikipedia.org/wiki/Blue_Banana#/media/File:Blue_Banana.svg*

Cadwallader, M. 1996. *Urban geography: An analytical approach.*Upper Saddle River, NJ: Pearson Prentice Hall

Central Intelligence Service. [On-line]. *https://www.cia.gov/index.html*

____. *http://www.romaniajournal.ro/charlottenburg-the-only-circular-village-in-romania*

____. The circular villages of the Wendland region. [On-line] *http://www.germany.travel/en/towns-cities-culture/traditions-and-customs/circular villages*

____.*https://en.wikipedia.org/wiki/List_of_cities_in_the_Americas_by_year_of_foundation*

____. *http://www.beyondblighty.com/best-colonial-cities-in-latin-america/*

____. *https://www.google.com/search?q=compact+village+maps&tbm=isch&tbo=u&sourc/*

____. *http://en.wikipedia.org/wiki/Compact_city*

____. *https://www.google.com/search?q=Indian+compact+villages+photos*

Dickenson, J., Gould, B., Clarke, C., Marther, S., Siddle, D., Smith, C., and Thomas-Hope, E. 1996. *A geography of the Third World*. Second edition. New York: Routledge

____. *https://en.wikipedia.org/wiki/Dispersed_settlement*

____. *https://www.google.com/search?q=Erbil+Iraq+photos*

Green, R. and Pick, J. 2006. *Exploring the urban community: A GIS approach.* Upper Saddle River, NJ: Pearson Prentice Hall

Johnston, R. J., Gregory, D., Pratt, G. and Watts, M. 2003. *The dictionary of Human Geography.* Malden, MA: Blackwell Publishing

Knox, P. and McCarty, L. 2005. *Urbanization: An introduction to Urban Geography.* Second Edition. Upper Saddle River, NJ: Pearson Prentice Hall

Knox, P. and Marston, S. 2010. *Human Geography: Places and regions in global context.* Fifth Edition. Upper Saddle River, NJ: Pearson

Knox, P. and Marston, S. 2013. *Human Geography: Places and regions in global context.* Sixth Edition. Upper Saddle River, NJ: Pearson

____. Kraal. [On-line]. *https://en.wikipedia.org/wiki/Kraal*

____. *https://www.google.com/search?q=circular+villages+in+Africa+Kraal*

Levis, K. Rural settlement patterns. [On-line]. *http://www.lewishistoricalsociety.com/ article/aruralsettlements.html*

____. *https://en.wikipedia.org/wiki/Linear_settlement*

____. *https://www.historicplaces.ca/en/rep-reg/image-image.aspx?id=13161#i1*

Marston, S., Knox, P., Liverman, D., Del Casino, V., and Robbins, P. 2017. *World regions in global context: People, places, and environments.* Sixth Edition. Upper Saddle River, NJ: Pearson

Means, B. K. 2007. Circular villages of the Monongahela tradition. [On-line]. *http://www. uapress.ua.edu/product/circular-villages-of-the-Monongahela-Tradition*

____. *http://www.ancientmesopotamians.com/cities-in-mesopotamia.html*

____. *https://en.wikipedia.org/wiki/Megalopolis Northeast_Megalopolis*

Population Reference Bureau. 2016 World population data sheet. [On-line]. *www.prb. org/pdf16/prb-wpds2016-web-2016.pdf*

Pulsipher, L. 2000. *World regional geography.* New York. W. H. Freeman and Company

____. Romania: Settlement patterns. Encyclopaedia Britannica. [On-line]. *https://www. britanica.com/place/Romania/Settlement.patterns*

Rowntree, L., Lewis, M., Price, M., and Wyckoff, W. 2006. Diversity amid globalization: World regions, environment, development. Third Edition. Upper Saddle River, NJ: Pearson Prentice Hall

Rubenstein, J. 2013. *Contemporary Human Geography.* 2e. Upper Saddle River, NJ: Pearson

Rubenstein, J. 2016. *Contemporary Human Geography.* 3e. Upper Saddle River, NJ: Pearson

____. Rundling. [On-line]. *https://en.wikipedia.org/wiki/Rundling*

_____. *https://en.wikipedia.org/wiki/smart-growth*

_____. *https://www.google.com/serach!q+Spanish and Portuguese Empires*

Todaro, M. 2000. *Economic development*. Seven Edition. New York: Addison-Wesley

United Nations, Department of Economic and Social Affairs, Population Division. World urbanization prospects: The 2011 revision. File 11a: The 30 Largest urban agglomeration ranked by population size at each point in time, 1950-2025. [On-line]. *http://esa.un.org/unpd/wup/CD-ROM_2011/WU*

United Nations, Department of Economic and Social Affairs, Population Division. World urbanization prospects: The 2014 revision. File 11a: The 30 Largest urban agglomeration ranked by population size at each point in time, 1950-2030. [On-line]. *https://esa.un.org/unpd/wup/Publications/Files/WUP2014-Report.pdf*

United Nations. The world's cities in 2016. [On-line]. http://www.un.org.en/development/desa/population/publications/pdf/urbanization/world-cities-in-2016_data_booklet.pdf/

_____. Degree of urbanization **(percentage of urban population in total population) by continent in 2016. [On-line].** *https://www.statista.com/statistics/270860/urbanization-by-continent*

_____. http://www.google.com/search?q=urbanization+world_maps

World Bank. World Development Indicators: Urbanization. [On-line]. *http://data.worldbank.org/indicator/SP.URB.TOTL.IN.ZS*

12.7 ENDNOTES

1. Population Reference Bureau. 2017 World Population Data Sheet. https://assets.prb.org/pdf17/2017_World_Population.pdf

2. Adapted from http://www.3dgeography.co.uk/settlement-patterns

3. Population Reference Bureau. www.prb.org. Accesses May 12, 2018.

4. Adapted from United Nations, Department of Economic and Social Affairs, Population Division. World Urbanization Prospects: The 2014 Revision. File 11a: The 30 Largest Urban Agglomeration Ranked by Population Size at each point in time, 1950-2030. http://esa.un.org/unpd/wup/CD-ROM_2014/WU

5. Adapted from https://urbandesigntheory.wordpress.com/2014/12/31/the-social-and-spatial-structure-of-urban-and-regional-systems/

6. www.prb.org/pdf16/prb-wpds2016-web-2016.pdf . Accessed June 10, 2017.

7. https://www.cia.gov/library/publications/resources/the-world-factbook/. Accessed 10 June 2017.

8. Population Reference Bureau. www.prb.org. Accesses May 25, 2017.

9. Population Reference Bureau. www.prb.org. Accesses May 25, 2017.

10. https://www.statista.com/statistics/270860/urbanization-by-continent. Accessed 25 May 2017.

13

Environment and Resources

Joseph Henderson

STUDENT LEARNING OUTCOMES

By the end of this section, the student will be able to:

1. Understand: the problems associated with the demand for nonrenewable energy resources and the available supply.

2. Explain: the issues associated with the pollution of air, land, and water.

3. Describe: the types of renewable and alternative energy resources and global initiatives to leverage these resources.

4. Connect: preservation efforts worldwide to anthropogenic pressures on the environment.

CHAPTER OUTLINE

13.1 INTRODUCTION

A discussion of resources makes an excellent capstone for this textbook, as most major topics that have been discussed previously relate to the consumption of resources in some way. Global *population* increase means a greater demand for resources. Job opportunities in fossil fuel mining and drilling are a significant pull factor for *migration* in many countries where fossil fuel extraction is an important part of the economy. *Cultures*, and their associated *political* and *economic* systems and *settlements,* affect and are affected by the availability of resources and how resources are used in a country. For example, as discussed in Ch. 9, the availability of coal in the United Kingdom helped bring about the *industrial* revolution in that country. Furthermore, as *developing* countries modernize, their energy needs for transportation and electricity grow, and there is a continued increase in the demand for such energy sources as fossil fuels.

13.2 NONRENEWABLE ENERGY RESOURCES

Fossil fuels are by far the most significant energy source for most countries worldwide, and these resources are **nonrenewable**, or in finite supply. Fossil fuels are burned in electric power plants, and provide energy to power vehicles, aircraft, and ships. While the technology exists to burn fossil fuels in power plants so that the emissions are not particularly significant in terms of air pollution, air quality reduction because of the combustion of fossil fuels is still considerable and is a cause for concern at the international level.

The major fossil fuels are coal, oil, and natural gas. **Fossil fuels** are formed from ancient living matter. In the case of coal, this fossil fuel was likely formed from primordial plant material in swamps that was buried and metamorphosed through time. The origin of oil and natural gas is attributed to microscopic organisms that sank to the bottom of prehistoric oceans and over time, transformed into these two fossil fuels.

The countries that hold the greatest share of fossil fuel **proven reserves**, or deposits that can be recovered with some certainty, are varied. For coal, the United States has the largest reserves, followed by Russia and China. In the case of oil, the vast majority of proven reserves are located in the Middle East in countries like Saudi Arabia (with the largest reserves) and Canada. Other countries that hold substantial reserves include Iraq and Venezuela. Lastly, for natural gas, Russia and the Iran have the largest reserves. These statistics do not take into account **potential reserves** of fossils fuels, or fossil fuels that may be exist but have not been definitively verified. One example of potential reserves in the United States is the Green River Formation in parts of Colorado, Utah, and Wyoming. This large deposit of shale rock may contain up to three trillion barrels of oil, only half of which is recoverable (7). 1.5 trillion barrels of oil, though, is roughly equivalent to the entire world's proven oil reserves. The Green River Formation is not likely to be tapped because of complications associated with extracting oil from the rock, as the process would require heating the rock and using significant quantities of water in a rather arid region. Another example of potential reserves is the deposits of oil and gas in the Arctic, a region that has not been extensively explored, but the melting of sea ice caused by rising temperatures in the Arctic is increasing opportunities for experimental drilling.

The countries that have the largest reserves are not necessarily the countries with the leading amount of fossil fuel production. **Production** refers to the extraction of fossil fuels from the Earth by drilling and mining. The leading coal producers are China, India, and the United States, largely because of significant demands for electric power. Saudi Arabia, Russia, and the United States currently lead the world in oil production. Natural gas is extracted most heavily in the United States and Russia.

13.3 RENEWABLE ENERGY SOURCES

Fossil fuel resources are finite, so efforts to seek other types of energy resources are ongoing in many countries around the world. Because fossil fuels are not limitless, some researchers are concerned that supply will eventually exceed demand, particularly as developing countries increase economically and technologically. Moreover, as we will see in a later section on pollution, the emissions from the burning of fossil fuels are a serious concern for the global atmosphere, and clean energy is seen as a better alternative to coal, oil, and natural gas.

The shift away from fossils fuels involves a number of different types of renewable energy and alternative energy sources. **Renewable resources** are those that are in infinite supply. Major types of renewable energy include solar, wind, hydroelectric power, biofuels, and geothermal energy. Although the risks are considered to be significant, an alternative energy source is nuclear power, but nuclear power is not renewable as it depends on the supply of uranium. Significant efforts worldwide to harness these energy sources can be seen in many regions of the world.

For example, in Europe, a concerted effort to fund renewable energy projects has been underway since the 1990s. In 2016, wind energy overtook coal as the second largest form of power production (after natural gas) in Europe, and 80 percent of new energy projects were from renewable energy sources. France, Ireland, Lithuania, the Netherlands, and Finland all set records for windfarm installation, while Germany continues to lead in wind power. Norway and Sweden are the leading countries overall in terms of renewable energy consumption in Europe, as more than half of their energy consumption is from renewables (**Figure 13.1**).

In Asia, while China continues to build many coal-powered plants, it is investing hundreds of billions of dollars on renewable energy, more money than the United Kingdom, the United States, and Japan, combined. This development is important as China is the world's largest emitter of greenhouse gases, and this pollution may contribute to global warming. Moreover, China is already the world leader in wind

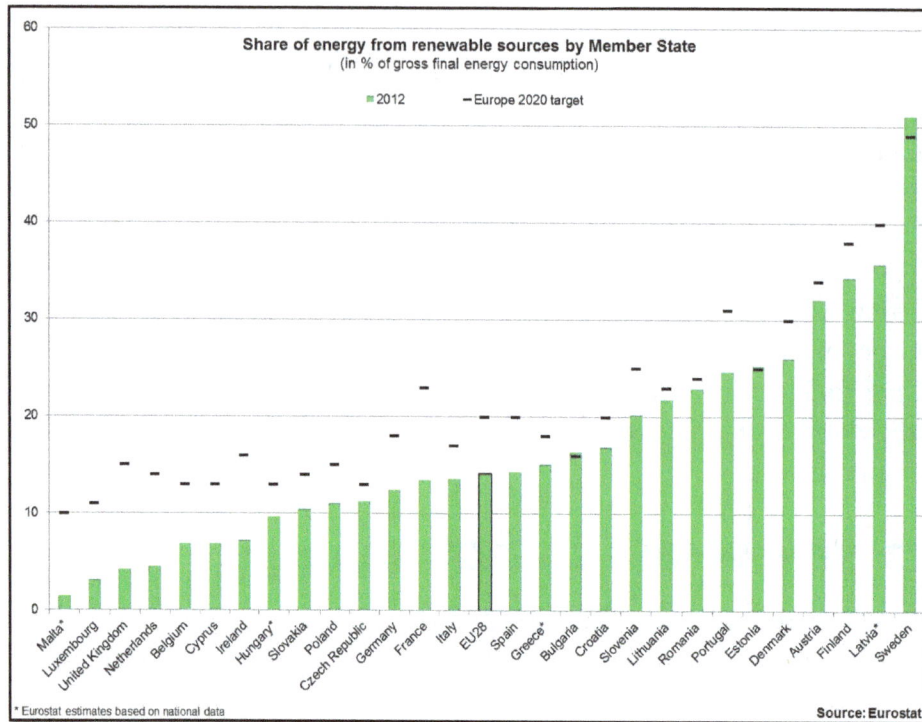

Figure 13.1 | Renewable energy for selected countries in Europe
Author | Eurostat
Source | Flickr
License | CC BY SA 2.0

power and plans to build the largest solar farm on the planet. India's goal is to produce over 50 percent of its power from renewable energy by 2027. All of these efforts require a considerable amount of capital investment to help cover the costs of the energy transformation.

In the Middle East, even the world leader in oil production, Saudi Arabia, is looking toward renewable energy as they are expected to be a net importer of oil by 2038 as their domestic supplies dwindle. The Middle East is an excellent location for solar power because of the number of cloud-free days in a year.

In the United States, only about 12 percent of energy production is from renewables, and much of that production is from hydroelectric power. However, America has some of the largest wind and solar farms in the world. Texas leads in wind power production, and California has the most solar energy generating capacity (**Figure 13.2**). Efforts to promote electric car use have also been promoted in California, as the state government has set a goal of having 1.5 million electric cars on the

Figure 13.2 | The Smoky Hills Wind Farm in Kansas
Author | User "Drenaline"
Source | Wikimedia Commons
License | CC BY SA 3.0

road by 2025. Although using a battery to power a car instead of fossil fuels is generally environmentally friendly, electricity generated primarily from fossil fuels is still used to charge the vehicles.

One of the more controversial types of renewable energy is **biofuels**. Biofuels produce energy from living matter, and much of the biofuel used in the world is from a variety of crops such as corn, cassava, sweet potato, sugar cane, and sorghum. The United States and Brazil lead the world in biofuel production, followed by China. In the United States, most gasoline has a certain percentage of ethanol, derived from corn. The major controversy with biofuels stems from the fact that many of the plants used for biofuels could also be used to feed people, so whether or not using scarce land to produce fuel as opposed to food is a good idea is hotly debated. However, other biofuel sources exist besides agricultural crops, and these biofuels include bio-methane and even bacteria such as E. coli. In the United States, India, and Germany, methane derived from cow manure and decaying plant matter is used to power electric plants. In Gwinnett County, Georgia, methane extracted during the wastewater treatment process is used to provide electric power for the wastewater treatment process (**Figure 13.3**).

Figure 13.3 | Egg-shaped "digesters" at the Newton Creek Wastewater Treatment Plant
Author | Jim Henderson
Source | Wikimedia Commons
License | Public Domain

13.4 POLLUTION

13.4.1 Air Pollution

While the technology exists to burn fossil fuels so that the emissions are not particularly significant in terms of air pollution, air quality reduction because of the combustion of fossil fuels is still considerable and is a cause for concern at the international level. Another pollutant that has received worldwide attention is chlorofluorocarbons (CFCs) and their impact on the ozone layer. Water pollution and the disposal problem of solid wastes are also important environmental concerns.

One of the major issues internationally is the suspected anthropogenic, or human-caused, warming of the Earth's atmosphere because of carbon dioxide in fossil fuel emissions. The issue is complex because "climate change", as global warming has been dubbed, is affected by both natural and anthropogenic processes. A number of scientists are concerned with warming temperatures in the lower atmosphere because this warming could contribute to rising sea levels, an increase in the frequency of storms and drought, a greater number of heat injuries, and the spread of tropical pests into higher latitudes. Many climatic factors operate to affect global temperatures, so carbon dioxide produced by the burning of coal, oil, and natural gas is not the only important element that dictates temperatures worldwide. Ocean currents, shifts in large warm and cold pools of the Oceans such as El Ñino/La Ñina, changes in sunspot activity, photosynthesis from global plant life, the extent of global sea ice, the type and extent of global cloud cover, the natural cycling of carbon, and volcanic activity are but a few of the climate controls of global temperatures.

Despite the complexity of the Earth's climate system, many scientists attribute the slight warming (degrees C) of the average temperature of the Earth since the Industrial Revolution to increased levels of carbon dioxide, a natural component of the Earth's atmosphere. Scientists have been measuring atmospheric carbon dioxide in Hawaii since the 1960s, and the trend has been upward for virtually every year on record. The increase in carbon dioxide levels has been linked by many scientists to the burning of fossil fuels. So, how does this increase in carbon dioxide relate to warming temperatures around the globe?

Scientists suspect that a natural phenomenon called the **greenhouse effect** is related to this warming. The greenhouse effect is the trapping of longwave radiation (heat) by certain greenhouse gases in the lower atmosphere, as these greenhouse gases absorb and reradiate the heat from the Earth, essentially forming a blanket in the Earth's lower atmosphere. Without the greenhouse effect, the Earth would be over 35° C cooler, and life on Earth would likely not exist. Global warming is thought to be related to the greenhouse effect because carbon dioxide is one of the many greenhouse gases that traps heat from the Earth, and by humans increasing the concentration of greenhouse gases in the atmosphere, the greenhouse effect is enhanced, leading to increased warming above and beyond the "natural" effect (**Figure 13.4**).

International efforts to curb emissions of anthropogenic carbon began in earnest with the Kyoto Protocol in 1997. This agreement, signed and ratified by many countries around the world, sought to establish goals whereby countries would reduce their greenhouse gas emissions. The U.S. never ratified the treaty, particularly because India and China, two of the world's biggest producers of carbon dioxide, were not required to reduce their emissions. However, the U.S. and other countries have continued their dialogue about lowering emissions, and the recent Paris Climate Summit is 2016, signed by 194 countries (5), is the newest plan for global cooperation in carbon dioxide reductions and with a goal to keep global average temperatures at less than 2° C above preindustrial levels.

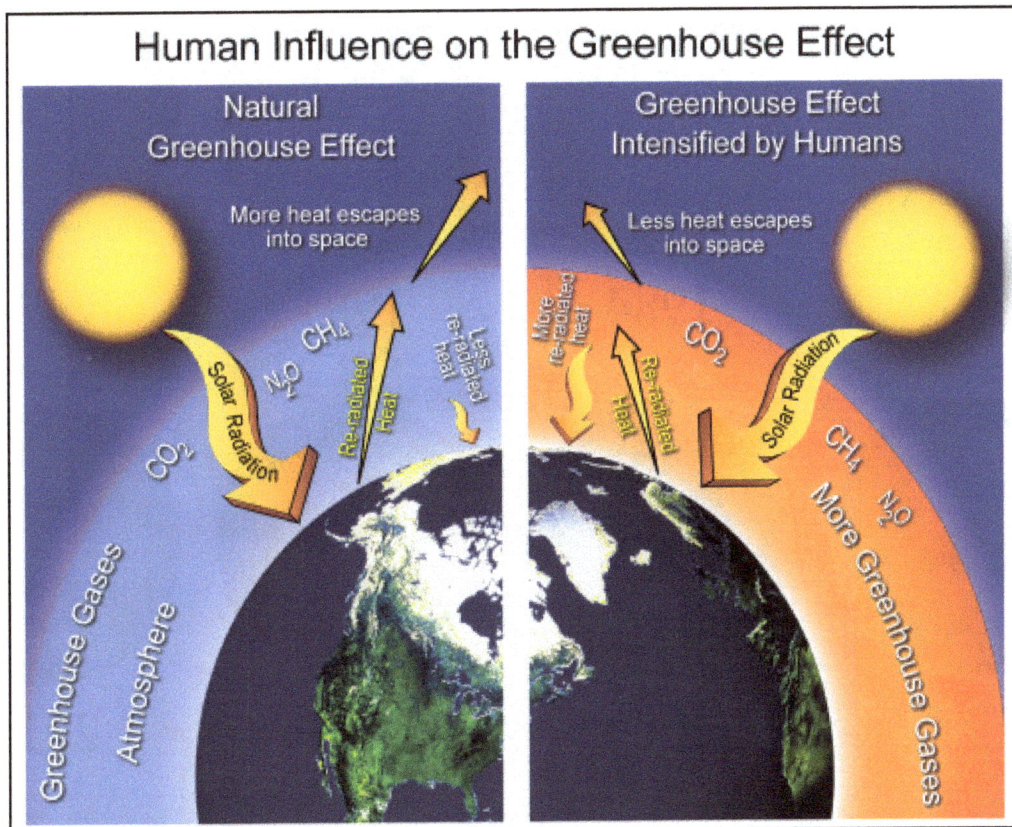

Figure 13.4 | Natural and anthropogenically-forced greenhouse effect
Author | National Park Service and U.S. Global Change Research Program
Source | U.S. Global Change Research Program
License | Used with Permission.

Another example of international cooperation to decrease atmospheric pollutants involves the ozone layer. The ozone layer is in the stratosphere or upper atmosphere from 10–50 kilometers above the Earth's surface. This important layer of ozone gas protects the Earth from harmful ultraviolet radiation that can be harmful to skin by causing sunburn and skin cancer. Scientists believe that chlorofluorocarbons (CFCs), used in refrigeration equipment and aerosols, have damaged the ozone layer and caused it to become dangerously thin over some parts of the globe. This thinning of the ozone layer has been named an "ozone hole", and is most pronounced around the South Pole during the Antarctic spring **(Figure 13.5)**.

In response to these scientific findings, in 1987, 105 countries signed an international agreement known as the Montreal Protocol. The participating countries agreed to reduce the consumption and production of CFCs and to cease producing CFCs by the year 2000. The Montreal Protocol process continues as countries work to find ever-better substitutes for CFCs. Although CFCs will linger in the upper atmosphere for many years, scientists have already seen some evidence of recovery in the ozone layer.

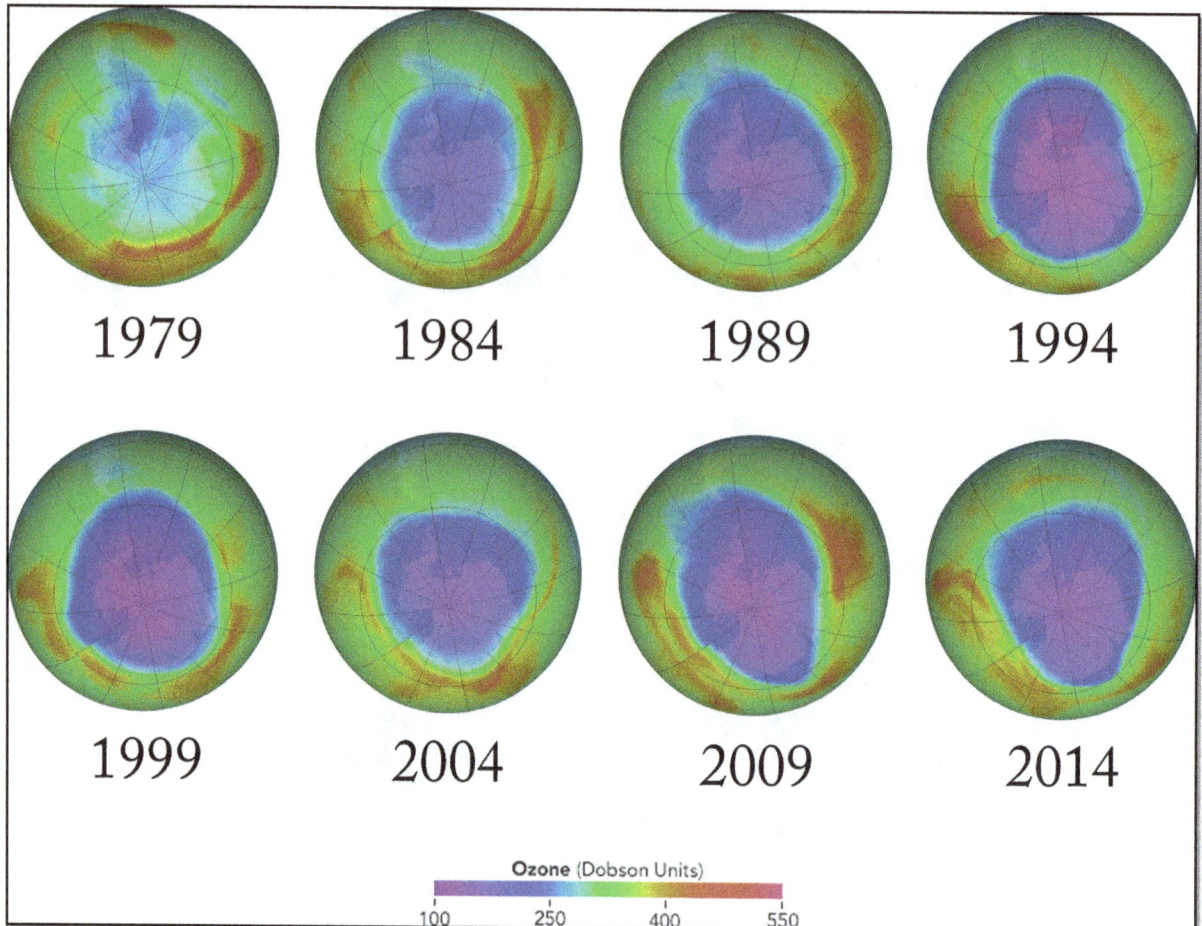

Figure 13.5 | Ozone depletion around Antarctica from 1979 to 2014
Author | NASA
Source | NASA
License | Public Domain

13.4.2 Water Pollution

While air pollution of the atmospheric global "commons" is an international issue, pollution of water and land areas is likewise problematic. The supply of clean, fresh water for drinking and other uses such as irrigation is under stress as population pressure continues to increase demand. Many regions with water deficits experience periodic droughts that strain available supplies. Unfortunately, both ground and surface waters are being polluted both in the United States and abroad. Pollutants such as industrial chemicals, runoff from roadways, and untreated sewage place a heavy toll on water supplies.

Not only is unclean water unsafe to drink and in some cases unfit for irrigation purposes, but pollution can also have a deleterious effect on aquatic life. Large industrial spills, discharge from paper mills, and other types of pollutants can render a lake or stream devoid of life for considerably long periods. Another type of pollutant that is of considerable concern in the United States and abroad is commercial fertilizers used in large-scale agriculture. Although it would seem that fertilizers would encourage plant growth and help aquatic ecosystems thrive, the opposite is, in fact, true in many cases. Rainfall washes commercial fertilizers into surface lakes and streams, and these fertilizers increase the growth of algae. Through a process called **eutrophication**, the abundant algal blooms die, and in the process of decomposition of the dead plant material, large amounts of oxygen are consumed. The lack of oxygen can result in extensive fish kills. One stark example of the harmful effects of eutrophication can be seen in the Gulf of Mexico. The Mississippi River carries a significant load of commercial fertilizers, and where the river discharges into the Gulf, "dead zones" form at certain times of the year. The bottom and near-bottom portions of the Gulf can become almost devoid of marine life over several thousand square miles. As a result, the Gulf of Mexico has the second largest "dead zone" of depleted oxygen area in the entire world, with the largest being in the Baltic Sea. (**Figure 13.7**).

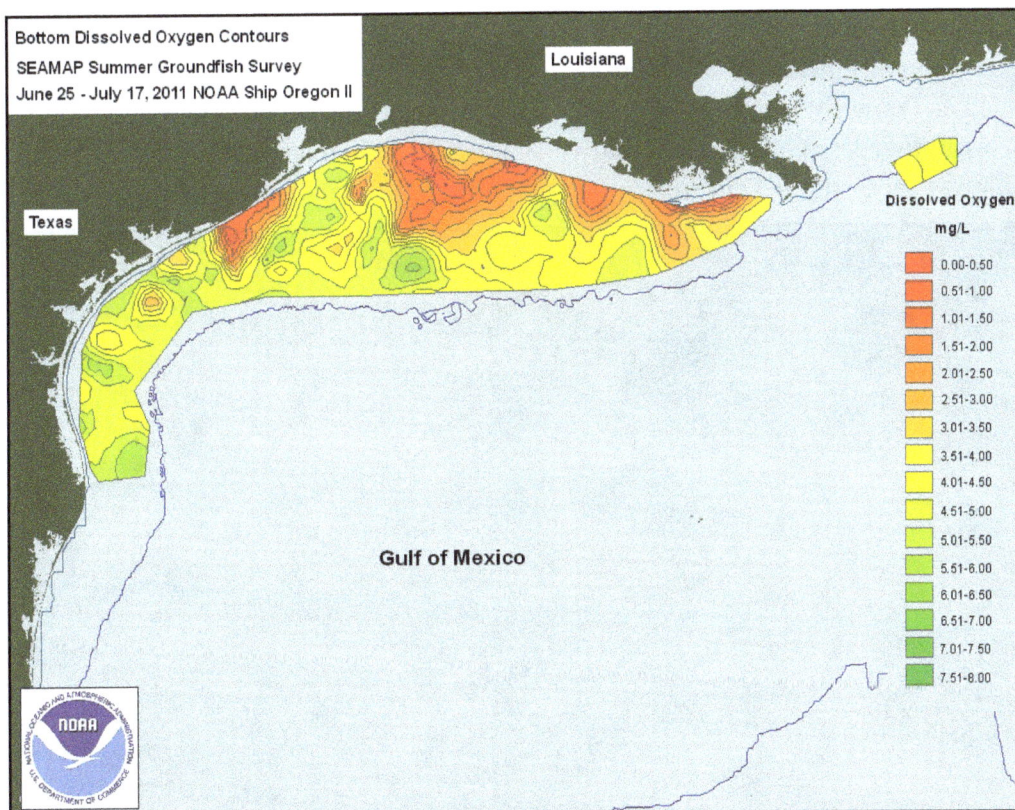

Figure 13.6 | The 2010 Dead Zone in the Gulf of Mexico
Author | NASA NOAA
Source | NASA NOAA
License | Public Domain

13.4.3 Solid Waste

As countries struggle to maintain clean water sources, a related problem with pollution occurs with solid waste and its disposal. Solid waste that is buried beneath the ground can, overtime, begin to pollute the groundwater below it. Water percolating downward through the rubbish can pick up harmful pollutants and deposit them in subsurface water sources. These groundwater reservoirs feed surface streams and are also tapped for well water. In the state of Florida, for example, the pollution of groundwater reservoirs, known as aquifers, is a significant issue because the state relies heavily on groundwater for its water supply.

Three primary solutions to the problem of solid wastes is disposal in landfills, incineration of the trash, or recycling. In the United States, solid waste collectors pick up trash in cities and neighborhoods and transports the waste to **landfills** where it is buried daily to decrease the pervasive odor and protect it from marauding pests. This is the most common method of dealing with solid waste in the United States, but unfortunately, as urban sprawl expands towards areas where landfills were on the periphery of populated areas, the matter of residents having to live close to landfills is problematic. Moreover, landfills require extensive areas of land, and in some states with heavy population concentrations such as New York City, government officials have resorted to shipping their refuse into adjacent states such as Pennsylvania where space for landfills is more available.

Another less common alternative to landfills is incineration, a process made famous in Disney's *Toy Story 2* (**Figure 13.7**). At incineration facilities, trash

Figure 13.7 | The "Claw" moving plant waste and rubbish to incinerator
Author | Senior Airman Sam Fogleman
Source | U.S. Air Force
License | Public Domain

is dumped into an enormous pile, and a claw-like machine picks up the trash and continually feeds the fire of the incinerator. The advantage of incineration is the vast reduction in the amount of trash that must be disposed of in landfills, and these facilities also generate electricity from the combustion of the burning trash. Fortunately, because of advanced technology in the smokestack systems, the emissions from modern incinerators are not particularly harmful in the atmosphere. Because of its enormous output of daily solid waste, New York City takes advantage of incineration through a facility on the Hudson River just north of the city.

The last method to deal with solid waste, recycling, is the most sustainable. **Recycling** involves the conversion of waste or other unwanted material into something that is reusable. Common wastes that are recycled include paper products, metals, plastics, and glass. In the United States, efforts to improve the percent of waste material that is recycled instead of disposed of in trash disposal units are many. For example, *Recyclemania*, started in 2001, is a nation-wide competition amongst colleges and universities to promote student knowledge and participation in recycling. Georgia Gwinnett College in Lawrenceville, GA, is one of the many competitors in *Recyclemania* every spring. Despite efforts such as these, the United States only recycles about 30 percent of solid waste, even though approximately 75 percent of the waste is recyclable. In contrast, Germany recycles about 65 percent of its municipal waste and leads the world, followed closely by South Korea at 59 percent. Germany's recycling emphasis reflects the generally strong emphasis on recycling in most European countries, thanks to rather strict recycling laws.

13.5 PRESERVATION OF NATURAL RESOURCES

Given the tremendous impact that human activities and settlements have on the environment, the question of how to preserve and protect the Earth's land areas and its resources is particularly important, especially as the global population continues to grow. Two major concepts in this effort are conservation and preservation. **Conservation** involves using the Earth's resources sustainably, which means they will be available to future generations. Ways in which conservation can be applied include not overfishing, replanting trees when they are logged, and protecting soil in farming areas from erosion. **Preservation** is the idea of protecting natural areas and trying to keep them as close as possible to their original, unspoiled state. As such, human impacts should me minimal, and resources in preserved areas are not for human use.

One of the foremost examples of preservation is the national park. The governments in nearly 100 countries worldwide have established these parks in wilderness areas where visitors can come and enjoy the scenery and the flora and fauna. The United States is a world leader in national parks, and in fact, the first national park in the world, Yellowstone, was created in the United States in 1872. However, the United States only has 59 national parks, much less than the 685 parks in Australia, the world leader in total parks. In Asia, China boasts the largest number of parks at over 200.

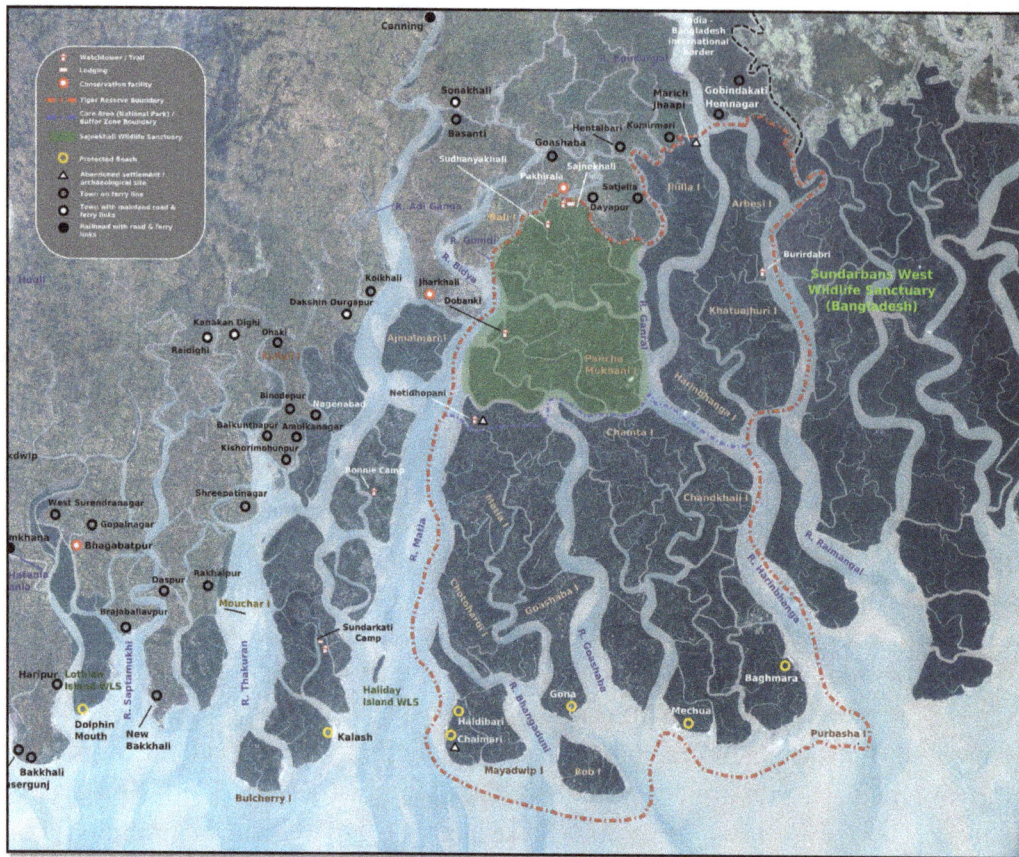

Figure 13.8 | Sundarban National Park, India
Author | User "Pradiptaray"
Source | Wikimedia Commons
License | Public Domain

These parks help protect some of the most vulnerable plant and animal species in the world. For example, in the coastal mangrove forests of eastern India, the Bengal tiger, an endangered species, is protected in the Sundarban National Park (**Figures 13.8 and 13.9**). Endangered species are species that are at risk of becoming extinct, and the International Union for the Conservation of Nature is an international organization aimed at protecting habitats, such as forests, and species around the globe (website at: https://www.iucn.org/). Deforestation is one of the greatest threats to endangered species, particularly in tropical rainforests in places like the Brazilian Amazon, as depicted in Chapter 1, but also in the United States in the forests of the Pacific Northwest, for

Figure 13.8 | Sundarban National Park, India
Author | User "Pradiptaray"
Source | Wikimedia Commons
License | Public Domain

example. Thousands of endangered species exist worldwide, and in the United States, the U.S. Fish and Wildlife Service oversees the identification, protection and restoration of endangered species. The Endangered Species Act of 1973 protects vulnerable plant and animals in the United States, but the needs and demands of human development often clash with the preservation of habitat. With the world in the midst of one of the highest rates of extinction in the history of the Earth, the necessity to protect the remaining species from disappearing is a cause for alarm, both in the United States and abroad.

13.6 KEY TERMS DEFINED

Biofuels – energy sources from living matter.

Conservation – using natural resources in a sustainable way so that they are preserved for future generations.

Eutrophication – the process by which nutrient-rich waters promote the growth of algae, and when the abundant algal blooms die, the decomposition of the dead plant material consumes large amounts of oxygen.

Fossil Fuels – energy sources such as coal, oil, and natural gas, derived from ancient plant and animal matter.

Greenhouse effect - the trapping of longwave radiation (heat) by certain greenhouse gases in the lower atmosphere; greenhouse gases absorb and reradiate the heat radiated from the Earth, increasing global temperatures by 350 C compared to an atmosphere with no greenhouse effect.

Landfill – An area where solid waste is deposited and buried to reduce odor, vermin proliferation, and unsightly trash.

Nonrenewable resource – a resource that is in finite supply and is depleted by humans.

Potential reserves – estimates on available energy in deposits that are thought to exist but have not been completely verified.

Production – the extraction of fossil fuels from the ground.

Proven reserves - state in which the territorial boundaries encompass a group of people with a shared ethnicity.

Preservation – setting aside areas so that resources are essentially untouched with as little human impact as feasible.

Renewable resource – a resource that is in infinite supply such as solar and wind energy.

13.7 WORKS CONSULTED AND FURTHER READING

"China Aims to Spend at Least $360 Billion on Renewable Energy by 2020." The New York Times. December 22, 2017. Accessed April 23, 2018. https://www.nytimes.com/2017/01/05/world/asia/china-renewable-energy-investment.html.

"Home." ETIP Bioenergy-SABS. Accessed April 23, 2018. http://www.biofuelstp.eu/global_overview.html.

Kirk, Ashley. "Paris Climate Summit: Which EU Countries Are Using the Most Renewable Energy?" The Telegraph. November 30, 2015. Accessed April 23, 2018. http://www.telegraph.co.uk/news/earth/energy/renewableenergy/12021449/Renewable-energy-in-the-EU-Which-countries-are-using-the-most-renewable-energy.html.

Kleven, Anthony. "Powering Asia's Renewable Revolution." The Diplomat. January 07, 2017. Accessed April 23, 2018. https://thediplomat.com/2017/01/powering-asias-renewable-revolution/.

"Paris Agreement." Wikipedia. April 23, 2018. Accessed April 23, 2018. https://en.wikipedia.org/wiki/Paris_Agreement.

Torgerson, Ross. "Untapped: The Story behind the Green River Shale Formation." Bakken.com. November 06, 2014. Accessed April 23, 2018. http://bakken.com/news/id/225072/untapped-story-behind-green-river-shale-formation/.

Vaughan, Adam. "Almost 90% of New Power in Europe from Renewable Sources in 2016." The Guardian. Accessed April 23, 2018. https://www.theguardian.com/environment/2017/feb/09/new-energy-europe-renewable-sources-2016.